Peter Comba, Trevor W. Hambley

# Molecular Modeling of Inorganic Compounds

# Further Reading from Wiley-VCH
## and John Wiley & Sons

H.-D. Höltje / G. Folkers
**Molecular Modeling. Basic Principles and Applications**
1997. 206 pages. Wiley-VCH.
ISBN 3-527-29384-1

F. Jensen
**Introduction to Computational Chemistry**
1998. 454 pages. Wiley.
ISBN 0-471-98425-6

W. Koch / M. C. Holthausen
**A Chemist's Guide to Density Functional Theory**
2000. 308 pages. Wiley-VCH.
ISBN 3-527-29918-1

K. B. Lipkowitz / D. B. Boyd (Eds.)
**Reviews in Computational Chemistry, Vol. 13**
1999. 384 pages. Wiley.
ISBN 0-471-33135-X

M. F. Schlecht
**Molecular Modeling on the PC**
1998. 763 pages. Wiley.
ISBN 0-471-18467-1

P. von Schleyer (Ed.)
**Encyclopedia of Computational Chemistry**
1998. 3580 pages. Wiley.
ISBN 0-471-96588-X

J. Zupan / J. Gasteiger
**Neural Networks in Chemistry and Drug Design**
1999. 400 pages. Wiley-VCH.
ISBN 3-527-29779-0 (Softcover), 3-527-29778-2 (Hardcover)

Peter Comba, Trevor W. Hambley

# Molecular Modeling of Inorganic Compounds

Second, Completely Revised and Enlarged Edition

Includes CD-ROM
With Tutorial

**WILEY-VCH**

Weinheim · New York · Chichester
Brisbane · Singapore · Toronto

Prof. Dr. Peter Comba
Anorganisch-Chemisches Institut
Im Neuenheimer Feld 270
69120 Heidelberg
Germany
e-mail: comba@akcomba.oci.uni-heidelberg.de
homepage: http://www.uni-heidelberg.de/
institute/fak12/AC/comba/

Prof. Dr. Trevor W. Hambley
School of Chemistry
University of Sydney
Sydney, NSW 2006
Australia
e-mail: t.hambley@chem.usyd.edu-au
homepage: http://www.chem.usyd.edu.au/
~hamble_t

Shown on the cover are the calculated structures and energetics of two conformations of a figure-of-eight-shaped dicopper(I) compound, together with the observed $^1$H-NMR spectra (J.Chem.Soc., Dalton Trans. (1999), 509; Eur.J.Inorg.Chem. (1999), 509).

Library of Congress Card No.: applied for

A catalogue record for this book is available from the British Library.

Die Deutsche Bibliothek – CIP Cataloguing-in-Publication Data
A catalogue record for this publication is available from Die Deutsche Bibliothek

ISBN 3-527-29915-7

Composition: ProSatz Unger, D-69469 Weinheim
Printing: Strauss Offsetdruck GmbH, D-69503 Mörlenbach
Bookbindung: Wilhelm Osswald & Co., D-67433 Neustadt

Printed in the Federal Republic of Germany

# Preface

Over the past thirty years molecular-mechanical modeling of organic molecules has developed to the point where comprehensive models are now available and the structures and energetics of most simple molecules can be reliably calculated. More recently there has been a rapid expansion in the application of molecular mechanics and dynamics to biological macromolecules such as proteins and DNA. Highly sophisticated commercial packages are available that combine molecular mechanics with computer-graphical construction, manipulation and graphical output.

Proceeding in parallel with these developments has been the application of molecular mechanics to inorganic and coordination compounds. Initially, simple metal complexes were modeled, but recently the field has been extended to include organometallic compounds, catalysis and the interaction of metal ions with biological macromolecules. The application of molecular mechanics to coordination compounds is complicated by the number of different metals and the variety of coordination numbers, coordination modes, geometries and electronic states they can adopt. For this reason the existing models used for metal containing compounds are more complex than those available for organic molecules, and only few of the commercially available packages are able to reliably deal with even a small subset of the possible metal-based systems. The difficulties encountered in modeling inorganic and coordination compound systems have deterred many from making use of the method.

The goals of this book are to provide an understanding of molecular mechanics, to show that it can be applied successfully to a wide variety of inorganic and coordination compound based systems and to show how to undertake such a study. This book should give the reader the ability to judge the reliability of published data, to evaluate reported interpretations and to judge the scope and limitation of the various models for her or his own studies.

The book is aimed at students and scientists who have a basic understanding of inorganic chemistry. No prior knowledge of theoretical chemistry, sophisticated mathematics or computing is assumed. The basic concepts of molecular mechanics are developed and discussed in Part I. Examples of applications and the difficulties encountered are reviewed in Part II. In Part III a practical guide to undertaking a molecular modeling study of a new system is presented and the problems and pitfalls likely to be encountered are outlined. The three parts of the book can be read and used separately.

We are grateful for the help of Sigrid Rieth, Brigitte Saul, Volker Licht and Dr Norbert Okon in the preparation of the manuscript, to Dr Thomas Kellersohn for a strain-free collaboration with VCH, to our coworkers for allowing us to report unpublished data and for their helpful hints on scientific, didactic and linguistic aspects. A special thank is due to Prof. Marc Zimmer for his invaluable help during his sabbatical leave as a Humboldt Fellow in Heidelberg. Many interesting discussions with and suggestions by him have been of importance to this book. Finally, we owe our families a huge debt for their love and patience.

Peter Comba                                                  Trevor W. Hambley

# Preface to the Second Edition

The field of inorganic molecular modeling has developed in the past five years to an extent that it has led us to add some chapters and rewrite others. The division of the book into three parts; I Theory, II Applications and III Practice that can be read and used separately is retained. Our emphasis is still on empirical force field calculations. Quantum-mechanical calculations have undergone an enormous development in recent years, and techniques such as DFT and combined quantum mechanics/molecular mechanics (QM/MM) are now routinely used by theoreticians and experimentalists to predict and interpret structures, stabilities, electronic properties and reactivities of metal-containing compounds. Where appropriate, we have included results derived from such methods in this second edition of our book, without going into detailed discussion of the theoretical background, since this is given in many recent textbooks and review articles.

We have made only a few changes to Part I and most examples from Part II have remained unchanged, since our goal is to cover the types of application rather than to provide a comprehensive review of inorganic molecular modeling. New developments and some new examples have been added.

Part III has been rewritten completely. Important rules for molecular modeling and for the interpretation of the results, possible pitfalls and guidelines for the publication of molecular modeling studies are given in an introduction. This is followed by a tutorial, based on software included in this book, where the reader experiences in 20 lessons, how inorganic molecular modeling works in reality. The appendices have been updated and lists of molecular modeling books and inorganic molecular modeling reviews have been added (see Appendix 4). Regularly updated lists and comments related to the field appear also on our homepages.

We are grateful for comments by colleagues on the first edition. We are also grateful for the help of Marlies von Schoenebeck-Schilli, Karin Stelzer and Brigitte Saul in preparing the manuscript, to Dr Norbert Okon for the setup of MO-MEClite and to Dr Roland Wengenmayr for an excellent collaboration with Wiley-VCH. The continuing support of our families has allowed us to complete this project and is greatly appreciated.

Peter Comba                                                    Trevor W. Hambley

# Contents

# Part I: Theory

In Part I of this book we describe methods for molecular modeling with special emphasis on empirical force field calculations. Molecular mechanics is an interpolative procedure, and its justification is that it works. However, there is a theoretical basis for force field calculations, and it will be given in this Part of the book. The fact that molecular mechanics is a rather simplistic method implies that there are a number of dangers and limitations, and these have to be discussed in detail to give the reader the ability to judge where these methods are applicable and what quality the predictions might be expected to have. Since the aim of Part I is to show explicitly these problems we might create a rather pessimistic view in terms of the accuracy, reliability, general applicability and scientific basis of the methods presented. This certainly is not our aim. The applications discussed in Part II clearly reveal the potential of molecular modeling, enabling useful predictions to be made in many areas of inorganic chemistry.

Part I: Theory

# 1 Introduction

## 1.1 Molecular Modeling

Advances in computing, and particularly the ready availability of high resolution graphics, have greatly increased the interest in computer-based molecular modeling. Molecular modeling is now widely used as an aid in the interpretation of experimental results and in the design of new materials with desirable properties. Examples drawn from the area of inorganic chemistry include the study of the interaction of metal ions with proteins and DNA, the design of new metal-based drugs, metal-ion-selective ligands and stereospecific catalysts.

The basis of molecular modeling is that all important molecular properties – such as stabilities, reactivities and electronic properties – are related to the molecular structure (Fig. 1.1). Therefore, if it is possible to develop algorithms that are able to calculate a structure with a given stoichiometry and connectivity, it must be possible to develop algorithms for the computation of the molecular properties based on the calculated structure and vice versa. There are many different approaches and related computer programs, including ab-initio calculations, various semi-empirical molecular orbital (MO) methods, ligand-field calculations, molecular mechanics, molecular dynamics, QSAR, neural networks and genetic algorithms that can be used to calculate structures and one or more additional molecular properties.

Before any computational study on molecular properties can be carried out, a molecular model needs to be established. It can be based on an appropriate crystal structure or derived using any technique that can produce a valid model for a given compound, whether or not it has been prepared. Molecular mechanics is one such technique and, primarily for reasons of computational simplicity and efficiency, it is one of the most widely used technique. Quantum-mechanical modeling is far more computationally intensive and until recently has been used only rarely for metal complexes. However, the development of effective-core potentials (ECP) and density-functional-theory methods (DFT) has made the use of quantum mechanics a practical alternative. This is particularly so when the electronic structures of a small number of compounds or isomers are required or when transition states or excited states, which are not usually available in molecular mechanics, are to be investigated. However, molecular mechanics is still orders of magnitude faster than ab-initio quantum mechanics and therefore, when large numbers of

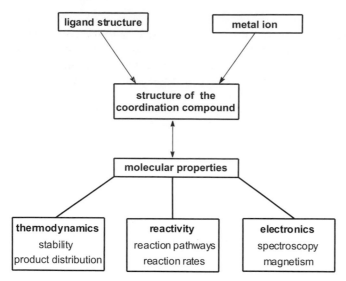

**Figure 1.1:** The relationship between the ligand and metal ion preferences, the resulting molecular structure, and the molecular properties.

compounds or isomers are to be investigated, molecular-mechanical methods are still preferred. Also, because of the speed of molecular-mechanics calculations it is possible to highly optimize the force field parameterization against a large set of compounds. With slower quantum-mechanical calculations the optimization and testing of basis sets, functionals and other variables against more than a few compounds is less feasible and, perhaps as a consequence, the balance between bonded and nonbonded forces is often better in molecular mechanics than in quantum mechanics.

Molecular mechanics can be considered to arise from the Born-Oppenheimer approximation, which assumes that the motions of the nuclei of a molecule are independent of the motions of the electrons. In molecular-mechanics calculations the arrangement of the electrons is assumed to be fixed and the positions of the nuclei are calculated. The basis of many quantum-mechanical calculations is, in contrast, that the electronic states can be calculated if the nuclei are assumed to be in fixed positions.

The basis of the molecular-mechanics method is that a good estimate of the geometry of a molecule can be obtained by taking into account all the forces between the atoms, calculated using a mechanical approach. For example, bonded atoms are treated as if they are held together by forces that behave as mechanical springs, and nonbonded interactions are taken to be made up of attractive and repulsive forces that together produce the typical van der Waals curve. The parameters that define the strength of the springs or the steepness of the van der Waals curves are derived, in the first instance, from experimental observables such as infrared vibrational frequencies and gas compressibility data. However, the parameters are usually modified empirically to enhance the reproduction of experimentally determined geometries. To optimize the geometry of a molecule, the total energy that arises from these forces, or stresses, is minimized by computational methods. The minimized total energy is taken to be an indication of the strain present in the molecule. It is frequently referred to as the "strain energy" or "steric energy" and is related to the molecule's potential energy and stability.

Some of the potential energy functions used to calculate the total strain energy of a molecule are similar to the functions used in the analysis of vibrational spectra. Because the parameters used to derive the strain energies from these functions are fitted quantities that are based on experimental data (e.g., X-ray structures or vibrational spectra), molecular mechanics may be referred to as "*empirical* force field calculations" (more often the simplification "force field calculations" is used). The quality of such calculations is strongly dependent on the reliability of the potential energy functions and the corresponding parameters (the force field). Thus, the selection of experimental data to fit the force field is one of the most important steps in a molecular mechanics study. An empirical force field calculation is in essence a method where the structure and the strain energy of an unknown molecule are *interpolated* from a series of similar molecules with known structures and properties.

Molecular modeling of transition metal compounds is complicated by the partially filled d-orbitals of the metal ions that are responsible for the multifarious structures of coordination compounds with a large variety of possible coordination numbers and geometries. The coordination geometry of a metal complex is always a compromise between the size and electronic structure of the metal ion, and the type, size, geometry and rigidity of the coordinated ligands (see Fig. 1.1). The fact that ligand-metal-ligand angles vary over a much larger range than corresponding parameters of organic molecules indicates that the competition between the ligand and metal ion in terms of coordination geometry is generally dictated by the ligand. Thus, the structure of a coordination compound, and therefore its thermodynamics, reactivity and electronics, is strongly influenced by the ligand structure. Since empirical force field calculations have been shown to be a powerful tool for estimating the structures of organic molecules, there is reason to expect that molecular mechanics can be a viable tool for modeling coordination compounds.

For a molecular-modeling technique to be useful and to achieve widespread application it must readily and reliably reproduce molecular properties that closely resemble experimentally determined data. The molecular-mechanics method has been successfully applied to a wide variety of problems in inorganic chemistry and many of these are outlined in detail in Part II of this book. However, the varied chemistry, particularly of the transition metal elements, greatly complicates the molecular-mechanical analysis of such systems, and in some cases molecular mechanics alone is unable to predict the geometry of a metal complex. For example, the assumption that the nature of the bonding does not change with the structure may not be valid when there is π-bonding between the metal and the ligand or when there is an equilibrium between two spin-states with similar energies. Coupling of the molecular-mechanics method with quantum-mechanical or ligand-field calculations has led to new models that can overcome some of these restrictions. The limitations of the classical molecular mechanics method as applied to metal complexes are discussed further in all three parts of this book.

## 1.2   Historical Background

Chemists in the 19$^{th}$ century were aware of the connectivity and the basic geometries of their molecules and therefore of structural formulae, but they were not able to quantify the structures of molecules on a metric basis. In addition to chemical bonds, they were aware of van der Waals interactions, electrostatic interactions, steric hindrance, Kekulé conjugation and donor-acceptor interactions. However, detailed information on electronic and molecular structure was lacking.

The 20$^{th}$ century brought two important advances. With the development of diffraction techniques, the arrangement of atoms could be determined on a metric basis. Depending on the size of the molecule, the quality of the crystal and the accuracy of the experiment, well-defined bond lengths, valence and torsional angles, as well as nonbonded contacts, can be determined.

The other development with far reaching consequences was the Schrödinger equation ($H\Psi = E\Psi$). The problems encountered when solving the Schrödinger equation for complex molecular systems have resulted in the development of various approximations. The most important one is the Born-Oppenheimer approximation, whereby the total energy of the molecular system is related to the coordinates of the nuclei. A quantitative description of a Born-Oppenheimer potential energy surface may solve many, if not most, chemical problems. However, there is not an efficient theoretical way to compute such a surface. The alternative developed in recent years is to use empirical models, based on experimental data, to calculate potential energy surfaces, and molecular mechanics is the least computationally intensive method that is able to compute energetic and structural information (Fig. 1.2).

The application of molecular mechanics to metal complexes developed in parallel with its application to organic molecules. Indeed, the earliest report that considered the importance of nonbonded interactions in determining the relative stabilities of isomeric molecules was a 1944 study of the six isomers of the coordination compound [Co($(S)$−pn)$_2$(NO$_2$)$_2$] (pn = propane-1,2-diamine)[1]. A number of similar, though more detailed, studies in the 1950s and 1960s[2−4] led to the full application of molecular mechanics to metal complexes. The common theme

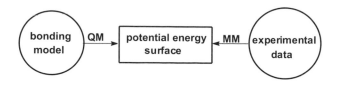

**Figure 1.2:** Computation of the potential energy surface.

in these early studies was the measurement of nonbonded contacts, from physical (Dreiding) models, or their determination by vector analysis. The van der Waals energies of the shortest nonbonded interactions for a series of isomers were calculated and compared, the hypothesis being that the isomer with the fewest short, high-energy contacts would be the most stable. It was realized in this early work that this was a rather crude approach. No account was taken of how other internal coordinates, such as bond angles, might adjust to accommodate and avoid close contacts, nor of the energy cost associated with the deformation of these internal coordinates.

In a series of more advanced studies this problem was partially addressed by systematically adjusting a limited number of internal coordinates (bond lengths, valence angles and torsion angles) to find the geometry of metal chelates with the lowest energy[5-8]. However, due to the computational limitations of the time, the approach was limited in that only a small number of internal coordinates could be adjusted simultaneously. It became clear that methods for calculating the energy costs associated with deforming all of the possible internal coordinates (bond lengths, valence angles, torsional angles, nonbonded contacts), and for finding the geometry with the lowest deformation or strain energy were required.

At that time, the first applications of the currently employed molecular mechanics techniques, to organic molecules, were being reported[9-12]. In particular, the first report of the use of the Newton-Raphson method for strain-energy minimization appeared[13]. Subsequently, new force fields for modeling cobalt(III) complexes were developed and used with the Newton-Raphson method to determine the strain energies and minimum energy geometries of a number of complexes[14-16]. Since then molecular mechanics has been used to model increasingly diverse metal containing systems, and, numerous reviews of these studies have been published (see Appendix 4).

# 2 Molecular Modeling Methods in Brief

## 2.1 Molecular Mechanics

In the early molecular mechanics studies in both inorganic and organic chemistry the strain energy $U_{total}$ is defined as arising from four principle energy terms (Eq. 2.1),

$$U_{total} = \sum_{molecule} (E_b + E_\theta + E_\phi + E_{nb}) \tag{2.1}$$

where $\Sigma E_b$ is the total bond deformation energy, $\Sigma E_\theta$ the total valence angle deformation energy, $\Sigma E_\phi$ the total torsional (or dihedral) angle deformation energy and $\Sigma E_{nb}$ the total nonbonded (van der Waals) interaction energy. The individual energy terms are calculated using simple functions. Bonds are modeled as springs that obey Hooke's law (Eq. 2.2),

$$E_b = \frac{1}{2} k_b (r_{ij} - r_0)^2 \tag{2.2}$$

where $k_b$ is the force constant or spring 'strength' and $r_0$ is the ideal bond length or the length the spring wants to be. Valence angles are modeled in a very similar way (Eq. 2.3),

$$E_\theta = \frac{1}{2} k_\theta (\theta_{ijk} - \theta_0)^2 \tag{2.3}$$

where $k_\theta$ is the strength of the 'spring' holding the angle at its ideal value of $\theta_0$. Torsion or dihedral angles cannot be modeled in the same manner since a periodic function is required (Eq. 2.4),

$$E_\phi = \frac{1}{2} k_\phi (1 + \cos(m(\phi_{ijkl} + \phi_{offset}))) \tag{2.4}$$

where $k_\phi$ is the height of the barrier to rotation about the torsion angle $\phi_{ijkl}$, $m$ is the periodicity and $\phi_{offset}$ is the offset of the minimum energy from a staggered arrangement. Nonbonded interactions are calculated using a function that includes a repulsive and an attractive (London dispersion) component (Eq. 2.5),

$$E_{\text{nb}} = Ae^{-Bd_{ij}} - Cd_{ij}^{-6} \tag{2.5}$$

where $d_{ij}$ is the distance between the two nuclei and $A$, $B$ and $C$ are atom based constants discussed later in this book.

More recently a number of additional components have been added to the calculation of the strain energy. Out-of-plane deformation terms $E_\delta$ have been included in models of aromatic or sp$^2$ hybridized systems (Eq. 2.6),

$$E_\delta = \frac{1}{2} k_\delta \, \delta^2 \tag{2.6}$$

where $\delta$ is the angle between the plane defined by three atoms and the vector from the center of these atoms to a fourth bonded atom, and $k_\delta$ is the corresponding force constant. Modeling the interaction of metal complexes with biological systems has necessitated the inclusion of electrostatic and hydrogen bonding interaction terms. Electrostatic interactions are modeled based on the Coulomb law (Eq. 2.7),

$$E_\varepsilon = \frac{q_i \, q_j}{\varepsilon \, d_{ij}} \tag{2.7}$$

where $q_i$ and $q_j$ are the partial charges on atoms $i$ and $j$, $\varepsilon$ is the dielectric constant and $d_{ij}$ is the interatomic separation. Hydrogen bonding interactions are generally modeled using a function of the type given in (Eq. 2.8),

$$E_{\text{hb}} = Fd_{ij}^{-12} - Gd_{ij}^{-10} \tag{2.8}$$

where $F$ and $G$ are empirically derived constants that reproduce the energy of a hydrogen bond and $d_{ij}$ is the donor-acceptor distance. The addition of these terms gives rise to the revised definition of $U_{\text{total}}$ given in Eq. 2.9.

$$U_{\text{total}} = \sum_{\text{molecule}} (E_{\text{b}} + E_\theta + E_\phi + E_{\text{nb}} + E_\delta + E_\varepsilon + E_{\text{hb}}) \tag{2.9}$$

The set of functions together with the collection of terms that parameterize them ($k_{\text{b}}$, $r_0$, etc.) is referred to as the force field. In some cases force field parameters can be related to experimentally determinable values. For example, the bond stretching force constant $k_{\text{b}}$ is approximately equivalent to the vibrational force constant derived from an infrared spectrum. However, in general the force field terms are derived empirically with the target of reproducing experimental structures and energy distributions.

Once a model and a force field have been chosen for a particular problem, the goal of molecular mechanics is to find the geometry with the minimum strain energy. This can be achieved by a variety of mathematical techniques described elsewhere in this book. The value of the strain energy is dependent on the force field and therefore has little meaning in absolute terms. However, because isomers have the same bond, bond angle and torsional angle types, strain energies of isomers can be compared to each other and differences correlated with experimentally determined isomer populations. This has formed the *raison d'être* of many molecular mechanics studies. In recent studies of more complex systems the primary goal has been to produce reasonable models that allow the investigator to visualize the interactions of metal ions with large molecules. Also, methods where molecular mechanics is used in combination with experimental data to determine molecular structures, for example in solution, are receiving increasing attention.

## 2.2  Quantum Mechanics

Empirical approaches such as molecular mechanics are able to deal with very large systems. With current technologies, a few thousand atoms, extensive conformational searching and long molecular dynamics trajectories can be handled, and these limits are steadily being extended due to the continuous improvement of hardware and software. The limits of empirical methods are that, since they are based on high degrees of parameterization, they are restricted to certain classes of molecules and to specific properties (vibrational spectra, thermodynamic properties and ground state geometries). These limits can be pushed back by the extension of existing force fields and the development of new ones (see Appendix 3); the refinement of generic force fields (see Section 3.3); quantum-mechanically driven molecular mechanics, e.g., for transition states (see Section 3.3); the development of tools that refine parameter sets based on data banks, including genetic algorithms, neural networks or more conventional techniques (see Sections 3.3 and 16.3).

The advantage of ab-initio quantum-mechanical methods is their ability to handle any element of the periodic table and ground states as well as excited and transition states. The cost is a heavy consumption of computing resources and this limits the size of systems that can be treated. These limits can be overcome by using combined QM/MM methods (see Section 3.3) or the thorough investigation of simplified models of the molecular systems of interest, and approximations to simplify ab-initio quantum mechanics, where certain quantities are neglected or replaced by parameters fitted to experimental data.

$$H\Psi = E\Psi \tag{2.10}$$

An important modification of the general Schrödinger equation (Eq. 2.10) is that based on the Born−Oppenheimer approximation[17], which assumes stationary nuclei. Further approximations include the neglect of relativistic effects, where they are less important, and the reduction of the many-electron problem to an effective one-electron problem, i.e., the determination of the energy and movement of each electron in the potential field created by the nuclei and the sum of the other electrons. The one-electron Hamiltonian in the Born-Oppenheimer approximation is restricted to the kinetic energy of the electrons ($K_e$) and their potential energy, with terms describing the electrostatic interactions between electrons and the nuclei ($V_{ne}$), between the electrons themselves ($V_{ee}$) and an electron-electron exchange and correlation term ($V_{ex}$). The latter is essentially responsible for the quality of a particular ab-initio method.

### 2.2.1   Hartree–Fock Calculations

Many ab-initio quantum-chemical methods are based on the Hartree-Fock (HF) approximation and describe electronic structures with sets of doubly occupied and vacant molecular orbitals (MO). Open shell systems need special methods to treat singly occupied orbitals with electrons in α or β spin (UHF or ROHF vs RHF). The MO's are defined as linear combinations of atomic orbitals (LCAO). The number and mathematical description of AO's is known as the basis set, and the orbital energy (and structure) of a molecular system depends on the extent of the basis set. The ground state electronic configuration is described mathematically by a single Slater Determinant (antisymmetrical product of N one-electron wave functions). Therefore, only averaged electron-electron correlation is included, and this is not well suited to transition metal systems. While structural parameters of organic and main group compounds are often reproduced satisfactorily, vibrational frequencies and ionization energies may have systematic errors that can be corrected with empirical scaling factors.

Most of the problems of single determinant HF calculations are due to the inadequate treatment of electron correlation. In post Hartree-Fock methods, this is generally remedied by the explicit inclusion of configuration interaction (CI) or by perturbation theory. Especially popular are perturbation theory expressions developed by Møller and Plesset[18], and the corresponding second-order treatment (MP2) leads to significant improvements. In systems, where multiple configurations are of importance, more sophisticated models are required. These include various self consistent field (SCF) approaches, such as the multi configuration SCF (MCSCF) and the complete active space SCF (CASSCF) methods. Due to the computational expense of these approaches (approximate scaling factors (n nuclei, N electrons): MM, $n^2$; HF, $N^4$; CI, $N^5$–$N^8$) large molecules are often geometry-optimized at a single determinant HF level with subsequent higher-level fixed-geometry computation of the molecular energy.

### 2.2.2   Semi-Empirical Approaches

Ab-initio quantum mechanics calculations are, because of the computational cost, impractical for large transition metal compounds. In semi-empirical methods, some of the quantities of ab-initio calculations are neglected or replaced by parameterized terms, derived from experimentally observed data (structures, ionization energies, heats of formation). A popular range of approaches is based on the neglect of differential overlap (NDO), ranging from complete neglect of differential overlap (CNDO)[19], intermediate neglect of differential overlap (INDO)[20], modified intermediate neglect of differential overlap (MINDO)[21] and neglect of diatomic differential overlap (NDDO)[19]. A number of INDO-based methods have been parameterized and used successfully for transition metal compounds[22–25]. Extended Hückel molecular orbital (EHMO) and Fenske-Hall (FH) approaches are even more approximate but can still provide important insights in the area of transition metal compounds, when applied with care[22,24,26,27].

### 2.2.3   Density Functional Theory

For transition metal systems, DFT methods generally lead to more accurate structures and vibrational energies than single determinant HF methods[22,28], and often they are similar in quality to high-level post-HF methods. Since, in addition, DFT calculations are less computationally expensive (approximate scaling factor: $N^3$) they have become the method of choice for routine applications in the area of transition metal compounds[22,28–30].

The basis of DFT is that the ground state energy of a molecular system is a function of the electron density[31]. The Kohn-Sham equations provide a rigorous theoretical model for the all-electron correlation effects within a one-electron orbital-based scheme[32]. Therefore, DFT is similar to the one-electron HF approach but the exchange-correlation term, $V_{ex}$, is different: in DFT it is created by the functional $E_{xc}(\zeta)$ and in real applications we need approximations for this functional. The quality of DFT calculations depends heavily on the functional. The simplest approximate DFT approach is the $X_\alpha$ method which uses only the exchange part in a local density approximation (LDA, local value of the electron density rather than integration over space)[33,34]. The currently available functionals for approximate DFT calculations can, in most cases, provide excellent accuracy for problems involving transition metal compounds. Therefore, DFT has replaced semi-empirical MO calculations in most areas of inorganic chemistry.

## 2.3   Other Methods

### 2.3.1   Conformational Searching

The analysis of potential energy surfaces may be of importance for both molecular-mechanical and quantum-mechanical computations. However, due to the fact that thousands of structures instead of only one need to be optimized, the methods briefly described here are only routinely used with force field calculations[35].

Grid or deterministic searches that cover the entire conformational space are very time consuming. Less computationally expensive are stochastic or Monte-Carlo searches and molecular dynamics calculations. Other methods not discussed here explicitly are simulated annealing, which mimics the process of cooling a molecule from high temperature, where it adopts many different conformations, to low temperature, where it adopts only one conformation, and genetic algorithms, where the starting geometries evolve towards the global minimum energy structure.

**Stochastic Methods**

Random searching of the conformational space is usually done by randomly changing the most flexible internal coordinates, e.g., dihedral angles or the cartesian coordinates. Using internal coordinates can be more efficient since this reduces the number of degrees of freedom[36,37]. For combinations of ring systems, which often occur in coordination compounds, they may also be more effective. Usually, the Metropolis[38] method is used in Monte Carlo searches, and this involves a weighting based on potential energies. Therefore, Monte Carlo searches are in general not completely random, and an ensemble of structures with a Boltzmann distribution can be obtained. For efficient stochastic searches it is important that the starting structure for the random perturbation is not always the same, and a number of ways are used to select the starting structure for each Monte Carlo cycle.

**Molecular Dynamics**

Molecular dynamics involves the calculation of the time dependent movement of each atom in a molecule[39]. Generally, the forces determined by empirical force fields are used in combination with Newton's laws of motion (Eqs. 2.11, 2.12).

$$F = m \cdot a = -\frac{dE}{dr} = m \frac{d^2r}{dt^2} \tag{2.11}$$

$$\frac{d^2r_i}{dt^2} = a_i = \frac{F_i}{m_i} \; ; \quad F_i = \frac{\delta E}{\delta r_i} \tag{2.12}$$

In contrast to molecular mechanics, in molecular dynamics these forces are not minimized but used to calculate changes in the inter- and intramolecular velocities of the investigated system. The changes in velocities and coordinates with time are recorded in trajectories. The time steps of molecular dynamics runs should be an order of magnitude smaller than that of the largest vibrational frequency found in the system. Since this is often that of the C−H stretching, which is about $10^{14}$ s$^{-1}$, typical time steps are a few femtoseconds. Unfortunately, mainly due to limits in computing power and storage, the length of molecular dynamics simulations is limited, typically to hundreds of picoseconds or nanoseconds at most. It follows that molecular dynamics is often only efficient at exploring local conformational space and is not effective for crossing large energy barriers or for searching globally. To achieve faster and more complete searching high temperatures can be used. However, sampling (and refining) structures during a high temperature molecular dynamics search might lead to relatively large fractions of high-energy and therefore irrelevant structures.

### 2.3.2 Database Searching

Experimental structures are often the basis for computational studies; they are used as input structures for structure optimizations and conformational searches, for the parameterization and validation of force fields and for analyzing the effects of crystal lattices. More than 200,000 experimental structures have been reported, and the majority are found in the Cambridge Structural Data Base (CSD, small molecular structures which include carbon atoms); the Inorganic Crystals Structure Database (ICSD); and the Protein Data Base (PDB; this database includes X-ray as well as optimized structures based on NMR data).

### 2.3.3 Cluster Analysis

Conformational searching and data mining produce large amounts of data which need special techniques for their analysis. A commonly used method is cluster analysis in which all elements that are similar to each other, in terms of a specific property such as a set of torsion angles, are grouped in a cluster. The basis on which the elements of a clusters are separated, how this property is measured quantitatively and what the clustering level is, i.e., how many elements there are in average per cluster, are all important factors. An obvious property for clustering in coordination compounds are chelate ring conformations[40], but it could also be a particular valence angle, a bond length ratio or a symmetry measure[41,42]. An important step is the measure of this property and the grouping of the compounds into different clusters. Methods to quantify the difference between all elements and group them into separate clusters include various rms techniques (cartesian coordinates, torsional angles, as described in Section 2.3.1). The distribution of all elements between various clusters depends on the relative energy of the clusters, and this can be used to compute the over-all isomer distribution.

### 2.3.4 Free Energy Perturbation

In cases where substrates bind selectively to enzymes or catalysts, and in the area of selective host-guest interactions, changes in the free energy are of importance. These are usually not available computationally, except by ab-initio quantum mechanics for very small systems. Free-energy-perturbation techniques are an interesting alternative for large systems, when one assumes that the two systems, for which a free-energy difference is required, can be related by a perturbation describing the mutation of one system into the other. The two systems X and Y, described by the Hamiltonians $H_x$ and $H_y$, are related by the perturbation $\Delta H$, which contributes to the free energy difference between the two systems (Eqs. 2.13, 2.14).

$$H_y = H_x + \Delta H \tag{2.13}$$

$$\Delta(\Delta G) = \Delta G_y - \Delta G_x = -RT \ln\langle \exp\{-\Delta H/RT\}\rangle_x \tag{2.14}$$

The exponential in Eq. 2.14 represents the average over the system described by the hamiltonian $H_x$, and the corresponding series of conformers and configurational isomers is usually created by molecular dynamics or Monte Carlo methods. When the two systems X and Y are very similar, the exponential term vanishes, leading to a very slow convergence of the average in Eq. 2.14. A number of techniques have been described to overcome this problem[43,44]. One of the few applications of this method to coordination compounds is the investigation of $O_2$ and CO affinities to iron porphyrins[45].

### 2.3.5   QSAR

Quantitative structure-activity relationships are primarily used for drug design. The underlying principle is that the shape and noncovalent interactions are the main contributors to the selectivity of the binding of substrates to an active center. Therefore, it must be possible to correlate structural properties of substrates with their activity. The assumptions on which QSAR methods are generally based are that all substrates bind to the same site, that structurally related compounds bind with a similar orientation and that dynamic effects can be ignored.

For a QSAR analysis a training set of compounds with known descriptor properties (e. g. $pK_a$-values, surface areas, dipole moments etc.), including the property of interest, is required. The Hansch Analysis[46] is a statistical method to analyze and correlate these data in order to determine the magnitude of the target property (Eq. 2.15).

$$\log(1/C) = k_1 \log(P) - k_2 (\log P)^2 + k_3 \sigma + k_4 \tag{2.15}$$

Recent developments include 3D QSAR methods which relate regions of the binding site with complementary properties[47]. The conformation of each molecule then needs be computed and the descriptor property determined. Another interesting development is the Electron Topological (ET) approach in QSAR methods[48]. Molecular compounds are described by quadratic matrices ($n^2$, n: number of atoms), where elements close to the diagonal represent electronic parameters while the other elements are related to the structure.

# 3 Parameterization, Approximations and Limitations of Molecular Mechanics

## 3.1 Concepts

The fundamental assumption underlying the molecular mechanics (MM) method is that the positions of the atoms of a molecule, ion, solvate or crystal lattice are determined by forces between pairs of atoms (bonds, van der Waals interactions, hydrogen bonding and electrostatic interactions), groups of three atoms (valence angles) and groups of four atoms (torsional angles, planes; Fig. 3.1).

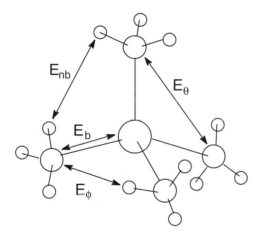

**Figure 3.1:** The molecular mechanics model.

The energies $E_i$ resulting from these forces are related to the positions of the nuclei in a molecule and therefore enforce the entire molecular structure. The energy lost by moving atoms away from their ideal positions is related to the strain or steric energy, $U_{\text{total}}$ (Eq. 3.1) as a function of the nuclear geometry.

$$U_{\text{total}} = \sum_{i=1}^{n} U_i \qquad (3.1)$$

Minimization of the strain energy $U_{\text{total}}$ by rearrangement of the nuclei leads to an optimized structure and a value for the minimized strain energy.

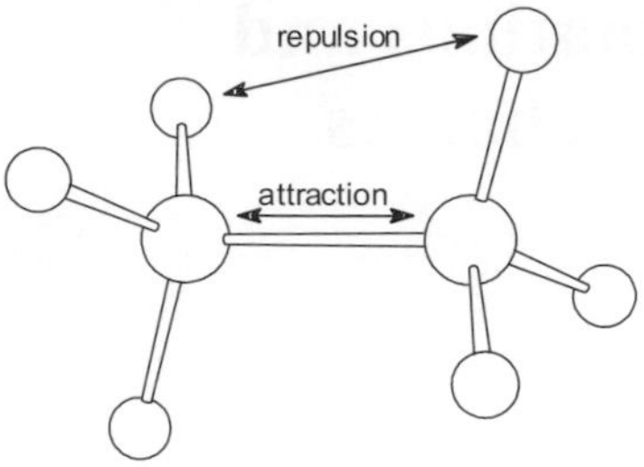

**Figure 3.2:** The relationship between the ideal and observed C–C distance in ethane.

It is important to realize that for any arrangement of more than two atoms the strain energy minimized structure does not have ideal (zero strain) distances and angles. This is demonstrated in the case of ethane (Fig. 3.2), where, due to the repulsion of the protons, the experimentally determined C–C distance in ethane of 1.532 Å, which is well reproduced by empirical force field calculations, is slightly longer than the ideal C–C separation of 1.523 Å used in the MM2 force field[49,50]. Further examples are presented in Table 3.1. With increasing substitution of the carbon atoms the C–C separation increases up to 1.611 Å in tris-*t*-butylmethane.

**Table 3.1:** Experimentally determined and calculated C–C bond distances of some simple alkanes[a].

| Compound | C–C [exp.] [Å] | C–C [calc.] [Å] |
|---|---|---|
| $H_3C–CH_3$ | 1.532 | 1.531 |
| $CH_3CH_2–CH_3$ | 1.534 | 1.534 |
| $(CH_3)_2CH–CH_3$ | 1.535 | 1.537 |
| $(CH_3)_3C–CH_3$ | 1.539 | 1.541 |
| $(CH_3)_3C–C(CH_3)_3$ | 1.582 | 1.574 |
| $((CH_3)_3C)_2–CH(C(CH_3)_3)$ | 1.611 | 1.620 |

[a]  The experimental data are from [49], the calculated values were obtained with HyperChem[51], using the MM+ force field (MM2[49,50]; $k^{CC}$ = 4.4 mdyn/Å; $r_o^{cc}$= 1.523 Å).

Similar effects are observed in coordination compounds. In Table 3.2\*, calculated and experimentally determined Co–N distances for various cobalt(III) hexaamines, including the calculated distance for the hypothetical $[Co(N)_6]^{3+}$ cation, are assembled. Analysis of these data shows that the repulsion of the amine substituents is lowered by an elongation of the cobalt(III)-amine bond. The energy gained by decreasing the repulsive forces is balanced by a concomitant increase in strain energy due to stretching of the cobalt-amine bond. Thus, the resulting structure is a compromise between maximum bonding energy and minimum repulsion. Ignoring inductive effects by alkyl substituents and realizing that even

---

\*  Momec[52] is a molecular mechanics program which was designed specially for coordination and inorganic compounds. It has been adapted for use with HyperChem[51].

**Table 3.2:** Experimentally determined and calculated Co–N bond distances of some cobalt(III)hexa-amines[a].

| Compound | Co–N$^{exp.}$ [Å] | Co–N$^{calc.}$ [Å] | Mean E$_b$ (kJ mol$^{-1}$)[b] |
|---|---|---|---|
| [Co(N)$_6$]$^{3+}$ [c] | – | 1.933 | 0.414 |
| [Co(NH$_3$)$_6$]$^{3+}$ | 1.961 | 1.955 | 1.320 |
| [Co(NH$_2$CH$_3$)$_6$]$^{3+}$ | 2.013 | 1.983 | 3.229 |
| [Co(en)$_3$]$^{3+}$ [d] | 1.956 | 1.963 | 1.355 |
| [Co(tmen)$_3$]$^{3+}$ [e] | 1.994 | 1.971 | 2.338 |

[a] The references for the experimental data are as follows: [Co(NH$_3$)$_6$]$^{3+}$[53] [Co(NH$_2$CH$_3$)$_6$]$^{3+}$[54], [Co(en)$_3$]$^{3+}$[55], [Co(tmen)$_3$]$^{3+}$[56]. The calculated values were obtained with MOMEC[52] using a published force field[57,58]; k$^{CoN}$ = 1.75 mdyn/Å; r$_o^{CoN}$ = 1.905 Å).
[b] Averaged strain energy of the Co-N bonds.
[c] Hypothetical cobalt(III) complex with six bare N-donor atoms.
[d] Most stable conformer; en = ethane-1,2-diamine.
[e] Most stable conformer; tmen = 2,3-dimethylbutane-2,3-diamine.

the hydrogen atoms in ammonia lead to some repulsion (this is nicely shown by the Co-N distances in the calculated structure of the hypothetical [Co(N)$_6$]$^{3+}$ cation), it is clear that ideal distances and angles are not experimentally observable quantities. Generally, just like force constants, values for the ideal distances and angles are derived empirically in order to reproduce experimental data. They depend on the molecular mechanics model used, and each parameter is dependent on all of the others used in the force field.

Two important messages emerge:

- Isolated force field parameters are generally of little value. A force field parameter set is only reliable as a whole and together with a specified set of potential energy functions.
- A molecular mechanics model is not directly related to physical reality. It is best described as an "as if" model since we assume, for example, that the bonds behave "as if" they were springs.

These points will be discussed further in Sections 3.3 and 3.5.

It is common practice to represent the total strain energy, $U_{total}$, of a molecule by a set of potential energy functions, including bonding ($E_b$), valence ($E_\theta$) and torsional angle ($E_\phi$) interactions as well as nonbonded ($E_{nb}$) and electrostatic interactions ($E_\varepsilon$) (see Eq. 3.2).

$$U_{total} = \sum_{molecule} (E_b + E_\theta + E_\phi + E_{nb} + E_\varepsilon + E_\delta + E_{hb}) \qquad (3.2)$$

Additional terms, including out-of-plane interactions, hydrogen bonding, cross terms etc., can also be included (see Section 3.2). The general approach of molecular mechanics, as shown in Figs. 3.1 and 3.3 and discussed above, need not be restricted to "physically reasonable" models. Parameterization schemes range from the representation of angle bending ($E_\theta$) by repulsion (points on a sphere or POS approach)[57,59] and the representation of all interactions by a set of two body central forces[60] to the development of generic force fields based on "first principles"[61]. In principle, any set of potential energy functions that are able to represent a molecule in the way shown in Fig. 3.1 can be established and parameterized on the basis of experimental data (see Fig. 3.3). The resulting force field (functions and parameters) can then be used to predict experimental observables of the type that were used to establish the force field.

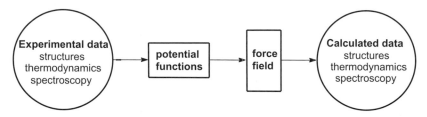

**Figure 3.3:** Development of a force field.

With a well chosen set of functions, a sufficiently high degree of parameterization, and parameter values that are based on high-quality experimental data of one molecule, e. g., a crystal structure, one must, in principle, be able to reproduce exactly this very data set. However, a fundamental goal of molecular modeling is to be able to predict the structures and energetics of unknown molecules. Since any prediction is only meaningful if the functions and force field used are applicable to the type of data and molecules analyzed, the parameterization has to be obtained by fitting a series of similar compounds. This is a typical requirement for a successful interpolation, and molecular mechanics clearly is an interpolative method.

Parameterizing a force field can be approached in two different ways. In order to get a generally applicable force field one can derive the parameters from as large and as widely varied sets of experimental data as possible. This has the obvious advantage that the force field used can be applied to a large variety of compounds without generating new parameters, as long as one stays within the limits of the interpolation. Alternatively, a force field can be derived to study a specific problem. In this case the parameters are based on as large a data set as possible, that can be found for a series of compounds that are similar to the desired compound[62,63]. In this approach the results should be more accurate because the force field is tuned for a very specific class of compounds. The disadvantage is that a new parameter set has to be derived and tested for each new class of compounds studied.

In view of the time and possible errors involved in fitting force field parameters to a large amount of experimental data, the degree of parameterization must be kept to a minimum. Contrary to this is the desire to optimize as large a range of compounds as possible with the same and constant set of force field parameters which requires a moderately high degree of parameterization.

From the discussion above it emerges that the total strain energies, which are often equated with enthalpy terms, are dependent on the molecular mechanics model used, and its associated parameter set. For this reason strain energies are generally taken to have significance only relative to one another. Even with the same model and force field, strain energies are generally of little value in absolute terms, especially since the energies are not normalized in terms of the stoichiometric composition of the molecules*.

Even for isomer and conformer distributions, where stoichiometric factors are not involved, it is questionable whether the calculated distributions based on force fields fitted exclusively with structural data can be expected to be accurate. This is because all parameters in a force field are interdependent and therefore all parameters have to be adjusted if one is changed. It follows that various combinations of parameters can lead to accurate (structural) results. For example, in the structures presented in Table 3.2 an increase in the ligand repulsion might have to be compensated by an accompanying increase in the metal-ligand bonding force constant. Doing this, the potential energy surface becomes steeper and the energy difference between various conformers might increase, leading to differences in the calculated conformational equilibria.

A number of force fields used in organic chemistry have been fitted to thermodynamic data, and most force fields used in inorganic chemistry are based on these. Also, initial guesses for parameters are often based on spectroscopic force constants and therefore related to physically meaningful parameters. Thus, in many cases, reasonably accurate thermodynamic data can be obtained with force fields where parts were fitted to structural data[65,66].

## 3.2 Potential Energy Functions

Conventionally, the model used in molecular mechanics involves the presentation of the total strain energy as a sum of through-bond and through-space terms (Eq. 3.3). The through-bond terms, $E_d$ (Eq. 3.4), are also, though not entirely correctly, called "electronic" or "directional" terms (specific electronic terms will be discussed in Section 3.4). Although all nonbonded interactions generally involve repulsive as well as attractive terms, the through-space or non-directional terms, $E_{nd}$ (Eq. 3.5), are also known as repulsive interactions.

$$U_{total} = \sum_{molecule} (E_d + E_{nd}) \tag{3.3}$$

---

* An exception is the calculation of formation enthalpies for organic compounds based on group increments and strain energies[64].

$$E_{\mathrm{d}} = E_{\mathrm{b}} + E_{\theta} + E_{\phi} + E_{\delta} \tag{3.4}$$

$$E_{\mathrm{nd}} = E_{\mathrm{nb}} + E_{\varepsilon} + E_{\mathrm{hb}} \tag{3.5}$$

The through-bond terms ($E_{\mathrm{d}}$) are comparable with the terms used in vibrational spectroscopy, and they consist of two-body interactions (bonding energy $E_{\mathrm{b}}$), three-body interactions (valence angle energy $E_{\theta}$) and four-body interactions (torsional angle $E_{\phi}$ and out-of-plane energy $E_{\delta}$) as well as additional terms discussed below. The through-space terms ($E_{\mathrm{nd}}$) consist of van der Waals interactions ($E_{\mathrm{nb}}$), electrostatic terms ($E_{\varepsilon}$), hydrogen bonding ($E_{\mathrm{hb}}$) as well as other possible interactions.

A separation of the through-bond interactions ($E_{\mathrm{d}}$) into bond stretching ($E_{\mathrm{b}}$), angle bending ($E_{\theta}$), out-of-plane deformation ($E_{\delta}$), torsional angle rotation ($E_{\phi}$) and other terms is only possible if these terms are not coupled, and this is most likely if the force constants are very different. This is generally true and leads to largely decoupled local oscillators. In cases where this requirement is violated correction terms have to be added. This can be done by the inclusion of cross-terms which will be discussed later.

Computer programs for empirical force field calculations that use other concepts have been tested, and some of these will be discussed in other parts of this book. Among these approaches are one based on a pure central force field model, used for simple organic compounds[60], an equipotential surface force field model, used for carbonyl cluster complexes[67,68], and one that includes ligand field terms in the CLF/MM model, developed for transition metal complexes[69] (CLF = cellular ligand field).

Typical functions for the various potential energy terms are given in Eqs. 3.6–3.12, with the

$$E_{\mathrm{b}} = \frac{1}{2} k_{\mathrm{b}} (r_{ij} - r_0)^2 \tag{3.6}$$

$$E_{\theta} = \frac{1}{2} k_{\theta} (\theta_{ijk} - \theta_0)^2 \tag{3.7}$$

$$E_{\phi} = \frac{1}{2} k_{\phi} (1 + \cos(m(\phi_{ijkl} + \phi_{\mathrm{offset}}))) \tag{3.8}$$

$$E_{\mathrm{nb}} = A e^{-B d_{ij}} - C d_{ij}^{-6} \tag{3.9}$$

$$A = 2014 (\varepsilon_i \varepsilon_j)^{\frac{1}{2}}$$

$$B = \frac{12.50}{vdw_i + vdw_j}$$

$$C = \frac{2.55 (\varepsilon_i \varepsilon_j)^{\frac{1}{2}} (vdw_i + vdw_j)^6}{144}$$

$$E_\delta = \frac{1}{2} k_\delta \, \delta^2 \tag{3.10}$$

$$E_\varepsilon = \frac{q_i \, q_j}{\varepsilon \, d_{ij}} \tag{3.11}$$

$$E_{\mathrm{hb}} = \frac{F}{d_{ij}^{12}} - \frac{G}{d_{ij}^{10}} \tag{3.12}$$

parameters and variables defined as shown in Fig. 3.4 (parameters and variables not defined in Fig. 3.4 will be given in the appropriate sections below).

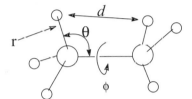

**Figure 3.4:** Parameters used in a molecular mechanics force field.

### 3.2.1   Bond Length Deformation

Bonding in a diatomic molecule can be described by the curve given in Fig. 3.5 which represents the potential energy ($V(r)$) as a function of the bond length ($r$). The bonding force constant, $k$, is given by the second derivative of the potential energy with respect to the structural parameter $r$, and corresponds to the curvature of the potential energy function. The anharmonicity can be described by higher order derivatives.

The experimentally observed parameters $r_e$, $D_e$ and $k_e$ are not directly related to the corresponding parameters $r_o$, $D_o$ and $k_o$, describing the theoretical curve

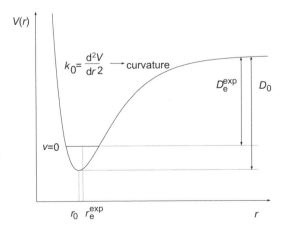

**Figure 3.5:** Bonding in a diatomic molecule.

(Fig. 3.5). For example, the ideal bond distance $r_0$ (minimum of the curve) and $r_e^{exp}$ (center of the zeroth vibrational level) are not identical because of the anharmonicity of the potential energy curve. For diatomic molecules there are acknowledged though complicated procedures for obtaining the theoretical values $r_o$, $D_o$ and $k_o$ from experiment, but for larger molecules this is largely impossible.

In molecular mechanics bond stretching is sometimes modeled using a Morse function (Eq. 3.13)[70] where $\alpha$ describes the curvature and $D$ the depth of the potential function (Fig. 3.6).

$$V(r) = D\left[1 - \exp\left(-\alpha\left(r - r_0\right)\right)\right]^2 - D \tag{3.13}$$

This treats the bond as a mechanical spring whose force constant is strong for small and weak for large interatomic distances. The disadvantage of using a Morse function in empirical force field calculations is that an exponential in addition to the square function and three parameters are involved, increasing the time requirement for the minimization process and the complexity of the force field parameterization.

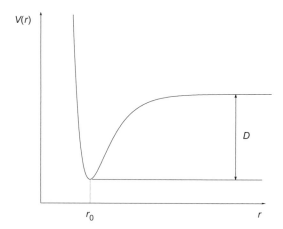

**Figure 3.6:** The Morse function.

Computationally, a Morse function can be mimicked by a Taylor expansion, where the first term (quadratic) describes a harmonic potential and successive higher order terms are included as anharmonic corrections (Eq. 3.14).

$$V(r) = D\left[1 - a^2\left(r - r_0\right)^2 + a^3\left(r - r_0\right)^3 - a^4\left(r - r_0\right)^4 + \cdots\right] \tag{3.14}$$

A harmonic function is a good approximation of the bond stretching function near the energy minimum (Fig. 3.7). Therefore, many programs use this approximation (see Eq. 3.6); however, the limits of the simplification have to be kept in mind in those cases where the anharmonicity becomes important. Harmonic potentials are usually satisfactory for normal coordinate analyses of vibrational spectra because the motions are small[65,71]. Parameters derived from vibrational spec-

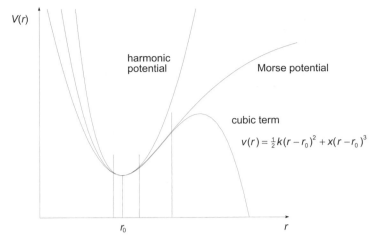

**Figure 3.7:** The bonding function $V(r)$ and its representation by a harmonic potential energy function and an additional cubic term.

tra are usually good starting points for an empirical force field. Apart from the possibility of including cubic terms to model anharmonicity (see second term in Eq. 3.14) which is done in MM2 and MM3 [49,50,72–75], the selective inclusion of 1,3-nonbonded interactions can also be used to add anharmonicity to the total potential energy function.

Bonding of ligands to alkali, alkaline earth and lanthanoid metal ions is mainly electrostatic. Consequently, these bonds can be described as a combination of electrostatic and van der Waals terms (see Chapter 15). Similar approaches have also been used for metallocene compounds (see Chapter 14).

### 3.2.2   Valence Angle Deformation

For organic molecules, valence angle bending, like bond stretching, is usually described by a harmonic potential (Eq. 3.7). Anharmonicity can again be introduced by the addition of higher-order terms of the corresponding Taylor expansion. In MM2 [49] a sixth power correction is used and in MM3 [72–75] all terms up to the sixth power are included. As with corrections to a harmonic bonding potential, these additional terms not only lead to slower optimization processes – which, with the increasingly fast and relatively cheap computers, is not too much of a problem for small molecules – but also to increased numbers of parameters which must be fitted to experimental data.

The fact that molecular mechanics is a well-developed tool for organic molecules, whereas coordination compounds have in the past been modeled less frequently, is partly related to the difficulty in reliably modeling the angles at transition metal centers. In organic compounds $sp^3$, $sp^2$ and sp hybrids lead to relatively stiff angles of $109.5°$, $120°$ and $180°$, respectively, which are conveniently modeled with functions of the type described above. In contrast, the open-shell electronic structures of transition metal centers lead to a variety of coordination geometries that cannot be modeled generally by simple harmonic valence angle functions.* For example, with d-metal ions four-coordinate (tetrahedral, 109.5 angles; square planar, 90° and 180° angles), five-coordinate (trigonal bipyramidal, 90°, 120° and 180° angles) and six-coordinate (octahedral, 90° and 180° angles) species are known, to mention just a few. Theoretically the lanthanoids and actinoids with their partially filled f-orbitals are even more complex, but the electronic effects in such complexes are generally minor.

The d-orbitals of the $e_g$ subset in an octahedral environment ($d_{x^2-y^2}$ and $d_{z^2}$) are directed along the axes of a cartesian coordinate system and, therefore, are more destabilized by σ-bonding ligand orbitals than are the orbitals of the $t_{2g}$ subset ($d_{xy}$, $d_{xz}$, $d_{yz}$), which lie in between the coordinate axes. Therefore, electronic factors dictate ligand-metal-ligand angles of 90° and 180° (i.e., square-planar and octahedral geometries) when strong ligand fields lead to preferential occupation of the $t_{2g}$ d-orbitals. The problem of modeling the angles around the metal in inorganic molecular mechanics is then reduced to two questions:

– How can simple functions with two or more minima (e.g., 90° and 180° for octahedral geometries) be derived?
– How can force constants for these functions be obtained that correctly model the balance between the metal ion preference (e.g. square planar) and the steric preferences of the ligand (e.g., tetrahedral)?

There are rather trivial solutions to the former problem[58]. Fortunately, the electronic influence exerted by the metal ion is most often a relatively small effect which can be added as a minor perturbation to the ligand dictated geometry. As discussed in the *Introduction* (Fig. 1.1) this emerges from experimental structural and spectroscopic data[65].**

There are a number of ways to model the geometry of transition metal centers. One promising treatment is based on the addition of a ligand field term to the strain energy function (Eq. 3.15)[69,77].

$$U_{\text{total}} = \sum_{\text{molecule}} (E_{\text{ligand}} + E_{\text{CLF}} + E_{\text{ML}} + E_{\text{nb}}) \qquad (3.15)$$

In this approach the metal-ligand interaction is modeled with a metal-ligand bonding interaction term, $E_{\text{ML}}$, approximated by a Morse function, a cellular ligand field stabilization energy term, $E_{\text{CLF}}$, (which is responsible for the coordination geometry) and a van der Waals term, $E_{\text{nb}}$, (ligand-ligand nonbonded interaction). It is necessary to use the Morse function, despite the associated problems (i.e.,

---

* Modeling of specific electronic effects due to the partly filled d-orbitals is discussed in Chapter 12.
** For simplicity, the discussion here is restricted to σ-bonding. Some aspects of π-bonding, including modeling of organometallic compounds, where stereoelectronic effects are more complex, are discussed in Chapters 12 and 14.

the number of parameters and exponentials required as discussed above), because high-spin and low-spin states must be modeled using a single function. The CLF/ MM approach has been applied to a number of problems involving transition metal coordination compounds, and the main highlights are that a single force field can be used for the accurate computation of high-spin and low-spin nickel(II) compounds, and no specific functions are necessary for Jahn-Teller distorted copper(II) compounds[77].

A very promising approach to modeling angular geometries, the VALBOND model[78,79], is based on Paulings 1931 paper[80] that established the fundamental principles of directed covalent bonds formed by hybridization. The VALBOND force field, which uses conventional terms for bond stretching, torsions, improper torsions and nonbonded interactions requires only two parameters per atom-pair for the angle bending potential. The hybridization of a compound is derived from its Lewis structure and the two required parameters, a hybridization weighting factor accounting for the hybridization preference of different substituents to the central atom and a scaling factor for the orbital overlap, are fitted to experimental data. The advantage of the VALBOND approach is that only few parameters are required since the two parameters used are atom based and therefore independent of the functional group. The accuracy of calculated structures and vibrational frequencies of organic and main group compounds (normal-valent molecules of the p-block) produced with the VALBOND force field is comparable with that of conventional force fields[78]. The VALBOND approach has also been applied successfully to hypervalent molecules of the p-block[81] and to transition metal hydrides and alkyls[82]. The obvious advantage of the VALBOND model – treatment of molecular shapes with a variable angle bending function defined by simple rules based on electronic structures – holds promise for the future.

A number of functions with multiple minima have been proposed for modeling the valence angles around metals. The molecular mechanics program DREIDING, based on a generic force field, uses a harmonic cosine function (Eq. 3.16)[83].

$$E_\theta = \frac{1}{2} k_\theta^c (\cos \theta_{ijk} - \cos \theta_0)^2 \tag{3.16}$$

The force constant $k_\theta^c$ of the harmonic cosine function is related to the force constant $k_\theta^h$ of the harmonic oscillator (Eq. 3.7) via Eq. 3.17.

$$k_\theta^c = \frac{k_\theta^h}{(\sin \theta_0)^2} \tag{3.17}$$

For linear geometries, DREIDING uses the angle bending function of Eq. 3.18.

$$E_\theta = k_\theta' (1 + \cos \theta_{ijk}) \tag{3.18}$$

The program and force field SHAPES, developed for transition metal complexes and tested for square-planar geometries, uses a single Fourier term (Eq.

3.19), which is similar to the torsional angle term in many molecular mechanics programs (see Section 3.2.3; periodicity = $m$, phase shift = $\psi$)[84].

$$E_\theta = k_\theta^F \left(l + \cos\left(m\theta_{ijk} + \Psi\right)\right) \tag{3.19}$$

The Fourier force constant $k_\theta^F$ is related to that of the harmonic potential ($k_\theta^h$, Eq. 3.7) by Eq. 3.20.

$$k_\theta^F = \frac{2\,k_\theta^h}{m^2} \tag{3.20}$$

where $m = \dfrac{\pi}{\pi - m\theta_0}$

Another approach for describing the geometry around metal centers is to use a harmonic sine function[52,58]. This function has minima at 0°, 90° and 180°, and based on Eq. 3.21 and 3.7 $k_\theta^s = k_\theta^h$.

$$E_\theta = \frac{1}{8}\,k_\theta^s \left(\sin\left(2\theta_{ijk}\right)\right)^2 \tag{3.21}$$

There is no need for a phase shift parameter if the deviation from orthogonality is modeled by a ligand-ligand repulsive term (see below). A similar approach is the use of a multiple harmonic function[52] and this has the advantage that it can easily be extended to other geometries such as trigonal bipyramids.

A quite different approach is to abandon all terms associated with deformation of angles about the metal center and include, instead, repulsions between the ligand donor atoms. The ligand-ligand-repulsion approach is based on the observation that the ligand-metal-ligand angles are generally easily deformed and require only small force constants if a harmonic potential is used. Thus, the coordination geometry is primarily dependent on the metal-ligand bond lengths, the ligand geometry and ligand-ligand repulsion. These ideas resulted in the conceptually simple ligand-ligand-repulsion based model[85−88] (Fig. 3.8). In this approach, the arrangement of ligands with fixed bite angles $\alpha$ or bite distances $b$ on a sphere around the metal ion with a fixed radius M-L is calculated by minimizing the ligand-ligand repulsion potential $U_r$ (Eq. 3.22).

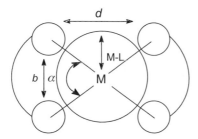

**Figure 3.8:** The ligand–ligand repulsion model.

$$U_r = f\left(\frac{l}{d^n}\right), \ (n \sim 6) \tag{3.22}$$

In the pure repulsion model the metal-ligand distances and ligand geometries are held constant, and the desire of the metal center in terms of bond directions, i.e., electronic effects, is neglected. The problems associated with this model are that the metal-ligand distances are fixed for mixed ligand complexes, where different bond lengths are obtained, and that rigid ligand geometries are an oversimplification for very strained molecules. However, the main limitation of this model is that the geometry of the ligand must be defined beforehand (bite angle $\alpha$, M-L distance). The model can therefore not be used for ligand design. Despite these problems, coordination geometries have been calculated rather accurately with the repulsion approach, indicating that the neglect of electronic effects in metal-ligand bond directionality is often an acceptable approach.

The success of the ligand-ligand-repulsion model prompted its adoption as an element of molecular mechanics.* In the resulting approach the valence angles around the metal ion are modeled solely by nonbonded interactions, using the usual van der Waals potential (i.e., Eq. 3.9 and $k_\theta = 0$ in Eq. 3.7)[59]. Again, the fact that the electronic effects responsible for the directionality of bonds are not explicitly modeled here may seem questionable but extensive tests have shown the model to be reliable[89]. An explanation for this apparent contradiction is that the electronic preferences of the metal ion are indirectly included in the bonding potential. A strong bond (strong ligand field) necessarily leads to relatively short metal-ligand distances and hence to strong ligand-ligand repulsion. This repulsion, together with the ligand constraints, defines the coordination geometry. For example, for hexacoordinate complexes with strong ligand fields and short bonds, such as for cobalt(III), the strong repulsion mimics the electronic preference of the metal center for an octahedral geometry.

An advantage of modeling the coordination sphere of transition metal ions by ligand-ligand nonbonded effects is that the parameterization of any new metal-ligand fragment is simplified because only the parameters for the bonding potential energy function have to be developed. Furthermore, by adding repulsive forces around the metal to the total energy, shortening of metal-ligand bonds is penalized more strongly than is the equivalent lengthening, which adds some anharmonicity to the potential.

Applications, where the accurate angular geometry of the chromophore is of importance (angular overlap model (AOM) calculations of ligand-field spectra, see Chapter 10) have uncovered the limits of a purely ligand-ligand repulsion based approach[90]. The problem of neglecting the electronic contributions to the metal-ligand bond directionality was solved by the addition of an angle bending term (Eq. 3.21), and tuning the contributions of repulsion (Eq. 3.9) and bond directionality (Eq. 3.21) by a ligand field dependent parameter[52,58].**

---

\* Generally, geminal effects are only modeled via valence angle terms, i.e., nonbonded interactions are only considered for 1,4- and higher interactions.

\*\* More details and examples of this approach are discussed in Chapter 12.

An additional problem probably arises from the use of standard nonbonded terms alone to model the repulsion between ligand donor atoms because at longer distances these terms become attractive. There is no doubt that these interactions occur but there may be an additional contribution to the repulsion arising from repulsion between electrons in the bonds (the electron pairs in the VSEPR model). This has led to the development of a VSEPR/POS or Gillespie-Kepert MM approach [91].

As discussed before, the coordination geometry depends largely on the strain induced by the coordinated ligands. Therefore, interligand angles are less well defined than intraligand angles. Accurate coordination geometries of complexes with monodentate ligands are therefore more difficult to obtain.

### 3.2.3 Torsion Angle Deformation

Torsional rotations about single bonds and multiple bonds are different processes. In a multiple bond a torsional rotation results in the transformation of one configurational isomer into another. In contrast, the rotation about single bonds leads to interconversion of conformational isomers (Fig. 3.9).

In both cases, repulsion of the substituents is modeled by van der Waals interactions (see below) and the torsional potential describes the additional electronic component, including distortion of the molecular orbitals and repulsion by the electron clouds.

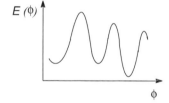

**Figure 3.9:** Rotation around double and single bonds.

It is common practice to describe torsional rotations about single bonds and those about multiple bonds with the same type of potential function but with very different force constants. The function must be able to describe multiple minima. Generally, a Fourier expansion of the torsional angle $\phi$ with only cosine terms is used (Eq. 3.23),

$$E_\phi = \sum_n \frac{1}{2} k_n \left(1 + \cos\left(m_n \left(\phi - \phi_{\text{offset}}\right)\right)\right) \tag{3.23}$$

where $m_n$ is the multiplicity and $\phi_{\text{offset}}$ is the phase shift (offset angle), which only has to be considered if, as often is the case, only one cosine term is included. Some examples of simple functions with common rotors are given in Table 3.3.

**Table 3.3:** Examples of potential energy functions for common rotors, using a single Fourier term (Eq. 3.23).

| Rotor | Example | Potential function |
|---|---|---|
| $C_{3v}-C_{3v}$ | $CH_3-CH_3$ | $E_\phi = \frac{1}{2} k_3 \left(1 + \cos 3\phi\right)$ |
| $C_{3v}-C_{1v}$ | $CH_3-OH$ | $E_\phi = \frac{1}{2} k_3 \left(1 + \cos 3\phi\right)$ |
| $C_{3v}-C_{2v}$ | $CH_3-NO_2$ | $E_\phi = \frac{1}{2} k_6 \left(1 + \cos\left(6\left(\phi + 30°\right)\right)\right)$ |
| $C_{2v}-C_{2v}$ | $CH_2=CH_2$ | $E_\phi = \frac{1}{2} k_2 \left(1 + \cos\left(2\left(\phi + 90°\right)\right)\right)$ |
| $C_{3v}-C_{4v}$ | $M^{oct}-LR_3{}^{a)}$ | $E_\phi = \frac{1}{2} k_{12} \left(1 + \cos 12\phi\right)$ |

### 3.2.4 Cross-terms

As mentioned earlier, the separation of bonding effects ($E_d$ in Eq. 3.3) into bond length, valence angle and torsion angle potentials is based on the assumption that these terms are only weakly coupled. Cross-terms can take care of exceptions to this approximation (see also Section 3.4). Generally, the distances $r_1$, and $r_2$ in Fig. 3.10 increase with a decreasing angle $\theta$ in large part because of repulsions between the atoms and, therefore, the bond lengths and angles are not independent.

In MM2 and MM3 [49,50, 72−75] stretch-bend cross terms are used (Eq. 3.24), but in many other programs this is neglected.

$$E_{\text{st/b}} = 2.51124 \left[k_{\text{sb}} \left(\theta - \theta_0\right)\left(\left(r_1 - r_{01}\right) + \left(r_2 - r_{02}\right)\right)\right] \tag{3.24}$$

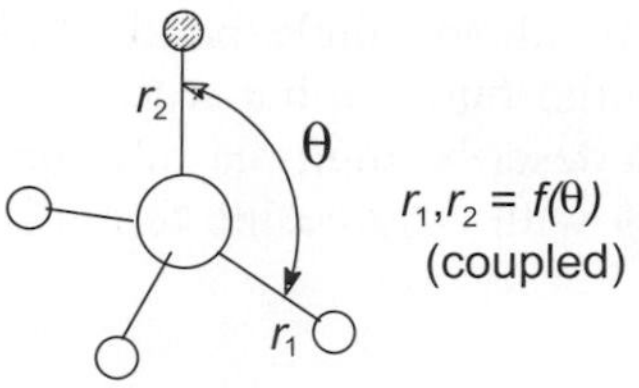

$r_1, r_2 = f(\theta)$
(coupled)

**Figure 3.10:** Coupling of bond length and bond angle potentials.

In MM3 a cross term involving torsional angles is also used[75] but in other programs it is neglected (exceptions are anomeric effects, e. g., with sugars). In other programs[52,58], 1,3-interactions can be selectively included, e. g., to model the geometry of the chromophore of metal complexes. In these cases, an additional correction via a stretch-bend cross term is probably redundant.

### 3.2.5   van der Waals Interactions

Various interactions, such as those of permanent electric dipoles, permanent multipoles and short lived multipoles, are assembled in the van der Waals term. The latter are responsible for some attraction (London dispersion) which is opposed by repulsion when the two nuclei approach one another. At distances below the van der Waals radius (defined here as the distance corresponding to maximum stabilization), the driving force emerging from the van der Waals potential (Fig. 3.11) is such that the atoms repel each other, above it, they attract each other. The significance of negative energy values resulting from a van der Waals term is that they contribute to the stability of a molecule. Therefore, even at interatomic distances smaller than the van der Waals radius, where repulsion becomes appreciable, the net effect might be that the molecule is stabilized by the van der Waals interactions.

The repulsion increases exponentially, and it is steeper than the bond length deformation potential. The attractive force is usually modeled by a $1/r^6$ term while

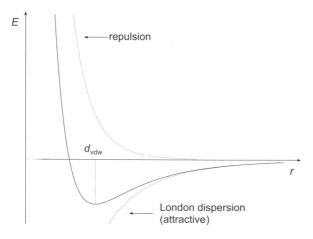

**Figure 3.11:** The van der Waals interaction.

various possibilities exist for the repulsion. The functions used in modern programs include, apart from the Morse potential (Eq. 3.13) the Lennard-Jones potential (Eq. 3.25)[92] (e. g., AMBER[93,94]), the Buckingham potential (Eq. 3.26)[95] (e. g., MOMEC[52,96]), or a modification thereof, the Hill potential (Eq. 3.27) (e. g., MM2, MM3[49,50,73]).

$$E_{\mathrm{vdw}} = \frac{A}{d_{ij}^n} - \frac{C}{d_{ij}^6} \ (\mathrm{n} \sim 12) \tag{3.25}$$

$$E_{\mathrm{vdw}} = A\mathrm{e}^{-Br} - \frac{C}{d_{ij}^6} \tag{3.26}$$

$$E_{\mathrm{vdw}} = k_{\mathrm{vdw}} \left( 2.9 \cdot 10^5 \exp\left(-12.5 \frac{d_{ij}}{d_{ij}^o}\right) - 2.25 \left(\frac{d_{ij}^o}{d_{ij}}\right)^6 \right) \tag{3.27}$$

The Lennard-Jones potential is simpler than the Buckingham potential since it has two rather than three parameters. Computations involving the Lennard-Jones potential are also faster as they do not involve any exponential terms. However, with the performance of the computers currently available the Buckingham potential, which gives a better description of short-range interactions, may be preferred.

The equations can be simplified by reducing the number of parameters required. This is demonstrated for the example of the Buckingham function (Eq. 3.28),

$$E_{\mathrm{vdw}} = \frac{\varepsilon}{1 - \left(\frac{6}{\alpha}\right)} \cdot \frac{6}{\alpha} \cdot \exp\alpha\left\{ \left(1 - 12.5 \frac{d}{d'}\right) \right\} - \left(\frac{d'}{d}\right)^6 \tag{3.28}$$

where $d'$ is the minimum of the function ($d_{\mathrm{vdw}}$, see Fig. 3.11) which is assumed to be the sum of the van der Waals radii of the two atoms concerned ($vdw_i + vdw_j$), $\varepsilon$ is the work that is needed to separate the pair of atoms to infinity, with $\varepsilon_{ij} = (\varepsilon_i \cdot \varepsilon_j)^{1/2}$, and $\alpha$ controls the steepness of the exponential repulsion term (see also Eq. 3.9). This approach only needs two parameters per atom type $i$ ($vdw_i$ and $\varepsilon_i$) and the global parameter $\alpha$, instead of three parameters ($A,B,C$) per atom pair type. van der Waals interactions involving the metal center are often omitted, usually with little obvious effect[65,97]. However, in some cases such interactions are clearly essential for the establishment of reasonable models[98,99]. This is particularly true for square-planar complexes, where the metal center is sterically accessible.

## 3.2.6  Electrostatic Interactions

In molecules with hetero atoms there is always some charge separation. In these cases, and more importantly with compounds that include charged metal ions, electrostatic effects may be critical. The problem is that, apart from some recent and, at least in the area of coordination chemistry, not yet fully tested methods there are no simple *and* accurate methods for calculating partial charges empirically. The methods used for the computation of charge distributions include ab-initio[100,101] and semi-empirical methods[102,103], and the charge equilibration approach which calculates charges analytically from the geometry of the molecule and a number of atomic parameters[104]. As quantum-mechanical methods for metal complexes become more readily available and more reliable, the determination of charges will become more routine. However, molecular mechanics will always be faster than quantum mechanics and, therefore, efficient semi-empirical methods for the determination of point charges are of particular interest[103], especially if the structural dependence is to be taken into account by recalculation of the charges in each cycle. Improvements in the way charges are extracted from quantum-mechanical models are also taking place, with ESP charges now preferred to Mulliken analysis[94].

One approach is to not include explicit electrostatic terms in the calculations. This does not mean that electrostatic interactions are neglected, but that they are absorbed into other functions. For example, the electrostatic attraction between say a metal ion and a directly coordinated anion is simply modeled as part of the bonding interaction between these two species. It is reasonable to not separate a bond into two types of interactions, especially since the amount of ionic character is unknown and is therefore difficult to parameterize.

Electrostatic interactions between nonbonded atoms are more of a problem, since they have a far reaching effect, decreasing only with $1/r$. Therefore, a parameterization scheme that uses only van der Waals terms, which decrease more rapidly might lead to inaccuracies. Since accurate charges are not readily available, electrostatic terms have been neglected in many calculations. The resulting inaccuracies are minimal, at least with relatively small molecules, where accurate parameterization schemes are available through extensive fitting. However, when modeling the interactions of metals with biomolecules such as proteins or DNA, the electrostatic effects can be dominant and they need to be included using reliable methods.

Most programs are able to include electrostatic effects through the function defined in Eq. 3.29. Other programs (e. g., MM2[50]) allow the computation of electrostatic interactions through dipole-dipole interactions (Fig. 3.12; Eq. 3.30).

$$E_\varepsilon = \frac{q_i\, q_j}{\varepsilon\, d_{ij}} \tag{3.29}$$

$$E_\mu = \frac{\mu_i\, \mu_j}{\varepsilon\, d_{ij}^3} \left(\cos\gamma - 3\cos\alpha_i\,\alpha_j\right) \tag{3.30}$$

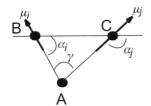

**Figure 3.12:** Dipole–dipole interactions.

### 3.2.7 Hydrogen Bonding Interactions

The function most commonly used for hydrogen bonding interactions is a two-parameter function with a repulsive term that decreases with $d^{-12}$ and an attractive term that increases with $d^{10}$ (Eq. 3.31).

$$E_{hb} = \frac{F}{d_{ij}^{12}} - \frac{G}{d_{ij}^{10}} \tag{3.31}$$

In some programs hydrogen bonding is not treated explicitly, but is included in the van der Waals and electrostatic terms.

There is some debate over whether or not angular terms should be included in the hydrogen bonding potential. The three deviations from linearity that have been discussed involve the angle around the hydrogen atom (D-H···A; $\theta$), the angle around the hydrogen acceptor (H···A-R; $\chi$, see Fig. 3.13) and the out-of-plane deformation around the acceptor A (improper torsion $\omega$, involving A, H and two selected atoms in the substituent R of A)[105]. The corresponding potential energy function is given in Eq. 3.32.

$$E_{hb} = \left( \frac{F}{d^{12}} - \frac{G}{d^{10}} \right) \cos^k (\theta) \cos^m (\chi - \chi_o) \cos^n (\omega - \omega_o) \tag{3.32}$$

In an extensive study of published crystal structures the average values of $\theta$ were, as expected, found to be 180°, while $\chi$ and $\omega$ depended on the hybridization of the hydrogen bond acceptor[105]. For example, $\chi$ was found to be 135° for carbonyl, carboxyl and sulfonamide acceptors, 109.5° for sp$^3$ and 120° for sp$^2$ hydroxyl acceptors, 126° for imidazole and 120° for pyrimidine type acceptors. These angular effects might imply some lone pair directionality in hydrogen bonds. However, the experimental observation that some bending occurs might simply be the result of repulsion of the substituents of the donor and acceptor, and does not necessarily imply that the hydrogen bond is other than isotropic (Fig. 3.13).

An important observation is that experimental structures can be reproduced satisfactorily with and without the angular terms involved in the hydrogen bonding interaction[105]. Thus, irrespective of the fundamental reason for the deviation from linearity the angular dependence can usually be modeled accurately either

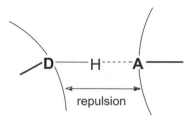

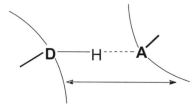

**Figure 3.13:** The angular dependence of hydrogen bonding.

implicitly by van der Waals repulsion or explicitly by including terms that mimic lone-pair directionality.

### 3.2.8   Out-of-Plane Deformation

The out-of-plane terms (Eq. 3.33) are used for $sp^2$ hybridized and aromatic systems such as carboxylates and phenyl groups. In metal compounds, out-of-plane terms are sometimes used for the modeling of square planar complexes (see Fig. 3.15, Section 3.6 below).

$$E_\delta = \frac{1}{2} k_\delta \, \delta^2 \tag{3.33}$$

## 3.3   Force Field Parameters

In order to set up a molecular mechanics model it is necessary to find mathematical expressions that are able to define the molecular structures and give the corresponding strain energies, and to find parameter values for these expressions so that the model can reproduce or predict the molecular structures and properties. These two parts of the molecular mechanics package have a direct influence on the optimized structure. The potential energy functions and the force field parameters are interrelated. Therefore, the parameters should not, in general, be transferred from one force field to another.

One of the primary applications of molecular mechanics is the prediction of structures of new molecules. The strategy adopted, when using molecular mechanics, is to parameterize a set of potential energy functions based on experimentally available data and to use this combination of functions and parameters (the force field) to predict the structure of new molecules. Ideally, the new molecules to be investigated should be similar to those used in the parameterization process. This not only applies to stoichiometries and connectivities but also, especially for metal complexes, to oxidation and electronic states. When parameterizing a force field, examples of both unstrained and highly strained molecules should be considered. For instance, the cobalt(III)-amine bond in hexaammine complexes should be parameterized with structures involving very short and very long Co-N bonds. Furthermore, the fact that the parameters for an M-N bond might differ depending on the other ligands cannot always be neglected. For instance an $M-N_{(pyridine)}$ bond could be affected by the presence of a $\pi$-acid or a $\pi$-base in the *trans* position[57,65]. In such cases the development of a *general* force field is difficult.

The parameterization of a force field can be based on any type of experimental data that is directly related to the results available from molecular mechanics calculations, i.e., structures, nuclear vibrations or strain energies. Most of the force fields available, and this certainly is true for force fields used in coordination chemistry, are, at least partially, based on structural data. The Consistent Force Field (CFF)[97,106,107] is an example of a parameterization scheme where experimentally derived thermodynamic data (e.g., heats of formation) have been used to tune the force field. Such data is not readily available for large organic compounds or for coordination complexes. Also, spectroscopic data have only rarely been used for tuning of inorganic force field parameters[13,74,108].

In theory, a properly developed force field should be able to reproduce structures, strain energies, and vibrations with similar accuracies since the three properties are interrelated. However, structures are dependent on the nuclear coordinates (position of the energy minima), relative strain energies depend on the steepness of the overall potential (first derivative), and nuclear vibrations are related to the curvature of the potential energy surface (second derivative). Thus, force fields used successfully for structural predictions might not be satisfactory for conformational analyses or prediction of vibrational spectra and vice versa. The only way to overcome this problem is to include the appropriate type of data in the parameterization process[107,109].

Most force fields used in coordination chemistry, in respect of the organic part of the molecules, are based on or are at least similar to the MM2[50], MM3[72–75] or AMBER[93,94,110,111] parameterization schemes or mixtures thereof. However, it is important to stress again that transferring parameters from one force field to another without appropriate checks is not valid. This is not only a question of the different potential energy functions that can be used, but is also a consequence of the interrelatedness of the entire set of parameters. Force field parameters imported from any source, whether a well established force field or experimental data should only be used as a starting point for further parameter refinement.

An important point that needs to be considered during the development of force fields used for coordination compounds is that, upon coordination of an organic molecule to a metal center, the bonds from the ligating atoms to their substituents are weakened through delocalization of electron density to the metal center. Therefore, force fields for organic ligands obtained from tuning with coordination compounds are *not* applicable to the metal-free ligands, i.e., to organic molecules[66,96].

The implementation of molecular mechanics of coordination compounds in existing modeling packages, particularly those designed for organic molecules, is only possible if

- there is the possibility of having more than four atoms attached to an atom,
- the coordination geometry can be modeled via ligand-ligand repulsion and/or a function with multiple minima or via substructures[112],
- new atom types (metal centers and ligand atoms) can readily be added and
- force field parameters can be changed and new ones added.

Programs that have been used for molecular mechanics modeling of coordination compounds and for which appropriate parameterization schemes exist include various versions of MM2[43–47] and MM3[118,119], the Consistent Force Field program (CFF)[97], various modifications of Boyd's program[15,52,57–59,120,121], SHAPES[84], HyperChem[122] and modified versions of MacroModel, BIOGRAPH, CHARMM, AMBER and MMPP[123]. The most extensive parameterization scheme is available with MOMEC[52], which employs pure ligand-ligand repulsion[57] or a combination of ligand-ligand repulsion with an angle function representing electronic effects[58]. The force field has been parameterized for coordination geometries for most of the first row and several second and third row transition metal ions and the lanthanoids[124], with a large number of common ligand systems[57,96].

A different and relatively new approach toward molecular mechanics modeling is the use of generic force fields. The driving force behind the development of these force fields is the fact that in molecular mechanics similar atoms in different environments need different force field parameters, resulting in large sets of parameters which are difficult and time consuming to generate. Generic force fields use simple algorithms that generate the parameters needed from atom-based parameters and the molecule's connectivity. Thus, once an atom type has been parameterized it can be modeled in any environment. A similar approach was discussed above in Section 3.2.5 on van der Waals interactions, where it was shown that general atom based parameters can be used successfully in the parameterization of the Buckingham and other similar potential functions.

The DREIDING force field, which has been developed for a range of small organic molecules, including the most common hetero atoms, and for main group inorganic compounds, uses a conventional set of potential energy functions with a set of parameters that are based on very simple rules[83]. The hybridization and the atomic number together fully define the atom type (to date the definition of the oxidation state and electronic structure of transition metal ions are not included). The parameters used are either derived from structural data of reference molecules (bond radii and angles), taken from the literature (van der Waals parameters, charges) or based on data fitting. Only a single force constant is used for bond stretching and angle bending, and six values are defined for torsional barriers. As might be expected, the accuracy of the model is inferior to those with more specialized force fields. However, new and unknown types of molecules can be considered with the same accuracy. Thus, similarly to model kits, DREIDING is an excellent tool for building new molecules that can then be optimized using more specialized force fields. Generic force fields are useful alternatives in situations where there are not enough experimental data to parameterize a force field for a given type of molecule. An attractive alternative is the parameterization based on structures and properties of molecules that have been modeled using quantum-mechanical methods. Note, that whether an experimentally or computationally derived basis set is used for the force field derivation, it should be as large and widely varied as possible.

The Universal Force Field (UFF) is a similar development[61]. Again, a conventional set of potential energy functions is used (angle bending is computed via a cosine Fourier expansion of the angle $\theta$) with atom based parameters. The force field has been extended over the entire Periodic Table, and has been derived from published atomic parameters and relatively simple rules found in the literature. The UFF has been tested for organic[125], main group[126] and transition metal compounds[127]. Again, the accuracy obtained with the UFF is respectable but inferior to that of conventional parameterization schemes, especially for main group inorganic, organometallic and transition metal coordination compounds. The future will show whether generic force fields will eventually lead to predictions of similar quality to those obtained with conventional parameterization schemes. However, at least at present, it seems that the original molecular mechanics philosophy, i.e., the computation of the coordinates of a potential energy surface based on *experimental* data of a series of similar molecules (interpolation) is less general but more reliable. The VALBOND force field[78,79] discussed in Section 3.2.2 uses a conventional parameterization for all terms except for angle bending, where general but fitted atom-pair-based parameters are used. Thus this model combines some of the advantages of conventional force fields with those of generic parameterization schemes.

What else can be done if no experimental data is available? This question arises for excited states, transition states and, to some extent, for fluxional molecules. This is clearly an important question, e.g., in the area of asymmetric catalysis, where the design of new catalysts leading to high enantiomeric excesses is a rewarding aim or where certain mechanistic aspects might be investigated. Ab-initio MO or DFT calculations can lead to accurate results but they are time consuming. Therefore, one promising approach is to calculate a number of key structures by thorough quantum-mechanical methods and then to use them to parameterize molecular mechanics force fields[128,129]. Alternatively, ab-initio quantum-mechanical techniques, used to refine the geometry around the metal center, can be directly coupled with the molecular mechanics refinement of the ligand structure. Substantial progress has already been made using this method[128,129] and it can be expected to become the basis of many future developments in molecular mechanics.

### 3.3.1  Bond Length Deformation

The function used to calculate the deformation energy associated with stretching or contracting a bond is given in Eq. 3.6. A first approximation to the force constant, $k_b$, can be calculated from the fundamental vibration frequency, $v$, of the X–Y bond, taken from the infrared spectrum of a representative compound by using Eq. 3.34, where $c$ is the speed of light and $\mu$ is the reduced mass of the two atoms involved in the bond.

$$v = \left\{ (k_b/\mu)^{1/2} \right\} / 2 \pi c \tag{3.34}$$

More accurate force constants for a number of transition metal complexes with ammine ligands have been derived by normal-coordinate analyses of infrared spectra[130,131]. The fundamental difference between spectroscopic and molecular mechanics force constants (see Section 3.4) leads to the expectation that some empirical adjustment of the force constants may be necessary even when these force constants have been derived by full normal-coordinate analyses of the infrared data. This is even more important for force constants associated with valence angle deformation (see below). It is unusual for bond-length deformation terms to be altered substantially from the spectroscopically derived values.

Preliminary values for ideal (undeformed) bond lengths are usually taken from crystal structures. However, the bonds in real compounds are necessarily deformed to some extent by the stresses present in all but diatomic molecules (see Section 3.1). There is rarely a substantial difference between ideal and observed bond lengths within ligands but metal-ligand bond distances can be substantially altered by the effects of steric stress. For example the Co–N bond length in a series of simple hexaaminecobalt(III) complexes varies from 1.94 to 2.05 Å[132,133] (see also Table 3.2). The ideal bond length used in molecular mechanics models of hexaaminecobalt(III) complexes is most commonly in the range 1.905 to 1.925 Å (see also Table 3.4), implying that all of these observed bonds are elongated with respect to the strain-free value.

Values for the undeformed M–L bond are arrived at empirically: a value slightly shorter than that observed in a relevant crystal structure is chosen. If, after strain energy minimization the experimental bond length is not reproduced then $r_0$ is altered and minimization repeated. Once the experimental bond length has been reproduced, the force field parameters are tested on a series of related compounds and small changes are made until a consensus value is obtained. If this proves to be impossible then the possibility of the force constant being incorrect needs to be considered as does the possibility that factors other than steric stresses are influencing the experimental bond lengths (see Chapter 12 and Sections 17.10–17.12).

It is important when carrying out an empirical development of the force field parameters in this way to model a large range of structures, i.e., those with bond lengths at the short end of the range and those with bond lengths at the long end. The value used for the strain-free bond length depends on the type of force field employed. For instance, if nonbonded interactions between the donor atoms are included in the model, then the stress on the M-L bond increases and it is generally necessary to use a lower value for $r_0$.

Care also needs to be taken in choosing values for the undeformed bond lengths in the ligand. Values of bond lengths and angles are often substantially altered as the result of coordination to a metal[96]. For example, a non-coordinated amine will have C–N bond lengths in the range 1.45–1.46 Å and C–C bond lengths in the range 1.52–1.53 Å. On coordination, C–N bond lengths, where the N is directly coordinated to the metal, typically increase to 1.49 Å and adjacent C–C bond lengths contract to 1.50 Å. For these reasons, it is often inappropriate to adopt the values used for ideal bond lengths in force fields for organic compounds for use in the modeling of metal complexes. Equally, force fields developed for transition metal compounds can not be used to accurately model purely organic molecules. Fortunately, the structural properties of organic ligands do not depend much on the metal center to which they are coordinated, i.e., a single set of force field parameters can be used, independent of the metal ion[96].

The modeling of bonds involving H atoms presents a particular problem which arises from the movement, on bond formation, of the single electron on the H atom away from the nucleus (proton) toward the bonded atom. This is readily observed in crystal structure analyses. X-ray diffraction analyses, which locate the center of the electron density, yield C–H bond lengths of 0.97 Å but neutron diffraction analyses, which locate the nucleus, give 1.08 Å. Since the nonbonded interactions arise in a large part from the interactions of the electron clouds it is appropriate that the repulsive center of the H atom should be moved away from the nucleus along the bond. In MM2 this is achieved by moving the center of repulsion by 20% of the bond length toward the heavy atom bonded to the H atom[49,72,134]. Elsewhere, the more simplistic but less accurate approach of using crystallographic bond lengths as ideal X-H bond distances has been adopted.

### 3.3.2 Valence Angle Deformation

The function used to calculate the energy associated with deformation of a valence angle is given in Eq. 3.7. A first approximation to the force constant, $k_\theta$, can again be derived from the infrared spectrum, though in the case of bond angles it is necessary to carry out a full normal-coordinate analysis in order to obtain accurate values. However, there are relatively few normal-coordinate analyses of metal complexes[131].

For the organic parts of the ligand unaffected by coordination, it is generally adequate to take force constants and undeformed angle values from appropriate organic force fields. For ligating and adjacent atoms the effects of coordination need to be taken into account and the force constants and the ideal angles will have to be developed empirically. In the absence of data from normal-coordinate analyses it is generally useful to take values from organic force fields as a starting point.

Force constants derived from normal-coordinate analyses might be over estimated by as much as a factor of two[49,72,134]. This occurs because, in most normal-coordinate analyses, nonbonded interactions other than 1,3- and 1,4-interactions are excluded. The forces arising from other nonbonded interactions are primarily taken up in the normal-coordinate analysis by the valence angle terms and hence they are over estimated. Also, normal-coordinate analyses relate to real, strained, molecules. Thus, the vibrations are about a mean that does not correspond to the ideal bond angle. In fact, the implementation of nonbonded interactions in molecular mechanics force fields is necessary to obtain molecule-independent force fields (see also Sections 1.2 and 3.4). The energy curve is steeper away from the minimum and therefore the spectroscopic force constant will generally be larger than that which should be used for molecular mechanics modeling. The same is undoubtably true for bond stretching terms but in that case the deformations are generally much smaller and hence the effect is not so significant. In the MM2 force field valence angle terms with force constants reduced by about 45% from the experimental values are used [49,72,134], and some force fields for metal complexes have similar reductions in these force constants [57–59,96,132].

Valence angles that involve a metal atom need to be considered more closely. These fall into two categories: angles in which the metal atom is not the central atom (M-L-X angles) and angles where it is the central atom (L-M-L' angles). In choosing force constants for M-L-X angles it is often assumed that the force constant is independent of the metal center [65,89,135], and then a value can be taken from an existing parameter set in the same force field. Alternatively, the force constants should be derived from an infrared spectrum and subsequently modified empirically because they can have a substantial effect on the strain energy and geometry that is obtained following minimization.

A number of fundamentally different approaches have been used in modeling the forces associated with distorting L-M-L' angles [15,65,120,123] (see Section 3.2.2). An often used approach is to treat them as Hookes Law vibrators, as for other angles. Force constants can be derived as outlined above, and the ideal angles can generally be predicted on the basis of the expected geometry, e.g., 90° and 180° for an octahedron and 90°, 120° and 180° for a trigonal bipyramid. A number of force constants have been used in force fields for hexaaminecobalt(III) complexes but their origin has not always been clear [15,120,123]. The most frequently quoted reference reports values of 0.68 [131] and force constants used range from 0.05 to 0.68 [76,120]. Some other methods developed recently involve more complex valence angle functions with multiple minima around the metal center [58,61,83,84]. For alkali, alkaline earth and lanthanoid metal ion compounds the angular geometry is usually modeled with an electrostatic potential, an approach similar to that involving 1,3-nonbonded interactions discussed below (see also Chapter 15).

An alternative approach to modeling the L-M-L' angles is to set the force constants to zero and include nonbonded 1,3-interactions between the ligand atoms. In most force fields, 1,3-interactions are not explicitly included for any atoms, instead they are taken up in the force constants for the valence angle terms. This is an approximation because the 1,3-interactions are most often repulsive and thus the function used to calculate the strain energy arising from valence angle deformation should be asymmetric. It was shown that the nonbonded 1,3-interactions around the metal atom are in many cases a major determinant of the coordination geometry[85–88,136]. The approach of including these interactions in molecular mechanics force fields for metal complexes was found to substantially improve the agreement between differences in minimized strain energies and experimental isomer distributions[59,89,135].

The use of nonbonded 1,3-interactions in place of force constants also offers a number of practical advantages. First, it is not necessary to determine the values of L-M-L′ force constants. If the assumption that the M-L-X force constants are independent of the metal center is also made, then replacement of the L-M-L′ force constants with 1,3-interactions results in a force field that is largely independent of the metal ion. Second, the coordination geometry is not predetermined by the choice of the undeformed bond angles. For instance, in the case of a five-coordinate complex the choice of angles will, if force constants are applied, predetermine whether a trigonal-bipyramidal or square-pyramidal geometry results. This can limit the generality and value of the modeling exercise if there are not strong electronic effects favoring one of the possible coordination geometries.

Whether the use of 1,3-interactions in place of L-M-L′ force constants is a valid approach depends on the metal ion being considered. If it is a metal for which the M-L bonding is primarily electrostatic, such as an alkali, alkaline earth or lanthanoid metal, then 1,3-interactions are definitely preferable. In such cases it may be important to include an electrostatic component in the 1,3-interactions in addition to the usual van der Waals term. If, however, the metal ion is one that has a clear preference for a particular coordination geometry then inclusion of at least a component of L-M-L′ force constants may be indicated (see Section 12.1). Recently, a model which includes both 1,3-interactions and force constants for the L-M-L′ angles has been described[58]. Other approaches are discussed in Section 3.2.2.

### 3.3.3 Torsion Angle Deformation

The function generally used to calculate the strain energy associated with rotation about a torsion angle is given in Eq. 3.8. When rotation about a bond occurs there are two sources of strain energy. The first arises from the nonbonded interactions between the atoms attached to the two atoms of the bond (1,4-interactions), and these interactions are automatically included in most molecular mechanics models. The second source arises from reorganization of the electron density about the bonded atoms which alters the degree of orbital overlap. The values for the force constants can be determined if a frequency for rotation about a bond in a model compound can be measured. For instance, the bond rotation frequencies of ethane and ethylamine have been determined by microwave spectroscopy. From the temperature dependence of the frequencies the barriers to rotation have been determined as 12.1 and 8.28 kJ mol$^{-1}$ respectively[137]. The contribution to this barrier that arises from the nonbonded 1,4-interactions is then calculated using the potential energy functions employed in the force field.

The component that remains is assumed to arise from the rearrangement of the electron density and it is this component that is calculated using Eq. 3.8. Care needs to be taken when converting the 'electronic' energy barrier to a force constant because some programs calculate rotation about a bond with a single term and others include all possible torsion angles. For example, in ethane there are nine distinct H–C–C–H torsion angles. The barrier can be modeled with a single term or with nine terms and in the latter case the force constant will be 1/9th of that in the former case. Force constants have been determined in this way for rotation about alkane C–C bonds and alkylamine C–N bonds[15]. However, it is important to remember that the contribution taken to arise from the nonbonded potential depends on the functions used to model the nonbonded interactions. Consequently, these terms must be recalculated any time there is a change in the way the nonbonded interactions are modeled. Force constants for rotation about other types of bonds are less well established and have often been estimated or derived empirically. In the development of most force fields for modeling metal complexes it has been assumed that torsion angle force constants employed for modeling organic compounds are satisfactory for the ligand backbone. This is an approximation, because, as discussed above, coordination often results in substantial changes in the intra ligand bonding. However, in the absence of an experimental determination of these barriers in coordinated groups it is probably sufficient to use the values that have been tested by application to organic molecules. Use of such values has not revealed any obvious problems in the modeling of metal complexes.

Force constants for rotation of torsion angles that have a metal involved need to be considered more closely. If the metal is one of the peripheral atoms (e. g., M–N–C–C) then it is generally assumed that the force constant is the same as for rotation about any other N–C bond. This is again an approximation that needs to be closely monitored but it has proven to be acceptable in the past. If the metal is one of the central atoms in the bond (e. g., N–M–N–C) then the practice has generally been to assign a force constant of zero for that torsion. That this approach is appropriate in the case of hexaaminecobalt(III) complexes was justified[15] on the basis that in $[Co(NH_3)_6]^{3+}$ there is rapid rotation about the Co–N bonds, even at 20 K[138]. Zero force constants for rotation about M–X bonds have been used extensively and successfully, and in some cases it has been shown that use of a non-zero force constant would lead to a substantially worse model[139]. However, there are situations where a zero force constant would not be appropriate. For example, where there is extensive π-bonding or backbonding between a ligand and the metal ion, then this will often confer a strong directionality on the preferred ligand orientation. In such a case it is necessary to derive the force constants empirically, unless experimental data on the barrier to rotation is available.

In some torsion angles, the energy minimum occurs when the attached groups are eclipsed rather than staggered. The most common examples are aromatic and delocalized systems, and in these cases the torsional barrier acts to keep the group planar. For many such groups (e. g., phenyl rings) highly refined force fields have been developed for organic molecules. There are, however, many common ligands that have conjugated bond systems that have not previously been modeled using molecular mechanics. The acetylacetonato family of ligands is a prominent example that forms a large range of complexes. In these cases initial estimates of force constants might be obtainable from the infrared spectrum of a model compound and then, as usually is the case, the parameters would need to be fitted to experimental structures.

### 3.3.4  Out-of-plane Deformation

The function used to calculate the strain energy associated with deformation of an atom out of the plane defined by three other atoms is given in Eq. 3.10. For common groups such as phenyl and ethene moieties a number of force fields have been reported. More of a problem are sp$^2$-hybridized groups and conjugated molecules that are directly coordinated to the metal atom. For example, the carboxylate group is a common ligand and it is necessary to employ an out-of-plane term to prevent the group from becoming excessively nonplanar. In theory, it should be possible to extract the spectroscopic force constants from infrared spectra, but in reality it is often difficult to do so because the absorptions are weak and are not easily assigned. In such cases the force constants should be taken from related force fields for uncomplexed molecules and adjusted empirically as necessary. Our experience is that the out-of-plane force constants are not critical elements of the force field because there is generally little deviation from planarity and therefore any force constant that maintains planarity is adequate. Of course, in those cases where large deviations from planarity are observed it is important to refine the parameters more carefully.

Out-of-plane deviations are critical in square-planar metal complexes. An important example are the anti-cancer active Pt(II) complexes such as *cis*-[PtCl$_2$(NH$_3$)$_2$]. In modeling the interaction of these compounds with DNA it has been found that the substantial stresses present cause deviations from planarity and, therefore, it is essential that a good estimate of the energy cost of this deformation be available. One set of force constants has been developed empirically by modeling simple *bis*-nucleotide complexes of the type *cis*-[Pt(NH$_3$)$_2$(Nu)$_2$]. In a number of these complexes significant deviations from planarity have been observed experimentally. By empirical adjustment of the force constant it was possible to obtain a value that reproduces adequately the deviations in these complexes[135]. Other methods for modeling square-planar metal complexes are discussed in Section 3.6.

### 3.3.5  Nonbonded Interactions

In the past the parameterization and calculation of the nonbonded interaction energy terms varied substantially between force fields. However, the wide-ranging success of the MM2 model and its variants has led to a broad acceptance of the nonbonded parameters used in these programs. It is generally accepted that a Buckingham function is the best of the simple functions for calculating the energy. However, for reasons of computational efficiency a 6,12-function is often used, particularly in the modeling of macromolecules. Both functions have an attractive and a repulsive component and are described in detail in Section 3.2.5.

As a general rule, the best approach when developing a force field for a metal complex is to use, without modification, the nonbonded interaction terms developed for organic compounds. The best known and most popular is the MM2 force field and this has been extended and adapted for modeling metal complexes by a number of groups[113−116]. Force fields developed for organic molecules do not have parameters for some of the elements present in metal complexes but reasonable estimates are available for most of these[57−59,65,96,120,123,135].

An unresolved point of contention is how the nonbonded interactions of the metal itself should be handled. In many force fields for metal complexes no nonbonded interaction terms involving the metal are included. This may seem at first to be rather surprising. However, it has not been shown to lead to significant problems. There are, perhaps, two reasons for the apparent success of this rather crude approximation. Firstly, most molecular mechanics studies of metal complexes have dealt with metals in positive oxidation states such as cobalt(III), nickel(II) and copper(II). The van der Waals radii of such metal centers are likely to be small and their bond lengths are long so that even the closest metal-to-nonmetal contacts are probably in the range of weakly repulsive or weakly attractive forces. Secondly, most simple metal complexes are approximately spherical with the metal at the center and, therefore, the effect of a spherical field emanating from the metal is minimal and is probably absorbed into other energy terms. Clearly then, the approximation of omitting nonbonded interactions involving metal ions may be inappropriate if the metal ion is not highly charged, as in many organometallic complexes, or if the complex is not spherical. An example of the latter is the binding of a metal ion with square-planar coordination geometry to a large biomolecule. For example, it has been found that when platinum binds to DNA via N7 of a guanine or adenine, it makes a number of close contacts with parts of the DNA. Electrostatic terms have in the past been used to model these interactions and this has proven successful in reproducing the structures of small Pt/nucleotide complexes[140,141]. However, force fields including van der Waals terms for Pt have now been reported[98,99].

One of the main reasons why nonbonded interactions involving metal ions have not been included in most force fields is a lack of good estimates for the parameters. As discussed in Section 3.2.5, values for the van der Waals radius and the polarizability ($\varepsilon$) are required. In the case of metal complexes it is difficult to obtain estimates for the van der Waals radius because the metal ion is generally buried inside an organic sheath and does not make close contacts with atoms free to move away from it. In a few cases close contacts to M(II) metal ions have been observed and these are consistent with van der Waals radii in the range of 1.5−1.7 Å. Where nonbonded interactions to the metal ions have been included in a force field the van der Waals radius used ranged from 1.7 to 2.44 Å[98,99,142].

### 3.3.6   Electrostatic Interactions

Electrostatic interactions are usually modeled by assigning point charges to the atoms and the contributions to the strain energy are then calculated using Eq. 3.11. In most molecular mechanics models for small molecules, organic and inorganic, the electrostatic terms have not been included explicitly without obvious detriment to the success of the model. However, the electrostatic terms are routinely included in models for biological macromolecules such as proteins and DNA because they make a substantial contribution to the stabilization of such molecules and to the interactions they make with other molecules[93,94,143]. Estimates for the point charges on the atoms have been obtained from quantum-mechanical calculations on fragments of the larger molecules, e. g., on the individual amino acids in a protein[93]. The accuracy of these point charges has been questioned by a number of workers[144] but they have been used successfully. Other approaches to the calculation of point charges have been discussed in Section 3.26.

When metal complexes are being modeled, the problem arises as to what point charge to assign to the metal atom itself. There are few efficient empirical or semi-empirical calculations on metal complexes that are sufficiently accurate to yield reliable estimates of the charge. As Pauling's electroneutrality principle shows[145], it is not appropriate to use the full formal charge and, therefore, partial charges have been used. For example, in the case of Pt(II) complexes an empirically assigned charge of +0.5 e on the Pt atom yielded reasonable results[140]. It is very difficult to judge whether such a value is appropriate because electrostatic energy terms, being dependent on $r^{-1}$, do not exert strong forces on even nearby atoms. It is, therefore, not surprising that in many cases the charge attributed to the metal center is not a critical parameter[63].

Assigning charges to atoms in the ligands is also a problem since these will almost certainly be modified by coordination. One approach is to use charges similar to those on groups that form similar strength hydrogen bonds. Thus, amine groups form hydrogen bonds similar to those seen between the bases of DNA and, therefore, charges similar to those on the amine groups of the DNA bases might be reasonable estimates[140].

A contentious issue in modeling electrostatic interactions in molecules of all types is what is the appropriate value for the dielectric constant. The value for a vacuum is 1 but this is not appropriate because it does not take account of the interference of the electron clouds of adjacent and intervening atoms. Since the number of intervening atoms increases with distance it has been suggested by a number of workers that the dielectric constant should be distance dependent[93,146]. A systematic molecular modeling study of the protein Crambin using a number of different forms for the dielectric constant showed that the best fit to experiment was obtained when a dielectric constant of $4\,d_{ij}$ was used[146]. The consequence of using such a term is that the equation for calculating the energy of an electrostatic interaction becomes Eq. 3.35.

$$E_{\varepsilon} = \frac{q_i\,q_j}{4\,d_{ij}^2} \tag{3.35}$$

Calculation of the electrostatic terms in this way is only an option when such a function has been coded into the molecular mechanics program. A perhaps unexpected consequence of using a distance dependent dielectric constant is that the energy surface arising from electrostatic energy terms becomes steeper, i.e., the force that these terms exert is stronger. The intention of using a dielectric force constant greater than 1 is to mute the electrostatic energy terms and, therefore, a distance dependent value may lead to unexpected or even incorrect results. The alternative is to use fixed values greater than 1 and values ranging from 2 to 8 have been proposed[147,148].

### 3.3.7   Hydrogen Bonding Interactions

Hydrogen bonds are only rarely an issue in the modeling of small metal complexes. There are, however, some cases where hydrogen bonds are particularly important. For example, when diamineplatinum(II) complexes bind to DNA, hydrogen bonds form between the H(amine) atoms and oxygen atoms on the DNA, and these interactions may be very important in determining the sequence specificity of Pt/DNA interactions[140]. Also, interactions between cationic and anionic complexes will inevitably involve hydrogen bonds and these terms will probably determine whether there is substantial stereoselectivity in the interactions.

In molecular mechanics force fields for biological molecules, hydrogen bonds are generally modeled using an empirical 10–12 function (Eq. 3.12). There is no theoretical justification for such a function but, as long as $F$ and $G$ are chosen to reproduce both the ideal hydrogen bonding distance and the energy of the hydrogen bond then they seem to work well. A possible angular dependence of hydrogen bonds has been discussed in Section 3.2.7. In modeling the hydrogen bonding of amine groups with oxygen atoms of DNA use of the same $F$ and $G$ values as those used to model the interstrand hydrogen bonds has successfully produced distances that closely resemble those in small molecule diamineplatinum(II)/nucleotide complexes[140].

## 3.4   Spectroscopic Force Fields

Force constants derived from normal coordinate analyses of infrared spectra have formed the basis for the parameterization of many molecular mechanics force fields. However, empirical adjustment of these molecular mechanics force fields has led to appreciable differences between the spectroscopic and molecular mechanics force constants. It is important to understand why this should be so and to appreciate the difference between spectroscopic and molecular mechanics force fields.

Let us consider the description of a simple three-atom molecule by the three most frequently used spectroscopic parameterization schemes, the general central

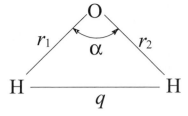

**Figure 3.14:** Parameters used for spectroscopic force field calculations of $H_2O$.

force field (GCF), the general valence force field (GVF) and the Urey-Bradley force field (UBF; Fig. 3.14).

While the GCF (Eq. 3.36) uses only interatomic distances ($\Delta r$, $\Delta q$)*, in the GVF the nonbonded interactions are replaced by valence angles (Eq. 3.37), and in the UBF (Eq. 3.38) both the interatomic distances and the valence angles are used ($r_e$ and $q_e$ in Eqs. 3.36-3.38, correspond to ideal distances)[71].

$$V_{GCF} = \frac{1}{2} k_r \left(\Delta r_1^2 + \Delta r_2^2\right) + \frac{1}{2} k_q \Delta q^2 + k_{rr} \Delta r_1 \Delta r_2 + k_{qr} \left(\Delta r_1 + \Delta r_2\right) \Lambda q$$
$$(3.36)$$

$$V_{GVF} = \frac{1}{2} k_r \left(\Delta r_1^2 + \Delta r_2^2\right) + \frac{1}{2} H_\alpha r_e^2 \Delta \alpha^2 + F_{rr} \Delta r_1 \Delta r_2 + F_{r\alpha} r_e \left(\Delta r_1 + \Delta r_2\right) \Delta \alpha$$
$$(3.37)$$

$$V_{UBF} = \frac{1}{2} k_r \left(\Delta r_1^2 + \Delta r_2^2\right) + k' r_e \left(\Delta r_1 + \Delta r_2\right) + \frac{1}{2} H r_e^2 \Delta \alpha^2 + H' r_e^2 \Delta \alpha$$

$$+ \frac{1}{2} F \Delta q^2 + F' q_e \Delta q$$
$$(3.38)$$

In each model the coupling of vibrations is taken into account by the addition of cross terms. Two important facts arise from this and a general appreciation of Eq. 3.36–3.38:

- The force constants are molecule specific.
  The force constants may not be transferred from one spectroscopic force field into another, i.e., they are not transferable physical quantities.

A molecule-independent, generalized force field for predictive calculations can be obtained by the inclusion of additional terms such as van der Waals and torsional angle interactions. This adds an additional anharmonic part to the potential (see below) but, more importantly, also leads to changes in the whole force field; thus the force constants used in molecular mechanics force fields are not directly related to parameters obtained and used in spectroscopy. It is easy to understand this dissimilarity since in spectroscopy the bonding and angle bending potentials describe relatively small vibrations around an equilibrium geometry that, at least

---

* An adaptation of the GCF to molecular mechanics [60] has been discussed in Section 3.1.

for strained molecules is far away from the ideal structure, whereas in molecular mechanics large deviations from the non-molecule-specific ideal geometry are involved. While the molecule-specific force constant $k_e$ at the molecule-specific equilibrium distance $r_e$ can be obtained by normal coordinate analysis of vibrational spectra, the corresponding molecular mechanics parameters have to be fitted empirically in order to reproduce experimental data.

## 3.5   Model and Reality

In this section we will discuss in some detail the relationship between molecular mechanics force field parameters and real physical parameters. As mentioned before, one fundamental difference between spectroscopic and molecular mechanics force fields is that the former are molecule specific while the latter are general. Empirical force field parameters can be used for the calculation of unknown structures and their strain energies, and for the prediction of vibrational frequencies of new compounds. However, the parameters themselves generally have limited meaning.

For example, the $r_0$ value for the Co−N bond in cobalt(III) amine complexes is smaller in parameterization schemes where 1,3-nonbonded interactions between the ligating atoms are included, than in force fields where only L-M-L′ angle bending functions are used. This is because the 1,3-nonbonded interactions in such complexes are highly repulsive, promoting an extension of the Co−N bonds. Thus, a smaller value for the ideal Co−N bond is required in order to reproduce the experimentally observed bond lengths. Values of $r_0$ for force fields that do and do not include 1,3-nonbonded terms are listed in Table 3.4 (see also [133] and Table 3.2).

**Table 3.4:**  Strain-free bond distances $r_0$ for the metal-ligand bonding interaction[a].

| Bond type | Strain-free value, $r_0$ [Å] | | |
|---|---|---|---|
| $Cr^{III}$-$N_{amine}$ | 2.045 * [57] | 2.050 [120] | |
| $Co^{III}$-$N_{amine}$ | 1.905 * [57] | 1.925 [15] | 1.950 [149] |
| $Co^{II}$-$N_{amine}$ | 2.120 * [151] | 2.180 [120] | |
| $Ni^{II}$-$N_{amine}$ | 2.090 * [57] | 2.100 [152] | 2.010 [153] |
| $Cu^{II}$-$N_{amine}$ | 1.970 * [57] | 2.030 [120] | 2.000 [154] |
| $Cu^{II}$-$S_{thioether}$ | 2.290 * [57] | 2.380 [150] | |
| $Cu^{II}$-$O_{carboxylate}$ | 1.900 * [57] | 1.950 [154] | |
| $Zn^{II}$-$N_{amine}$ | 2.220 * [57] | 2.170 [120] | |
| $Rh^{III}$-$N_{amine}$ | 2.050 * [57] | 2.040 [120] | |

[a]  parameters with an asterisk relate to a force field with 1,3-nonbonded interactions included.

As the bond lengths increase the difference between the $r_0$ values for the two types of force fields decreases because the 1,3-nonbonded interactions become less repulsive.

Another way to demonstrate this effect is shown in three sample calculations using the MOMEC force field[52,57,58,96,99,124,140,155]. In Table 3.5 the experimental M-N distances ($r$(obs)), the calculated bond lengths ($r$(calc)), the corresponding force constants ($k$) and ideal bond lengths ($r_0$) are shown for some sample calculations of some hexaamminemetal complexes.

**Table 3.5:** The influence of 1,3-nonbonded interactions on the ideal metal-ligand bond distance of chromium(III), cobalt(III) and nickel(II) hexaamines.

| Parameter[a) | $[Cr(NH_3)_6]^{3+}$ | $[Co(NH_3)_6]^{3+}$ | $[Ni(NH_3)_6]^{2+}$ |
|---|---|---|---|
| $r$ (obs) [Å] | 2.064 [156] | 1.959 [53] | 2.129 [157] |
| $r$ (calc) [Å] | 2.079 | 1.955 | 2.132 |
| $k$ [mdyn Å$^{-1}$] | 1.100 | 1.750 | 0.600 |
| $r_0$ [Å] | 2.045 | 1.905 | 2.090 |
| $r'$ (calc) [Å] | 2.063 | 1.942 | 2.118 |
| $r'_0$ [Å] | 2.063 | 1.930 | 2.115 |

[a) $r$ values are the metal-amine distances; the force field parameters ($k$, $r_0$) are from the MOMEC force field with the harmonic sine function[52,57]; the primed values are for calculations without the Urey-Bradley term (see text).

The calculated distances $r'$ (calc) were obtained with the original values for $k$ and $r_0$ after setting the 1,3-nonbonded terms to zero and with constrained angular geometries. Readjustment of the ideal M-N distances to $r'_0$ led to calculated M−N bond lengths in agreement with the experimental values. Thus, for these hexamine complexes the inclusion of 1,3-nonbonded terms is responsible for a reduction of the ideal bond length by 1−2% ($r_0$ vs $r'_0$), depending on the force constant and the actual bond distances.

Obviously, it would be of interest to know what the true ideal metal-ligand bond length is. Some studies, particularly some of those aimed at predicting metal ion selectivities based on hole size calculations, require accurate values for $r_0$[66,158] (see Section 9.2). The question is, whether the parameters determined with or without ligand-ligand repulsion are closer to physical reality. On the basis of chemical intuition, as well as the success of ligand-ligand repulsion based calculations, one has to assume that 1,3-nonbonded interactions are present in coordination compounds. Parameters derived without considering 1,3-nonbonded interactions have these interactions intrinsically included in their ideal metal ligand bond lengths. Therefore, the $r_0$ values obtained for the force fields that do not include nonbonded interactions are longer than those that include them and they are longer than the "true" ideal metal-ligand distance.

## 3.6   Electronic Effects

The arrangement of atoms in a molecule is based on attractive and repulsive forces (see Fig. 3.2), as well as the directionality of the bonds, which is determined by the orientation of the bonding orbitals and their desire for maximum overlap. At first glance, the simple mechanical model (see Fig. 3.1) does not explicitly included specific electronic interactions. However, in developing a model that reproduces experimentally derived structural and thermodynamic data, it is inevitable that electronic factors are included implicitly to account for electronic effects responsible for some of the structural and thermodynamic variation present in the data used in the parameterization. Depending on the model used, the electronic effects cannot be directly attributable to specific parameters.

In the conventional model, where the total strain energy is represented by bonding, valence angle, torsional angle, nonbonded and other interactions, some electronic effects are included in obvious ways. For example, strong orbital overlap may lead to short bonds and, concurrently, to relatively short ideal distances with strong force constants. Similarly, the directionality of bonds can be defined by valence angle functions (e.g. $sp^3$, 109.5°; $sp^2$, 120°; $sp$, 180°), where the strength of the geometric preference is represented by the respective force constant. However, since all force field parameters are interdependent, an accurate assignment of specific electronic effects is, even in these simple cases, not possible, i.e., as pointed out before, isolated force field parameters have dubious meaning.

This point can be illustrated by the examination of hexacoordinated transition metal ions with a common ligand system and variable electronic properties (d-electron occupancies)[65,89,159]. When modeling a strong ligand field metal center such as low-spin $d^6$ (e.g., cobalt(III)), which exhibits a strong preference for octahedral geometry, and one, where octahedral geometry is not preferred in terms of ligand field stabilization, such as high-spin $d^5$ (e.g., manganese(II)), the differences in geometric preferences can be reproduced using valence angle force constants. However, this need not be necessary since a strong ligand field leads to relatively short metal-ligand bonds and thus to an increase in ligand-ligand repulsion which induces a preference for octahedral geometry. Therefore, the electronically dictated directionality is, at least partly, included indirectly as a consequence of the bond deformation and repulsive terms. Such an approach is not possible for other stereoelectronic effects such as the *trans* influence in metal complexes, Jahn-Teller distortions, bonding of π-donor and π-acceptor ligands and conjugation in extended π-systems. Basically, all these and other possible effects can be, and most have been, included explicitly in an empirical force field calculation. Some specific examples are given in Chapter 12.

Electrostatic interactions can be used to account for the extra electron density transferred to the metal center from anionic π-donating ligands such as carboxylates. Alternatively, the ensuing effects might be modeled by redefining the force constants related to metal centers coordinated to donors such as carboxylates, since, effectively, the electronic properties of the metal center have changed, allowing it to be defined as an electronically different species[57,65]. As in this example, there is often a variety of possibilities to account for specific electronic effects. However, great care needs to be taken to define these as generally, and with as few parameters, as possible if they are to have wide ranging utility.

This is also demonstrated with square-planar nickel(II) complexes (low spin), where the planarity of the chromophore can be enforced in at least four different ways (Fig. 3.15).

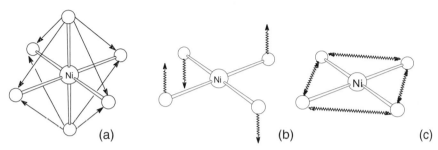

**Figure 3.15:** Three possible methods to enforcing a square planar geometry of low spin nickel(II).

Dummy atoms, defined on axial sites, can be used to impose a square planar arrangement via ligand-ligand repulsion (Fig. 3.15(a))[153]. The problems associated with this approach are (1) how to decide where exactly the dummy atoms are to be positioned and what van der Waals parameters they should be assigned, and (2) that the whole ligand including its carbohydrate backbone is repelled by the dummy atoms, which may lead to the incorrect calculation of ligand distortions. Alternatively, square-planar geometry can be enforced by an out-of-plane function of the type generally used for $\pi$-bonded systems (e.g., aromatic compounds and carboxylates) (Fig. 3.15(b)) or by using stiff ligand-metal-ligand angle functions (90° and 180°) (Fig. 3.15(c)). Finally, the planarity can be enforced using a tetrahedral twist function. This approach has not been tested extensively but preliminary results are promising (see Section 17.14)[109].

## 3.7 The Environment

Usually, force field parameters are developed on the basis of solid state data, e.g., crystal structural coordinates. It is therefore not entirely appropriate to refer to these molecular mechanics calculations as "gas phase calculations", even if the environment is not explicitly included in the structure optimization procedure. Environmental effects such as ion-pairing and hydrogen bonds to counter ions, co-crystallized solvent molecules and neighboring molecules are present in crystal lattices. Therefore, an averaged influence of these is implicitly included in the force field and at least partially mimics a non-specific environment, similar to that present in solution.

The fact that only "naked" molecules are modeled is based on the problem that for crystal lattices at least 27 unit cells would have to be included (with at least one unit per cell, including counter ions and solvents of crystallization), and in solution at least 200 molecules of water must be refined in the solvent sheath interacting with the compound to be modeled. Since CPU time $\sim f(m^2)$, where $m$ is the number of nuclei, the time required for a single optimization cycle increases dramatically under these conditions. Even more importantly, the initial configuration of the molecule and its environment is not easy to predict since the intermolecular contacts (crystal lattice, ion pairing and solvation) of a compound to be modeled are not known beforehand. Thus, inclusion of environmental effects in modeling studies, has necessitated the use of some severe approximations [160–163].

There is no doubt that environmental effects are of importance in coordination chemistry [65,164], e. g., in labile complexes exhibiting geometries dependent on the counter ion and solvent, in fractional crystallization processes, and where severe geometric distortions due to crystal packing occur. If these effects are neglected, as is the usual practice, the resulting inaccuracies must be recognized and reduced, where possible. This can be achieved by choosing appropriate experimental data for comparison with the calculation. That is, data to be used for the validation of the model should be obtained with solvents and counter ions that do not lead to strong and specific interactions with the complex ions.

Optimized structures of "naked" molecules do not generally differ extensively from crystallographic data. For small organic molecules, bond lengths are reproduced to within 0.005 Å and angles are usually accurate to approx. $1°$ [120,134]. Corresponding values for small coordination compounds are approx. 0.01 Å and $2-5°$ [57,65,96]. In part, this accuracy is because molecular mechanics is an interpolative process that uses crystal structures of related molecules for the parameterization. The largest difference between experimental and force field optimized structures is that the latter often have higher symmetry, and the approximate error limits mentioned above are related to average bond lengths and to average valence angles. If solution properties, where distortions by the crystal lattice are absent, are of interest, which is very often the case, the computed structures might be more relevant than experimentally obtained crystal structures, especially if the appropriate model is chosen and if solvation and ion-pairing are not selective. Combinations of strain energy minimization with experimental data, (usually spectroscopic) have been successfully used to determine solution structures. Examples of these techniques are given elsewhere in this book (see Chapters 5, 8 and 10) [90,164–166].

Minimized strain energies are used extensively to model relative stabilities in the design of stereoselective reactions and metal-ion-selective ligands (see Chapters 8 and 9) [65,120]. In these calculations it is not unusual to interpret energy differences of a few kilojoules per mole* as significant. Thus ion-pairing and solva-

---

* Based on a Boltzman distribution an energy difference of 10 kJ/mole between two conformers already leads to an almost exclusive population of the less strained species.

tion of charged complexes (hundreds of kJ mol$^{-1}$) have to be assumed to be non-selective if they are not included in the modeling procedures. Obviously, deviations from this assumption may lead to erroneous predictions, and therefore the experimental data have to be chosen carefully with respect to the selective associations formed by the solvent molecules and counter ions.

The solvation free energy is usually considered to consist of a solvent-solvent cavity term ($G_{cav}$), a solute-solvent van der Waals term ($G_{vdw}$) and a solute-solvent electrostatic polarization term ($G_{es}$) (Eq. 3.39).

$$G_{sol} = G_{cav} + G_{vdw} + G_{es} \tag{3.39}$$

In a recent approach that was successfully tested for small hydrocarbons the solvation was treated semi-analytically as a statistical continuum[167]. The method treats the sum of the solvent-solvent cavity ($G_{cav}$) and the solute-solvent van der Waals ($G_{vdw}$) terms by determining the solvent accessible surface[168], and the solute-solvent electrostatic polarisation term ($G_{es}$) is calculated by a modified version of the generalized Born equation[167,168].

The determination of the charge distribution in a molecule, needed here for the latter term, ($G_{es}$), has been a considerable problem in force field calculations, especially for transition metal compounds (see Sections 3.2.6 and 3.3.6). Most promising but not yet fully tested for transition metal complexes are semi-empirical quantum-mechanical methods[103,104]. Future studies might show whether a combination of approximate methods for the computation of charge distributions and solvation will lead to a reliable approximation of solvation parameters of co-ordination compounds.

## 3.8 Entropy Effects

A further complication associated with the application of molecular mechanics calculations to relative stabilities is that strain energy differences correspond to $\Delta(\Delta H)$ between conformers with similar chromophores (electronic effects) and an "innocent" environment (counter ions and solvent molecules), whereas relative stabilities are based on $\Delta(\Delta G)$. The entropy term, $T\Delta S$, can be calculated by partition functions, and the individual terms of $\Delta S$ include vibrational ($S_{vib}$), translational ($S_{trans}$) and rotational ($S_{rot}$) components, and in addition to these classical terms, a statistical contribution ($S_{stat}$). These terms can be calculated using Eqs. 3.40–3.43[121].

$$S_{vib} = R \sum_{i=1}^{3N-6} \frac{hv_i/kT}{\exp\left(\frac{hv_i}{kT} - 1\right)} - \ln\left(1 - \exp\left(-\frac{hv_i}{kT}\right)\right) \tag{3.40}$$

$$S_{\text{trans}} = R\left\{\frac{3}{2}\ln m + \frac{5}{2}\ln T - \ln p - 3.664 + \frac{5}{2}\right\} \tag{3.41}$$

$$S_{\text{rot}} = R\ln\left(\frac{8\pi^2 (8\pi^3 \text{ABC})^{\frac{1}{2}} (kT)^{\frac{3}{2}}}{\sigma h^3}\right) + \frac{3}{2}R \tag{3.42}$$

$$S_{\text{stat}} = R\ln\rho \tag{3.43}$$

Here, $v_i$ are the vibrational energy levels, $T$ is the temperature, $m$ the molecular weight, $p$ the pressure, $ABC$ the product of the moments of inertia, $\sigma$ the rotational symmetry number and $\rho$ the number of ways in which a given conformer can be formed.

The moments of inertia $A$, $B$, $C$ can be calculated from the optimized structures and the vibrational energy levels $v_i$, and these are available from the second-derivative matrix if a full-matrix Newton-Raphson refinement is used[13,75]. However, the approximations involved in the calculation of the entropies that have been used for the computation of the conformational equilibria of coordination compounds have led to considerable uncertainties[7,121,169,170].

Entropic factors are a major problem for relatively large molecules. For organic macromolecules, the simulation of the probability, $W$ ($S = k \cdot \ln(W)$), by molecular dynamics calculations or Monte-Carlo simulations, has been used to calculate the entropy from fluctuations of the internal coordinates[171–174]. For simple coordination compounds the corrections based on calculated entropy differences are often negligible in comparison with the accuracy of the calculated enthalpies[65,121,170]. Therefore, the relatively easily available statistical term ($S_{\text{stat}}$) is usually the only one that is included in the computation of conformational equilibria (see Chapters 8 and 9).

## 3.9   Summary

Molecular mechanics is a simple technique for scanning the potential energy surface of a molecule, molecular ion, crystal lattice or solvate. The model is based on a set of functions which may or may not be based on chemical and physical principles. These functions are parameterized based on experimental data. That is, the potential energy surface is not computed by fundamental theoretical expressions but by using functions whose parameters are derived empirically by reproducing experimentally observed data. Molecular mechanics then is, similar to a neural network, completely dependent on the facts that it has been taught. The quality of results to be obtained depends on the choice of the experimental data used for the parameterization. Clearly, the choice of potential energy functions is also of some importance. The most common model used is loosely derived from chemical principles but the simplifications that have been employed are rather extensive. Together with other approximations and omissions outlined in this chapter, including electronic contributions, entropy and the environment, the limits of molecular mechanics are, even for the novice, easy to comprehend. It is therefore obvious that one should be cautious about over-interpreting data produced by empirical force field calculations.

Molecular mechanics is an interpolative method. It thus follows that the strain energy and the structure of a very strained species may not be reliably computed based on a parameterization scheme fitted with a set of experimental data obtained from unstrained molecules since this amounts to extrapolation. Therefore, to get a generally reliable force field, extreme cases must be included in the fitting procedure of the force field. The speed with which structure optimizations are produced (seconds of CPU time on a simple personal computer for molecules with a few hundred atoms) does not place any restriction on the size of a data base for the parameterization of a class of compounds – the limit is usually the amount of experimental data available. The major appeal of molecular mechanics is the fact that unknown compounds can be modeled in much the same way as model kits are used, with the important difference that quantitative information becomes available, enabling the design of new materials.

In cases, where results obtained, be it computed structures or energies, are not in agreement with experimental data, it is very bad practice to simply adjust the force field. Rather, one should carefully consider possible reasons for the disagreement. Very often, this leads to the uncovering of novel aspects related to the chemical problem involved.

# 4    Computation

So far we have discussed the functional form of the energy terms and the parameters used to define these potentials in a molecular mechanics program. Clearly, these are primarily responsible for the accuracy of the computed structures, strain energies and molecular vibrations. However, they are not the only factors that have an influence on the calculation and the result. The type of input accepted by the program, the output produced, and the algorithms used to achieve energy minimization can impact on the accuracy of the results, limit the type of problems that can be handled and the type of constraints that can be imposed. It is, therefore, important to understand the advantages and disadvantages of different methods for carrying out energy minimization.

## 4.1    Input and Output

The type of input and output data that are available and the way in which they are handled not only has a strong impact on the aesthetics and ease of use of a molecular modeling package, but they also affect the accuracy of the interpretation of the results and the insights obtained. Ideally, molecular mechanics routines should be combined with molecular graphics packages that allow the importation of experimental structural data, the building of new molecules and the representation of these molecules. The manipulation of the calculated geometries in three-dimensional space should occur in real time. There are many such programs available but molecular graphics is beyond the scope of this book. The molecular mechanics programs listed on our homepage (see Appendix 3) are either available as packages with a graphical interface or they are easily adapted to other graphics packages.

Potential energy surfaces usually have many different minima. The starting structure, the algorithm used for strain energy minimization and the concurrent geometry optimization have a strong influence on the particular minimum to which the refinement converges and on the time required for the optimization. The type of minimum, whether it is a global or local minimum, can be of importance depending on whether a thorough conformational analysis or just a local energy minimum is the aim of the study. The time needed for a single optimization process becomes critical when a series of calculations must be performed. For instance, a full conformational analysis with, say, 100 conformers and 5–60 sec per energy minimization requires 1–2 hours to complete.

Coordinates of molecules can be represented in a global or in an internal coordinate system. In a global coordinate system each atom is defined with a triplet of numbers. These might be the three vectors $x_i$, $y_i$, $z_i$ in a crystal coordinate system defined by the three vectors $a$, $b$, $c$ and the three angles $\alpha$, $\beta$, $\gamma$ or by $a$, $b$, $c$, $\alpha$, $\beta$, $\gamma$ with dimensions of 1,1,1,90°,90°,90° in a cartesian or orthonormalized coordinate system. Other common global coordinate systems are cylindrical coordinates (Fig. 4.1) with the coordinate triples $r$, $\theta$, $z$ and spherical coordinates (Fig. 4.2) with the triples $\rho$, $\theta$, $\phi$.

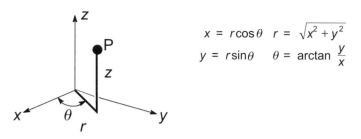

$$x = r\cos\theta \quad r = \sqrt{x^2 + y^2}$$
$$y = r\sin\theta \quad \theta = \arctan\frac{y}{x}$$

**Figure 4.1:** Cylindrial coordinates $r$, $\theta$, $z$.

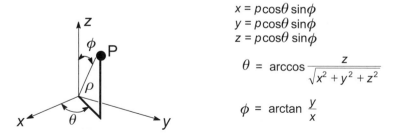

$$x = \rho\cos\theta\,\sin\phi$$
$$y = \rho\cos\theta\,\sin\phi$$
$$z = \rho\cos\theta\,\sin\phi$$

$$\theta = \arccos\frac{z}{\sqrt{x^2 + y^2 + z^2}}$$

$$\phi = \arctan\frac{y}{x}$$

**Figure 4.2:** Spherical coordinates $\rho$, $\theta$, $\phi$.

Internal coordinate systems include normal coordinates which are symmetry adapted and used in spectroscopy, and coordinate systems based on interatomic distances ("bond lengths"), three-center angles ("valence angles") and four-center angles ("torsion angles"). In the latter case a Z-matrix of the form shown in Table 4.1 defines the structure of a molecule. The input and output files of nearly all molecular mechanics programs are either cartesian coordinates or Z-matrices.

The connectivity of molecules can be defined in the process of drawing the molecule (model build function in the graphics program) or it might be recognized by a subroutine of the molecular graphics or molecular mechanics program. In the latter case the connectivity is defined by comparison with bonding radii. These are obtained from well-established tabulations but problems might occur in cases

**Table 4.1:** Z-matrix for methanol.

| Atom-number | Atom | d | α | ω | Definition[a] | | |
|---|---|---|---|---|---|---|---|
| 1 | $C_1$ | 0.00 | 0.0 | 0.0 | 0 | 0 | 0 |
| 2 | $O_2$ | 1.40 | 0.0 | 0.0 | 1 | 0 | 0 |
| 3 | $H_3$ | 0.94 | 107.0 | 0.0 | 2 | 1 | 0 |
| 4 | $H_4$ | 1.10 | 110.0 | −180.0 | 1 | 2 | 3 |
| 5 | $H_5$ | 1.10 | 110.0 | −119.0 | 1 | 2 | 4 |
| 6 | $H_6$ | 1.10 | 110.0 | 119.0 | 1 | 2 | 5 |

[a]  For example atom 5: $d(1\text{-}5) = 1.10$ Å, $\alpha(5\text{-}1\text{-}2) = 110°$, $\omega(5\text{-}1\text{-}2\text{-}4) = -119°$

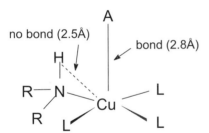

with strong anisotropy such as those associated with Jahn-Teller distortions, where some bonds are so long that they are not recognized as bonds (Fig. 4.3). These problems can be solved by defining anisotropic bonding radii.

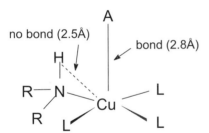

**Figure 4.3:** Anisotropic bonding and connectivity.

The parameterization of an interaction is dependent not only on the atoms involved but also on their environment. With the exception of generic force fields, each chemically unique interaction type has to be parameterized independently. For both, generic and conventional force fields, the atom types need to be specified in the input file. Most modern molecular modeling packages have the ability to recognize the atom types through the connectivity and assign them automatically.

## 4.2   Energy Minimization

The search for an advantageous condition of a system is a common problem in science and many algorithms are available to optimize objective functions. In this chapter we will discuss the methods commonly used in molecular mechanics. For a more detailed discussion we refer to specialist texts on computational chemistry[175–178].

Prior to minimization, little information is available about the high-dimensional energy surface (3N-6 dimensions with N atoms). Put simply, the program cannot "see the landscape". Ideally, the minimization process should adapt to the shape of the surface and the distance from the minimum. Also, the type of energy minimization procedure used should depend on whether a specific local minimum or any minimum, is sought. Most programs offer a choice of different optimization methods and the step size can often be chosen interactively.

Generally, an energy-minimizing routine will produce an optimized structure of the conformer most closely related to the geometry defined by the input coordinates. That is, the structure falls into the closest energy minimum. However, this cannot be guaranteed, and if one wishes to optimize a particular conformation a constrained optimization might be needed (see below). A significant problem in molecular mechanics is that of establishing a general method for finding the global minimum of a system. This will be discussed in some detail in Chapter 5.

The first derivatives of a potential energy function define the gradient of the potential energy surface and the second derivatives describe its curvature (Fig. 4.4). In most molecular mechanics programs the functions used are relatively simple and the derivatives are usually determined analytically. The second derivatives of harmonic oscillators correspond to the force constants. Thus, methods using the entire set of second derivatives result in some direct information on vibrational frequencies.

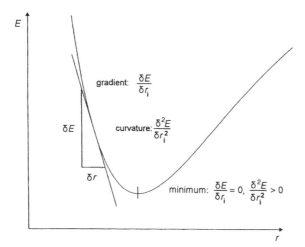

gradient: $\dfrac{\delta E}{\delta r_i}$

curvature: $\dfrac{\delta^2 E}{\delta r_i^2}$

minimum: $\dfrac{\delta E}{\delta r_i} = 0, \ \dfrac{\delta^2 E}{\delta r_i^2} > 0$

**Figure 4.4:** First and second deriatives of a potential energy function.

The energy minimizing routines discussed here include direct search methods where only the potential energy function is used (e.g., Simplex), gradient methods which involve the first derivatives (e.g., steepest-descent), including conjugate-gradient methods where the history of the search can influence the search direction and step size (e.g., Fletcher-Reeves, Polak-Ribière), second-derivative methods (Newton-Raphson and block-diagonal Newton-Raphson), as well as least-squares methods (e.g., Marquardt).

A problem common to all minimization programs is the choice of the step size. If the step size is too large the starting structure might be completely deformed, corresponding to a move to an area of the potential energy surface far away from the minimum. Conversely, excessively long optimization times result from step sizes that are too small. Generally, programs allow the starting step size to be chosen, and recent algorithms include loops to modify the step size during the optimization process. That is, the step size is increased when there is a decrease in strain energy relative to the previous cycle or reduced when the energy increases. The most effective method to find the energy minimum is to use an algorithm that starts with a crude minimization procedure far from the minimum (e.g., steepest descent) and changes energy-minimizing routines to more complex algorithms (e.g., Polak-Ribière and then Newton-Raphson) as a function of the energy gradient or the root mean square (rms) shift.

## 4.2.1   The Simplex Method

The simplex search algorithm[175–178] is a common pattern search method. A simplex is a geometrical figure that has one more vertex than the dimension of the space in which it operates, i.e., $n + 1$ for $n$-dimensional systems ($3N + 1$ for Cartesian coordinates with $N$ atoms) or 3 vertices (i.e., a triangle) in the two-dimensional case demonstrated in Fig. 4.5. The simplex (a triangle in this example) is moved downhill with certain rules (reflection of the least favorable point through the centroid of the other vertices; expansion or contraction of the new simplex depending on the improvement in terms of strain energy). Once the simplex has reached the valley it is allowed to shrink in order to reach the energy minimum. The simplex algorithm is relatively easy to implement in molecular mechanics

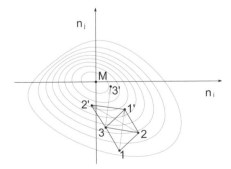

**Figure 4.5:** The two-dimensional simplex minimization method.

programs. It is efficient for preoptimization of structures far away from the energy minimum, particularly for small molecules, but becomes impractical as the molecular size increases.

### 4.2.2  Gradient Methods

The steepest-descent method[175–178] is a simple gradient method (Fig. 4.6). At each point the first derivative vector $A$ is used to find the direction of steepest descent (Eqs. 4.1 and 4.2),

$$A = \left( \frac{\delta E}{\delta x_1}, \frac{\delta E}{\delta y_1}, \frac{\delta E}{\delta z_1}, \frac{\delta E}{\delta x_2}, \cdots \frac{\delta E}{\delta z_n} \right) \tag{4.1}$$

$$X_{i+1} = X_i - \lambda_i \frac{A_i}{|A_i|} \tag{4.2}$$

where $\lambda_i$ is the step size. The main advantage of the steepest-descent method is that each cycle requires relatively little computational time. For starting points far away from the minimum, optimization is relatively rapid and reasonably good. However, close to the minimum the convergence slows because the shifts go to zero as the first derivative goes to zero. Therefore, the steepest-descent procedure is only recommended for initial optimization of very distorted geometries.

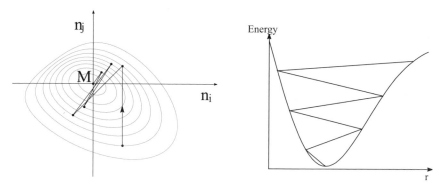

**Figure 4.6:** The steepest-descent minimization method.

### 4.2.3  Conjugate-Gradient Methods

In the steepest-descent method the gradient is calculated after each iteration. Therefore, depending on the surface, the search direction can change at each step if the molecule is moved through the minimum at each step. This can be avoided if the history of gradients is stored and used to modify subsequent steps. In a

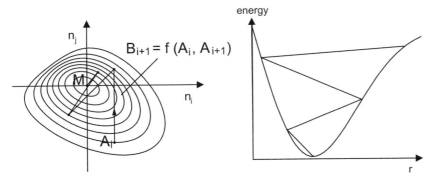

**Figure 4.7:** The conjugate-gradient minimization method.

sense, the history is a map of the curvature obtained without directly calculating and storing the second derivatives. An important difference between conjugate-gradient and steepest-descent algorithms is that each step in the former is one-dimensional, i.e., the minimization occurs along a line, and from each lower-energy point found on the surface the search direction is changed (Fig. 4.7).

The computation involved in each cycle is more complex and time consuming than for the steepest-descent method but convergence is generally more rapid. Two commonly used examples are the Fletcher-Reeves and the Polak-Ribière methods[175–178].

### 4.2.4   The Newton-Raphson Method

Similar information to that obtained during a conjugate-gradient refinement is obtainable from second derivatives (curvature)[13,64,175–178]. For a harmonic function the gradient (linear matrix of first derivatives, [**A**]) multiplied by the curvature (Hessian matrix of second derivatives, [**C**]) should lead directly to the shifts ($\Delta$**X**) to be applied in order to move toward the minimum (Eq. 4.3).

$$[\mathbf{C}]\Delta\mathbf{X} = -[\mathbf{A}] \qquad (4.3)$$

Because the potential energy surfaces are close to harmonic near the energy minimum, the Newton-Raphson technique converges very efficiently for molecules close to the optimum structure. However, for molecules far from the energy minimum, Newton-Raphson minimization can be unstable. Another disadvantage arises from the fact that the Hessian matrix has $(3N-6)^2$ elements*, which leads to large amounts of computer memory being required for large molecules.

In the block-diagonal Newton-Raphson minimization the generally small size of the off-diagonal terms is exploited and the matrix describing the curvature is re-

---

* If advantage is taken of its symmetry, the size is $(3N^2/2) + 3N/2$.

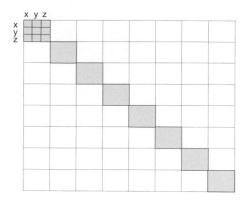

**Figure 4.8:** Arrangement of the blocks in the block-diagonal Newton–Raphson minimization produce.

duced to $N$ $3 \times 3$ matrices, i.e., to $9N$ elements (Fig. 4.8). Due to the approximations involved, convergence is slower but each cycle can be computed more rapidly. The choice of which of the two methods is more appropriate is mainly dependent on the size of the molecule. The block-diagonal Newton-Raphson method cannot readily yield vibrational frequencies and convergence cannot be confirmed.

### 4.2.5   Least-Squares Methods

Least-squares methods are usually used for fitting a model to experimental data. They may be used for functions consisting of square sums of nonlinear functions. The well known Gauss-Newton method often leads to instabilities in the minimization process since the steps are too large. The Marquardt algorithm[179] is better in this respect but it is computationally expensive.

### 4.3   Constraints and Restraints

A systematic analysis of a potential energy surface often requires minimization of the strain energy with one or more internal coordinates fixed. Successive variation (stepping) of fixed internal coordinates can be used in the analysis of activation barriers and energetic minima along a reaction coordinate. Applications include:

– the analysis of the mechanism and the determination of the energy barrier for the conformational interconversions of ring systems[180,181];
– the analysis of conformational preferences of ligand systems as a function of the metal centers to which they are coordinated and the prediction and analysis of coordination geometries as a function of metal-ligand bond distances[65,89,182];
– the prediction and analysis of metal ion selectivities of ligand systems[66,120];
– the calculation of activation barriers of outer-sphere electron transfer processes (see Chapter 11).

Some of these applications are discussed in detail in various chapters of Part II.

Two methods are used in molecular mechanics to fix internal coordinates: restraints and constraints. Applying restraints involves stiffening the corresponding potential energy function via excessively strong force constants in combination with a variation of the values for equilibrium distances and angles[183]. Constraints are applied by mathematical fixation of internal coordinates by using Lagrangian multipliers[183–185]. Although constraints offer a number of advantages compared with restraints (see also Chapter 9) they are not yet implemented in many programs[52] and are only available when full-matrix second-derivative methods are used. Convergence to a specified value is only accurately possible with a constraint. Also, the strain energy does not have to be adjusted for the additional strain induced by an artificially steep potential function. However, in practice, similar results can usually be obtained using constraints or restraints. Recently, a new method, based on Langrangian multipliers has been developed that allows the constraint of the sum of internal coordinates (see also Chapter 9)[186], and this is not possible in a general way using restraints.

# 5 The Multiple Minima Problem

In many applications of molecular mechanics it is important that the lowest energy structure, i.e., the global energy minimum, be found. This is particularly true when one would like to model the structures that are predominant in solution, or establish the distribution of all the conformers[65].

The energy-minimization routines employed by molecular mechanics programs usually refine the starting geometry to a local minimum which is not necessarily the global minimum. The aim of a conformational search is to find as many low energy minima as possible, which hopefully include the global minimum. In order to do this a large number of high energy starting conformations are generated, which are then minimized, compared with previously found conformers and stored if they have low strain energy and are unique. The probability of finding a new conformation is proportional to the number of undiscovered conformers, and thus the yield of new conformers decreases as the search progresses. This can be used as a qualitative indication of how far the search has progressed. Three methods can be used to generate the starting geometries: deterministic or grid searches that cover all areas of the potential energy surface systematically, stochastic or Monte Carlo methods, which use a random element to generate starting geometries, and molecular dynamics.

An excellent comparison of the methods employed in conformational searching of organic molecules has been published[187]. In the vast majority of published inorganic molecular modeling studies no stochastic or molecular dynamics conformational searches have been conducted. There are three reasons for this:

- Coordination of a ligand to a metal reduces the ligand flexibility and therefore the number of conformations available to the metal-ligand system is reduced.

- In many cases a type of deterministic search is conducted, where all probable structures are entered individually and minimized separately. The possible structures are generated by drawing all possible combinations of the chair, boat and skew boat conformers for six-membered rings, and λ and δ conformers for five membered rings[65,89,150,188,189].

- Methods using internal coordinate (torsion angle) frames, which are very common in organic chemistry, are not always applicable to inorganic systems.

There have, however, been a number of recent studies involving conformational searching of metal complexes. The geometries of flexible side-chains have been investigated using molecular dynamics[190], and conformational searching using Monte Carlo and molecular-dynamics methods have been carried out on transition metal complexes[37] and metalloproteins[191].


## 5.1   Deterministic Methods

For a thorough search, the crude starting geometries need to be spread over the entire potential energy surface. If only part of the surface is covered one cannot be sure that the global minimum will be found. The most reliable method for finding the global energy minimum is by systematically (deterministically) scanning the entire potential energy surface. It is well known that the bond angles and distances do not change much between different conformations of a molecule and that the major variations are in the torsion angles[192]. For this reason deterministic searches systematically vary all the torsional angles in a molecule. The step size of the torsional variation is critical in a search of this type. If the steps are too small, excessive computational time is required, but if they are too large, some conformations might be missed. The advantage of a systematic search is that it covers all areas of the conformational space; however, for large or flexible molecules the CPU time required can be prohibitive. The dimensions of the problem can be demonstrated by reference to the linear alkanes (Fig. 5.1).

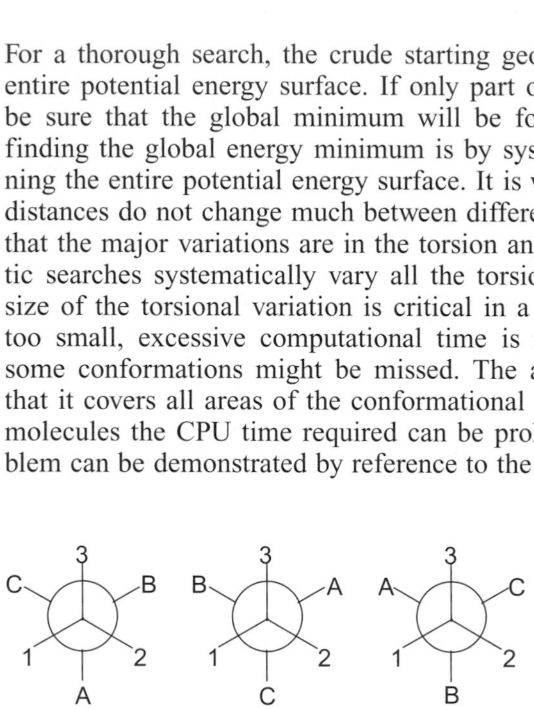

**Figure 5.1:** The conformation of alkanes.

With a chain length of $n$ the number of conformers is $3(n-1)$, thus 'hexane' has 243 conformations and 'decane' 19,683 (these are only non-degenerate if each carbon is uniquely substituted).


## 5.2   Stochastic Methods

Stochastic searches are not conducted, as suggested by the name, in a totally random fashion, as the variation from one starting conformation to another is limited in magnitude. Thus, one starting geometry will have some similarity to the next. The most effective way of generating starting geometries is to perturb the starting geometry that has been perturbed least often among all the conformations that meet the energetic requirements (e. g., 15 kJ mol$^{-1}$ from the energy minimum). In this way all the energetically acceptable conformations are used as starting geometries as often as each other. Always perturbing the conformation of the last optimized structure to generate a new starting geometry can lead to some areas of the potential energy surface being sparsely sampled.

The most commonly used stochastic methods are the torsional Monte Carlo method[193] and the cartesian stochastic (or random kick) method[37,191,194]. The two methods differ in the coordinate system in which they operate. The torsional Monte Carlo method uses internal coordinates, while the random kick method uses cartesian coordinates. The advantage of using internal coordinates is that the molecular degrees of freedom are reduced. The reason for choosing torsional angles as the variable internal coordinate is the same as that discussed earlier in connection with deterministic methods, namely that bond lengths and valence angles do not differ significantly between conformations, whereas dihedral angles do.

In each Monte Carlo step a random number of torsional angles are varied by a random amount, generating a new starting geometry that can be optimized. For cyclic systems the rings need to be cleaved at one point so that all the other torsional angles can be varied, and a ring closure constraint needs to be set to prevent the two ends of the cleaved ring from being too close or far apart from each other. When two or more rings adjoin each other, as is often the case for coordination complexes, the choice of the ring cleavage site becomes restricted and varying torsional angles is cumbersome. These types of systems are best handled with external coordinates or molecular dynamics.

In a cartesian stochastic search a random kick is applied to each atom in the molecule, generating a new starting geometry for energy minimization. In the conformational search of cycloheptadecane it was shown that the maximum kick size should be between 2.7 and 3.1 Å[187]. Larger kicks distort the molecule so badly that the average time for energy minimization was increased dramatically. The probability of falling back to the starting geometry increased when kicks smaller than 1.5 Å were applied[187]. The advantages of the kick method are that it can easily be applied to inorganic systems, and that it is easily implemented in force field programs.

## 5.3  Molecular Dynamics

Molecular dynamics involves the calculation of the time dependent movement of each atom in a molecule[39]. This is achieved by solving Newton's equations of motion. For this process the energy surface and the derivative of the energy in terms of the nuclear coordinates are required (Eqs 5.1, 5.2; mass $m$, acceleration $a$, potential energy $E$, coordinates $r$, time $t$).

$$F = m \cdot a = -\frac{dE}{dr} = m\,\frac{d^2r}{dt^2} \tag{5.1}$$

$$\frac{d^2r_i}{dt^2} = a_i = \frac{F_i}{m_i}; \quad F_i = \frac{\delta E}{\delta r_i} \tag{5.2}$$

The maximum time step depends on the highest frequencies (usually R–H bonds). One of the main problems in dynamics calculations is that the movements leading to appreciable conformational changes usually have lower frequencies (milliseconds, e. g., torsions). Depending on the size of the molecule the computation of such long time intervals is usually prohibitive due to the CPU time and storage space involved.

Structures for starting geometries are normally sampled as a function of time or geometry during a molecular dynamics run of at the most a few nanoseconds. Therefore, molecular dynamics is efficient at exploring local conformational space but it is not effective at crossing large energy barriers, and so it is not suited for global searches. To achieve faster and more complete molecular dynamics, searching at high temperatures can be used. However, sampling at temperatures that are too high (>1000 K) is not constructive because a large proportion of resulting minima are high-energy conformers. In order to prevent the molecular dynamics search from going to areas of the potential energy surface that have already been searched, a penalty can be assigned to sampled points[195]. In this way the molecular dynamics search can be made more global.

## 5.4   Practical Considerations

A good conformational search requires a method for generating starting geometries that cover all of the potential energy surface. Once generated, the starting geometries need to be optimized. Since a very large number of structures are produced in a conformational search, highly efficient methods for refining the structures, determining duplicates and high-energy conformers, are needed. The refinement process can be accelerated by using a pre-refinement cycle with rigid geometry or with the pure central force field approach which was developed and successfully tested for carbohydrates[60].

The stochastic and deterministic search methods are not influenced by high-energy barriers and are efficient at scanning large areas of the potential energy surface. On the other hand molecular dynamics, due to its difficulty in overcoming energy barriers, is predominantly a technique used for searching local conformational space. A common procedure for screening the potential energy surface therefore involves a combination of the three methods[196] (Fig. 5.2).

One obvious limitation of all of these methods is that the energy surface is not known and thus, there is no certainty of finding the global energy minimum. One

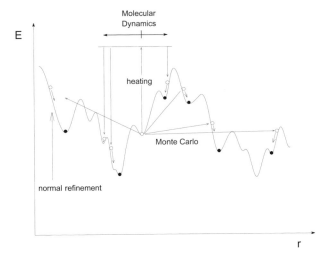

**Figure 5.2:** Conformational searching procedures.

never knows whether the lowest-energy structure found is indeed the global mini-mum of the system.

## 5.5 Making Use of Experimental Data

More direct and less time-consuming methods for searching conformational space may be used if the structure of an *existing* compound is to be determined. Be-cause electronic and other molecular properties are dependent on the molecular structure (see Fig. 1.1 in Chapter 1), experimentally determined data can be used in combination with the computation of molecular properties to find the solution structure[164,197]. Thermodynamic, kinetic and/or electronic properties may be used but spectroscopic data is usually the most informative. Methods that have been used to determine solution structures with this general approach include combinations of molecular mechanics with experimentally determined isomer dis-tributions[164,198,199], with NMR spectroscopy[164,200−203], with EPR spectroscopy and spectra simulations of dinuclear compounds[65,164,165,204−206], and with EPR and UV-vis-NIR spectroscopy in combination with Angular Overlap Model (AOM) calculations[65,90,164,166,201]. Some of these methods will be discussed in more detail in Part II.

# 6   Conclusions

Molecular mechanics is an empirical method based on simple elements of theory that every user can and should understand. With modern software the user is able to control the calculations in terms of the energy minimization routine, the potential energy functions and the force field parameters used. A significant advantage of molecular mechanics calculations is that they are relatively rapid and therefore that large series of calculations can be performed.

If molecular mechanics is to be a valid modeling tool for the design of new compounds and the interpretation of experimental results, the compounds under consideration must belong to a class for which the molecular mechanics model is verified. In other words, the accuracy of the results obtained depends critically on the parameterization of the force field and how this has been obtained (Fig. 6.1).

Here, it is of importance to stress that, ideally, all calculations falling in a given area should produce data of similar quality. There is no advantage in developing a model that generally produces results of high accuracy if a small percentage of the calculations fail to do so. This is clearly intolerable if the method is used for

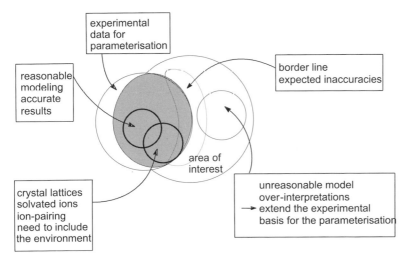

**Figure 6.1:** Liminations of molecular mechanics.

design and/or prediction of structures and properties. The philosophy is to accept a generally lower degree of accuracy or a more specific parameterization scheme.

As with any empirical model, there always exists the possibility of improvement. However, improvements should preferably not increase the complexity of the model since the main appeal of molecular mechanics is its simplicity. Instead, improvements should concentrate on the functional forms, the parameterization, and on the choice of experimental data used in developing approximations of the energy surfaces.

# Part II:  Applications

In Part II applications of molecular mechanics and related methods in inorganic chemistry are discussed. We have attempted to give a wide ranging coverage in terms of all the areas of interest to inorganic chemists. However, in each chapter, we have only selected the most instructive studies and these are discussed in detail. The aim is to show the types of problems that can be addressed using molecular modeling, and to indicate the directions being taken in an attempt to overcome existing limitations. Comprehensive reviews of each of the areas are not the aim of this book. In a review[123] covering the literature up to 1992, a table reporting molecular mechanics studies of coordination compounds had 613 entries. Many more have appeared since then, and there have been many reviews covering various aspects of modeling of inorganic systems (see Appendix 4).

# 7   Structural Aspects

Two fundamental types of information are obtained from any molecular-mechanics study, the minimum value of the strain energy and the structure associated with that minimum. Agreement between the energy-minimized and experimental (crystallographic) structures has often been used as the primary check on the validity of the force field and to refine the force field further, but often little predictive use has been made of the structures obtained. As force fields have become more reliable, the potential value of structure predictions increases. More importantly, when no unequivocal determination of a structure is available by experimental methods then structure prediction may be the only means of obtaining a three-dimensional model of the molecule. This is often the case, for instance, in metal-macromolecule adducts, and structures obtained by molecular mechanics can be a genuine aid in the visualization of these interactions. In this chapter we consider the ways in which structure prediction by molecular mechanics calculations has been used, and point to future directions.

## 7.1   Accuracy of Structure Prediction

The accuracy of molecular mechanics models in terms of reproducing known structures varies enormously, depending on the quality of the force field. The best-established force fields for metal complexes are for hexaaminecobalt(III) cations. Many such complexes have been modeled and the agreement is generally $\pm 0.01$ Å in terms of bond lengths, $\pm 2°$ in terms of valence angles and $\pm 5°$ in terms of torsion angles[57,65,120,123,150]. It is unlikely that this level of agreement can be improved on substantially because the disagreement is either close to the level of precision of the crystallographic experiment or close to the level of variation observed in crystallographic structures as a result of crystal packing and hydrogen bonding. The same is true for some other transition metal complexes with amine and a few other simple ligand systems, but in many other cases the reliability of structural predictions is lower.

The ability to predict structures can be important in the design of ligands. Chapter 8 deals with stereoselectivities predicted and interpreted by molecular mechanics. The use of molecular mechanics to design metal-ion-selective ligands is discussed Chapter 9. Spectroscopic and electrochemical properties are intimately related to the structure. For instance, the energy of the d-d transitions varies according to the metal-ligand bond length raised to an inverse power of 5 or 6. Thus, the ability to accurately predict, say, a Co$-$N bond length means that it should be possible to design a ligand with a given absorption spectrum (see Chapter 10). There is also the potential to design ligands that promote a particularly fast or slow electron transfer rate based on structural features, or reductants and oxidants with a given redox potential (see Chapter 11).

## 7.2   Molecular Visualization

High quality computer graphics is now widely available and is used as an aid in visualizing molecules, molecular processes, and intermolecular interactions. In some cases molecular graphics has replaced the building of physical (e.g., Dreiding) models but the potential applications are much greater. In order to visualize molecules they need to be constructed in a reliable manner and molecular mechanics is an ideal tool for doing so.

Potential applications of molecular visualization include structures and processes not accessible by experiment. For example, $[Co(en)_3]^{3+}$ (en = ethane-1,2-diamine) can adopt four conformational isomers, *lel₃*, *lel₂ob*, *ob₂lel* and *ob₃* (Fig. 7.1). Crys-

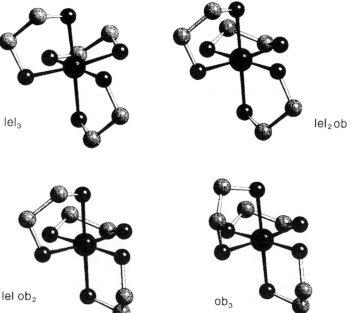

lel₃                                 lel₂ob

lel ob₂                    ob₃

**Figure 7.1:** The four nondegenerate conformations of $[Co(en)_3]^{3+}$.

tal structure analyses of the first three have been reported[208–210] but the fourth is sufficiently unstable not to have been observed. It has, however, been modeled by molecular mechanics[207,211] and therefore can be readily visualized.

Conformational interconversion is a process that cannot be observed experimentally. However, the energy profile associated with interconversion can be mapped using molecular mechanics. Then, representative structures along the minimum energy pathway can be predicted and visualized. For example, the mechanism of conversion from the $\lambda$ to the $\delta$ conformer of $[Co(en)(NH_3)_4]^{3+}$ has been calculated in this way[180].

Shown in Fig. 7.2 are the relevant structural parameters, schematic representations of the extreme, intermediate and transition state conformations of the chelate ring, and the calculated energy profile. Heavy equipotential lines are spaced by

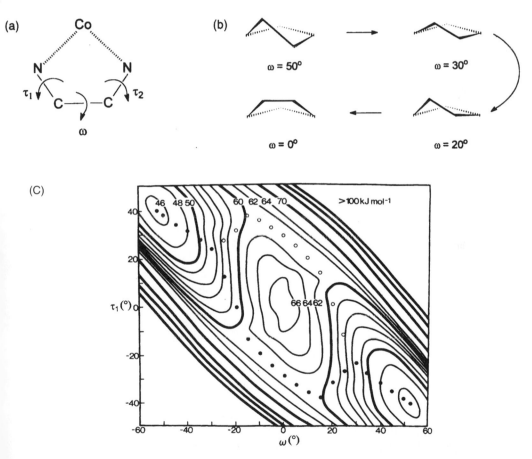

**Figure 7.2:** Conformational interconversion of the five-membered diamine chelate ring of $[Co(en)(NH_3)_4]^{3+}$. (a) Nomenclature of the torsional angeles. (b) Representation of the extreme, intermediate and transition state conformations. (c) Graphical representation of the potential energy surface (see text). Taken from [180]. © John Wiley and Sons, Inc, 1987.

10 kJ mol$^{-1}$, light lines by 2 kJ mol$^{-1}$ (see the labeling of some of the equipotential lines). Also shown in Fig. 7.2 are the two symmetrically related lowest-energy pathways for the $\delta$-$\lambda$ interconversion.

A more common use of molecular visualization is the interaction of metals with macromolecules. Examples of the use of this technique are given in Chapter 13. Here, we note that metal-macromolecule interactions are inevitably complex, and molecular modeling and visualization can be of enormous assistance in understanding the nature of the interactions and the factors that mediate them. However, the complexity of the interactions also leads to the expectation that any model generated is but one possible representation of the interaction and corresponds to one of the many possible energy minima on a complex potential energy surface. Force fields for macromolecules are less highly tuned than those for smaller molecules but, given the qualifications referred to above, this is probably not a serious restriction. Rather, in this field, the use and interpretation of the currently available molecular mechanics models is generally restricted to molecular visualization, and in most cases, little quantitative significance should be attached to the strain energies calculated.

## 7.3   Isomer Analysis

The energetics of isomer prediction using molecular mechanics is discussed in detail in Chapter 8. One of the results of such a study is the structure of each of the isomers. The archetypal studies in this field relate to the complex $[Co(dien)_2]^{3+}$ (dien = 3-azapentane-1,5-diamine; see Chapter 8). Other important studies include those on macrocyclic ligands (see also Chapter 9). Tetraaza macrocyclic ligands, for example, can adopt a series of configurational isomers, and these have been the subject of numerous molecular mechanics calculations. Consider an equatorially coordinated tetraaza macrocycle. Each of the amine groups can coordinate with the amine proton or substituent disposed above or below the coordination plane. How many isomers result depends on the symmetry of the macrocycle. For example, in the classic case of cyclam (cyclam = 14-aneN$_4$ = 1,4,8,11-tetraazacyclotetradecane) there are five *trans* isomers [212] and these are shown schematically in Fig. 7.3. It is not always possible to prepare or separate all of these isomers and, therefore, in many cases only a minority have been structurally characterized. Thus, the energy minimized structures represent the best available three dimensional representations of the other isomers. *Cis*-isomers are also possible and using conformational searching procedures it is now possible to generate these automatically (see Section 5.2) [37].

The nickel(II) complexes of the closely related macrocycle, tetra-*N*-methyl-14-aneN$_4$ (tmc), can adopt the same set of isomers. It has been noted that the isomer observed depends on whether nickel(II) is four-, five- or six-coordinate [213]. Molecular mechanics modeling of this system and prediction of each of the structures allowed an analysis of the specific interactions between axial ligands and the

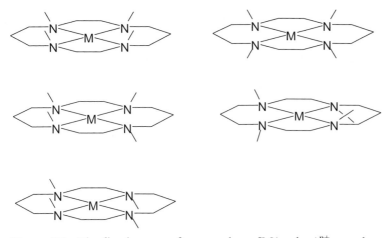

**Figure 7.3:** The five isomers of square planar [M(cyclam)$^{n+}$ complexes.

macrocycle to be carried out[153]. In this way a possible explanation for the experimental observations was arrived at. Specifically, it was concluded that the interactions between the methyl substituents on the ring and any axial ligands control the stabilities of the different isomers[153].

## 7.4 Analysis of Structural Trends

Cage complexes of the type shown in Fig. 7.4 have presented an excellent opportunity to study how the coordination geometry varies as a function of the metal ion when presented with a constant ligand geometry[89,90,159,214,215]. It has been noted that the trigonal twist angle $\phi$ varies from 28 to 55° through the first row transition metal elements (Fig. 7.4)[159]. The question arises as to whether this is a consequence of bond length variation or the electronically controlled preferences of the metal ion. Both, a ligand field[159] and a molecular mechanics analysis[89] of the series reproduced the twist angles well. It was concluded that simple steric factors predominate and that electronic factors exerted by the metal center are accounted for in the molecular mechanics analysis by the metal-ligand stretching function[65] (see also Sections 3.2.2, 3.6 and 12.1).

Where steric factors alone cannot account for observed trends, then electronic factors can be assumed to be playing a role. In this sense molecular mechanics is the ideal tool for factoring out the steric aspects of structural variability so that the electronic factors can be more readily analyzed. A good example is the analysis of transition metal complexes with ammonia, and with primary, secondary and tertiary amines. The increasing nucleophilicity of the amines (bond shortening) is paralleled by increasing ligand-ligand repulsion (bond lengthening). The two

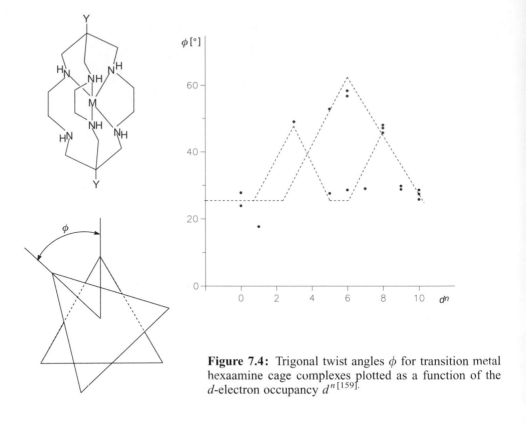

**Figure 7.4:** Trigonal twist angles $\phi$ for transition metal hexaamine cage complexes plotted as a function of the $d$-electron occupancy $d^{n}$[159].

structurally opposing effects can be analyzed by molecular-mechanics calculations and ligand-field analysis, respectively[90] (see also Section 10.2).

## 7.5  Prediction of Complex Polymerization

It is rare for a single complex to form as both dimers and monomers but just such a situation has been reported for copper(II) complexes of 1,8-diamino-3,6-diazaoctane-4,5-dione (Fig. 7.5)[216].

A molecular mechanics analysis of this system revealed that the dimer is more stable than the monomer and that the strain inherent in the ligand coordination in the monomer is primarily responsible for the energy difference. Selective formation of oligomers is an important part of the synthesis of large helical metal complexes and oligonuclear transition metal catalysts, and this result suggests that molecular mechanics could be used to design ligands that would increase the likelihood of oligomerization.

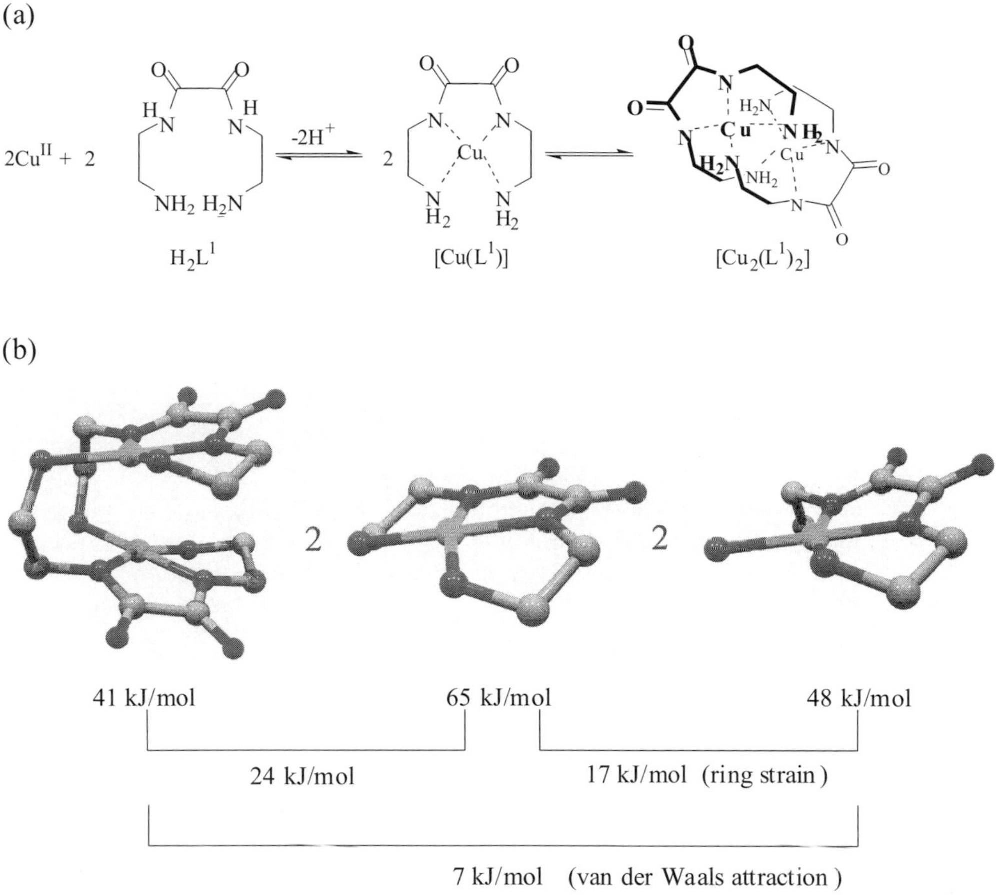
(a)
2Cu^II + 2
-2H+
H2L1
[Cu(L1)]
[Cu2(L1)2]
(b)
2
2
41 kJ/mol
65 kJ/mol
48 kJ/mol
24 kJ/mol
17 kJ/mol (ring strain)
7 kJ/mol (van der Waals attraction)

**Figure 7.5:** Dimerization of a copper(II) compound with a diaminodiamide ligand, (a) stoichiometry of the reaction; (b) reason for the dimerization process assumed on the basis of molecular mechanics calculations; reproduced with permission from reference [216].

## 7.6 Unraveling Crystallographic Disorder

Conformational disorder is a common phenomenon in crystal structure analyses of metal complexes of chelate ligands. For instance, en (ethane-1,2-diamine) rings can adopt two conformations, $\lambda$ or $\delta$ (see Fig. 7.1 and 7.2), and frequently both, or a smeared average of the two, are observed. If there is more than one such ring disordered, then it is often not possible to determine what overall conformations are adopted. For instance, in the case of [Fe(*trans*-diammac)]$^{3+}$ (Fig. 7.6), there are two five-membered chelate rings, *trans* to one another, and in the crystal structures, both show evidence of conformational disorder [182,207,217]. Thus, it is

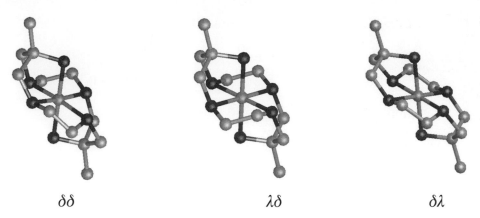

$$\delta\delta \qquad\qquad \lambda\delta \qquad\qquad \delta\lambda$$

**Figure 7.6:** The three conformers of [Fe(*trans*-diammac)]$^{3+}$.

not possible to determine which of the conformations $\delta\delta$, $\lambda\delta$ or $\delta\lambda$ (Fig. 7.6), are adopted, nor is it possible to determine the symmetry of the complex. A molecular mechanics analysis of this system revealed that the $\delta\lambda$ conformation had substantially higher strain energy than the other two [182,207] and this result was central to the interpretation of the solution EPR spectra [207] (see also Section 10.3).

A second example relates to the complex bis(1,5-diazacyclooctane)nickel(II) (Fig. 7.7). A crystal structure analysis of this complex revealed apparently planar NCCCN moieties with unusual bond lengths and angles [218]. A molecular mechanics analysis of the same complex showed that these chelate rings are puckered and concluded that the crystal structure result was a consequence of conformational disorder, the refined geometry corresponding to an average of two puckered conformations (see Fig. 7.7) [219].

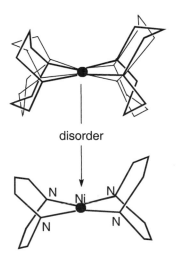

**Figure 7.7:** Structural disorder in bis(1,5-diazacyclooctane) nickel(II).

# 7.7   Comparison with Solution Properties

The use of molecular mechanics as an aid in the interpretation of spectroscopic data is outlined in more detail in Chapter 10. One of the most rapidly developing applications of molecular mechanics is the use of the structures to aid in the analysis of multi-dimensional NMR spectra[164,220]. This is particularly pertinent to the study of metal-macromolecule interactions where the spectroscopic data often has too low an observation/variable ratio to allow an unequivocal determination of the structure. Therefore, an additional source of structural information is needed. The number of studies involving metal ions has increased rapidly in recent years and examples are discussed in Chapters 10 and 13.

# 8 Stereoselectivities

## 8.1 Conformational Analysis

The pioneering applications of molecular mechanics to coordination compounds were isomeric and conformational analyses[1,2]. Recent applications involving the computation of conformer equilibria discussed in this chapter are studies of solution structure refinements[164,198,199], racemate separations[221-223] and the evaluation of conformer interconversion pathways[180,224,225]. The importance of conformer equilibria in the areas of electron transfer rates and redox potentials is discussed in Chapter 11, and many examples discussed in the other chapters of Part II indicate how important the prediction of conformational equilibria is in various areas of coordination chemistry.

The main problem that occurs when strain energies are used for the evaluation of relative stabilities have been discussed in Chapter 3, and they will be only briefly mentioned here:

- *Entropic* terms, other than statistical contributions, are often neglected.
- *Environmental effects*, such as solvation and ion pairing, are usually neglected.

Approaches related to overcoming these limitations will be discussed in the context of the examples presented ahead. Furthermore, the modeling of electronic effects is not usually addressed specifically. This can lead to additional uncertainties in terms of the computed structures and conformer distributions.

A carefully tuned force field is obviously critical, especially if energy differences as small as a few kilojoules per mole are the basis of an interpretation, as is often the case. Another problem is the choice of an appropriate experiment for comparison with the calculated data. This is especially true when inert compounds (for instance cobalt(III)amines) are studied, where equilibrated isomer mixtures have to be produced using special techniques[170,226]. Mixed ligand systems are especially difficult to handle since equilibration with inert metal centers might lead to preferential formation of mixtures of the homoleptic complexes[198,221], and for labile systems an excess of ligand needs to be used[223].

One of the most extensively studied systems is that of $[Co(dien)_2]^{3+}$ (Table 8.1).

**Table 8.1:** Experimentally determined and calculated isomer distributions of $[\text{Co(dien)}_2]^{3+\,\text{a)}}$.

| Experiment | *mer* [%] | *sym − fac* [%] | *unsym − fac* [%] | Ref. |
|---|---|---|---|---|
| (a) Experimentally determined distribution | | | | |
| $H_2O$, $X^{-\,\text{b)}}$ | 63 | 8 | 29 | 227 |
| $SO_4^{2-}$ (2M) | 37 | 25 | 38 | 227 |
| $PO_4^{3-}$ (0.08M) | 20 | 55 | 25 | 227 |
| $H_2O$, $CH_3COO^-$ (RT) | 66 | 7 | 27 | 227 |
| (80°) | 44 | 14 | 42 | 227 |
| $H_2O$ | 63 | 8 | 29 | 227 |
| MeOH | 53 | 18 | 29 | 227 |
| DMSO$^{\text{(c)}}$ | 80 | 6 | 14 | 227 |
| Acetone | 74 | 9 | 17 | 228 |
| (b) Molecular mechanics calculation | | | | |
| | 20 | 40 | 40 | 229 |
| | 34 | 34 | 31 | 230 |
| | 93 | 2 | 5 | 151 |
| | 66 (57)$^{\text{(d)}}$ | 2 (3)$^{\text{(d)}}$ | 32 (40)$^{\text{(d)}}$ | – |
| | 56 (48)$^{\text{(e)}}$ | 3 (4)$^{\text{(e)}}$ | 41 (48)$^{\text{(e)}}$ | – |

a) Structures

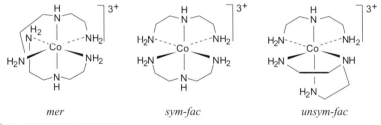

*mer*          *sym-fac*          *unsym-fac*

b) $X = ClO_4^-$, $Cl^-$, $Br^-$, $NO_3^-$
c) DMSO = dimethyl suloxide
d) force field [57]; 298K (353K)
e) force field ref [58,231]; 298K (353K) [57,58,231]

The *mer*-isomer is chiral ($C_2$) because of the two possible orientations of the proton at the secondary nitrogen atoms [232], the *unsym-fac*-isomer is also chiral ($C_2$) while the *sym-fac*-isomer is achiral ($C_i$). For the five-membered chelate rings all possible combinations of $\delta$ and $\lambda$ conformations (see Fig. 7.6 for the $\delta/\lambda$ nomenclature) have to be considered, leading to a total number of 40 isomers and conformers (some conformers are calculated to be unstable and can therefore be neglected [58,151]). The calculated distribution is based on partition functions (Eqs. 8.1, 8.2);

$$N_i\,[\%] = \frac{100 \exp\left(-E_i/RT\right)}{Q_{\text{tot}}} \tag{8.1}$$

$$Q_{tot} = \sum_i \exp\left(-E_i / RT\right) \qquad (8.2)$$

$Q_{tot}$ is the sum of the strain energies of all conformers considered, while $E_i$ and $N_i$ are the total strain energy and abundance of each individual conformer. The statistical factor, contributing to the energy term, is given by Eq. 8.3

$$\Delta S = RT \ln \rho \qquad (8.3)$$

($\rho$ is the number of ways in which a conformer can be formed). From Table 8.1a (experimentally determined data) it emerges that the isomer distribution is strongly dependent on the temperature, solvent, and counter ion.

The temperature dependence is described by the thermodynamic functions given in Eqs. 8.1–8.3. For two of the molecular mechanics calculations (last two entries in Table 8.1(b)) the abundances are given at the two temperatures where experimental values have been reported. While the expected and observed tendencies are well reproduced, the magnitude of the thermal effect is slightly underestimated. Similar results have been reported for other cobalt(III) hexaamines[170]. The small inaccuracy is presumably due to the neglect of entropic terms other than the statistical components. It is important to remember that the entropic terms are not only important in the prediction of the temperature dependence but also in the calculated abundances at each specific temperature. The calculated isomer ratios of [Co(dien)$_2$]$^{3+}$, especially the most recent ones (last two entries in Table 8.1(b)), are of satisfactory accuracy. Nevertheless, a more thorough treatment of the entropic terms might lead to an improvement. The problems and possible solutions associated with modeling entropy effects have been discussed in Section 3.8.

The influence of solvation and ion association on the relative stability of conformers is evident and quantitatively shown for the example of [Co(dien)$_2$]$^{3+}$ in Table 8.1. Why then are these effects neglected in force field calculations? Basically, because the inclusion of environmental effects is computationally intensive and is of questionable accuracy. Solvation and ion association are neglected to keep the modeling scheme as simple and as efficient as possible. In fact, if large or highly flexible systems are investigated, a thorough conformational analysis, including the computation of a large number of local minima, is only practical if some simplifications such as the neglect of environmental effects are adopted. A practical alternative is the use of molecular dynamics as it allows local minima to be investigated and facilitates the inclusion of solvent and ion interactions and entropy factors.

The question then is whether and in which cases meaningful results can be expected from the computation of "naked" molecules. It was pointed out that the relative abundance of an isomer may be related to its elution rate from an ion exchange column[233]. This is evidently so because both the abundance and elution rate are related to the stability of the solvated and ion-paired species. Aqueous solutions of chloride, nitrate and perchlorate do not lead to any chromatographic separation of the three isomers of [Co(dien)$_2$]$^{3+}$ [233]. These anions may therefore be regarded as "innocent", i.e., they do not become involved in any stereoselective interactions with any of the three isomers (first entry in Table 8.1a). Oxyanions such as sulfate and phosphate (the other two entries in Table 8.1a) are known to lead to enhanced separation in ion-exchange chromatography, and the energy differences obtained from the isomer distributions have been shown to be closely related to the elution rates of the three isomers from ion-exchange columns[151,233]. Thus, it seems that the effect of ion pairing and solvation can be neglected as long as "innocent" solvents and anions are used. Satisfactory results have been obtained in many systems when water was the solvent and halide, perchlorate and nitrate salts have been used to produce the experimental isomer ratios for comparison with the molecular mechanics calculations.

Despite the relatively good agreement between the experimentally determined (top line in Table 8.1a) and the calculated abundances (at least for the most recent studies which have involved a full conformational analysis and a well tested force field) for $[Co(dien)_2]^{3+}$, one should not forget that approximations, such as the neglect of entropic and environmental effects, lead to limitations in terms of the applications of molecular mechanics. It should be possible, with improved methods, to account for these effects, so that the remaining uncertainties can be removed without using prohibitive amounts of CPU time. An interesting development in this context is the treatment of solvation as a statistical continuum[167], in which the energy term and its derivative may be calculated analytically, resulting in only a moderate increase of computational time for the energy minimization.

Methods for conformational searching have been discussed in Chapter 5. Until recently there have been few applications of Monte Carlo searching and molecular dynamics to investigating the conformational isomers of metal complexes. The random kick method has been used to investigate bis(tridentate) and macrocyclic cobalt(III) complexes[37] and torsion angle searching has been used to investigate the geometry about the active site in urease[191].

## 8.2   Enantioselectivities

The production of enantiomerically pure chemicals is one of the more demanding and important challenges in chemical research, both at fundamental and at applied levels. Enantioselective synthesis and racemate separation rely on specific interactions between a chiral phase and a substrate. Thus, molecular modeling may be applied to the design of new, highly efficient, chiral phases that are able to selectively stabilize a specific adduct with the substrate, leading to the desired configuration of the product. Many enantioselective syntheses are induced by chiral transition metal compounds, and stereoselective ligand exchange on chiral metal complexes has been used for enantiomer separation. Application of molecular mechanics to both approaches will be reviewed in this section.

## 8.2.1   Racemate Separation

Racemate separation by stereoselective ligand exchange occurs when a chiral matrix complex has additional coordination sites that are capable of readily exchanging a racemic substrate ligand. The chiral induction, i.e., the efficiency of the matrix complex, is related to the product distribution which depends on the relative stabilities of the complexes with the two enantiomers of the racemic substrate (Fig. 8.1). The problem to be solved in the design of effective chiral matrix complexes for specific racemic substrates is therefore related to isomeric analyses of the type discussed in Section 8.1.

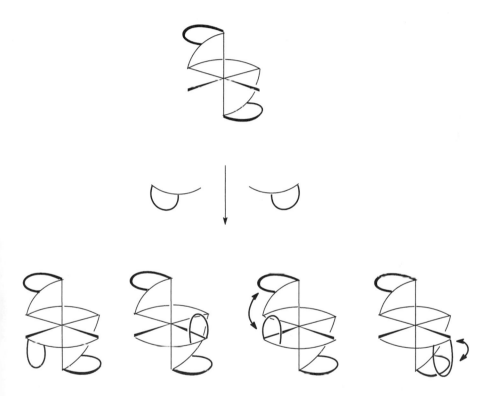

plus other possible isomers and conformers

**Figure 8.1:** Racemate separation by stereoselective ligand exchange.

An important condition for chiral matrices is that they need to form labile interactions with the substrate in order to facilitate both the recovery of the enantioselectively coordinated substrate ligand and the recycling of the chiral matrix. Usually copper(II) complexes have been used[234]. Due to the problems involved in the modeling of Jahn-Teller distorted copper(II) complexes (see Chapter 12 for a detailed discussion on Jahn-Teller effects) nickel(II) complexes have been used in the molecular mechanics design. They too are relatively labile. Some of the systems described in the literature are presented in Table 8.2.

**Table 8.2:** Racemate separation by stereoselective ligand exchange.

| Chiral matrix [a] | Racemic substrate [a] | Calculated (%) | | Observed (%) | | Reference |
|---|---|---|---|---|---|---|
| | | S | R | S | R | |
| Co$^{III}$((R)-trab) | pn | 56 | 44 | 54 | 46 | 221 |
| Co$^{III}$((S),(S)-ppm) | pn | 55 | 45 | 51 | 49 | 222 |
| Ni$^{II}$((S),(S)-ppm) | pn | 44 | 56 | 43 | 57 | 222 |
| | pam | 53 | 47 | 56 | 44 | 222 |
| | ala | 58 | 42 | 60 | 40 | 223 |
| | val | 69 | 31 | 75 | 25 | 223 |
| | leu | 67 | 33 | 68 | 32 | 223 |
| Ni$^{II}$((S),(S)-epm) | pn | 62 | 38 | 57 | 43 | 222 |
| | pam | 27 | 73 | 30 | 70 | 222 |

[a] Structures

The difference between calculated and experimentally determined distributions is generally not larger than 5%. This accuracy is clearly reasonable in the light of the approximations discussed above. More importantly, the error limit of ca. 5% seems to be general[221−223], i.e., the method is reliable. Two additional factors emerge from Table 8.2:

- As expected, the effectiveness of a chiral matrix is related to the substrate to be resolved, i.e., a specific matrix complex might be effective in the resolution of one substrate but it might not efficiently resolve another.
- Predictions based on simple molecular models can not replace a molecular mechanics study involving all possible conformers.

Instructive examples for cases where simple intuition might fail are the reversal of selectivity of $\{M^{n+}((S),(S)\text{-ppm})\}$ with respect to *rac*-pn for M = Co(III) and Ni(II) (see Table 8.2 for ligand abbreviations), the increased selectivity towards *rac*-pam of $\{Ni^{II}((S),(S)\text{-epm})\}$ compared to $\{Ni^{II}((S),(S)\text{-ppm})\}$, and a reversal of selectivity with the same matrix complexes towards *rac*-pn (see Table 8.2). All of these effects are accurately predicted with molecular mechanics.

The examples shown in Table 8.2 do not show a particularly high enantioselectivity. However, with an enantiomeric excess of 50% (75/25), material of acceptable optically purity can be obtained in five cycles, and for chromatographic resolutions, separation factors greater than one are sufficient for efficient resolution processes [222,223]. Nickel(II) complexes of the type shown in Table 8.2 have been used to modify ion-exchange resins that were used for racemate separations [222], and derivatives of $(S),(S)$-ppm with functional groups that may be fixed to supports are readily available [235].

Other methods for racemate separation include fractional crystallization of diastereomers with optically pure counter ions [236], conglomerate crystallization [237], and various chromatographic techniques incorporating both mobile and stationary chiral phases [238]. All of these techniques involve intermolecular interactions of some kind. While it is theoretically possible to model these, using molecular mechanics, the complexity of the problems involved has so far been prohibitive when the time required for a thorough design is compared with experimental trial-and-error studies.

## 8.2.2 Stereoselective Synthesis

The application of molecular mechanics to enantio- and diastereo-selective synthesis is less straight-forward, and publications in this area have only started to appear recently. In the case of the racemate separations described above, the isomer abundances of equilibrated solutions are taken to be related to the energy of all local minima. In contrast, in order to predict the enantiomeric excesses arising from chiral syntheses, the reaction mechanisms and the structures of relevant intermediates or transition states have to be known since their relative energies need to be calculated in order to predict the enantiomeric excesses. Thus, it is to be expected that quantum-mechanical methods such as DFT, in conjunction with molecular mechanics will provide the best insights into enantioselectivity [239] (see also Section 2.2).

Also, metal ion directed stereoselective syntheses often involve organometallic complexes. While, in terms of a molecular-mechanics description of the structures, there is no fundamental difference between metal-carbon and metal-heteroatom bonds, modeling π-bonded ligands is not trivial.* Given a known reaction mechanism (which is not possible for many catalytic reactions) the main problem is the parameterization of the potential energy functions for the intermediates and transition states. The problem is that force field parameters are generally carefully

---

* Various aspects of molecular mechanics of organometallic compounds are discussed in Chapter 14.

fitted to experimental results, i.e., structures or other data related to the output of force field calculations of the type of compound to be modeled have to be available. For short-lived transition states this is a considerable problem, but it can be addressed using quantum-mechanical calculations[239,240].

Possible applications and limitations of mechanistic interpretations based on molecular mechanics calculations are discussed later in this chapter. Here, some studies involving molecular modeling of stereoselective syntheses are discussed[241–244]. The two modeling studies of the asymmetric hydrogenation catalyzed by chiral bisphospine rhodium complexes (Fig. 8.2) both use a crude model based on rigid ligand geometries and the minimization of the van der Waals energy as a function of the angular geometry (P-Rh-O angle in Fig. 8.2)[241,242]. In this respect these studies are comparable to the early work in the area of coordination compounds[1,2]. As in those early studies, a cautious interpretation of the data allows some important conclusions to be drawn. From the eight possible reaction trajectories only two are energetically allowed, and these lead to intermediates that have been detected experimentally[241]. However, for quantitative interpretations, eventually leading to a prediction of the enantioselectivity, a more thorough method based on refinement of the entire structure with a well-established force field has to be awaited, especially since some of the structural features of the calculated molecules did not seem to be reasonable[241].

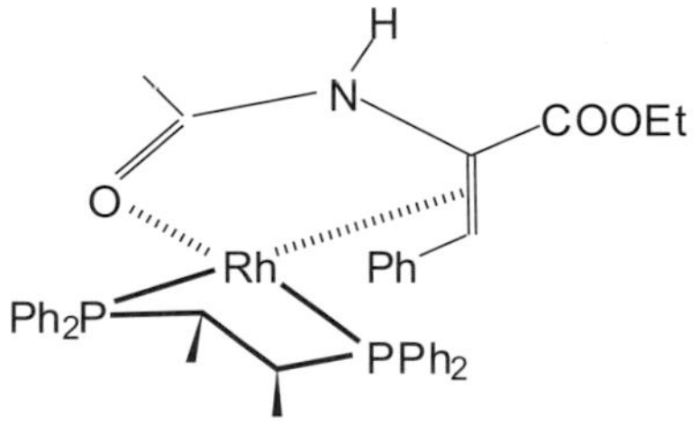

**Figure 8.2:** The chiral (phosphine)rhodium(I) catalyst used for asymmetric catalytic hydrogenation of an amino acid precursor.

Molecular mechanics was also used to model enantioselective metal-carbene transformations catalyzed by chiral dirhodium(II) compounds[243]. Here, a considerably more thorough approach was used, and the experimental structures of the catalysts were accurately reproduced. A difficulty encountered in this study was the parameterization of the metal-carbene intermediate. This might be part of the reason why in some cases the predicted enantioselectivities were opposite to those observed[243].

Metal-π-allyl complexes are important in a number of stereoselective catalytic reactions and therefore represent an interesting challenge to computational chemists (see also Section 14.2). An empirical force field study, based on the MM2 parameterization scheme, aimed at predicting stereoselective nickel(0)-catalyzed cycloadditions, was recently conducted[244]. As in a similar study[245], where a force field for the structure optimization of palladium allyl systems was developed, dummy atoms were needed to define the structural model. A significant improvement on this model has been achieved using an initial analysis with quantum-mechanical methods, followed by parameter refinement for a molecular-mechanics model[239]. Using these methods, models of the transition states accurate enough to produce useful estimates of selectivities were obtained.


## 8.3   Structure Evaluation


There is no other method that can determine molecular structures as accurately as an X-ray or neutron diffraction analysis of a single crystal. However, single crystals are not always available. This is especially true for large and highly flexible molecules (e. g. metalloproteins). Also, because of the poor quality of the crystals and the large size of the molecules the structures of biomolecules are often of low resolution. Furthermore, crystal structures represent only the product crystallized from a solution and this might not be the most abundant or important species but the least soluble part of a mixture of compounds. Another important aspect related to labile compounds, for instance complexes of labile metal ions, is that the structure of the crystallized compound might not be identical to the structure in solution. In view of the fact that the properties of compounds in solution are often the main point of interest (e. g., in preparative chemistry, homogenous catalysis, and biological systems) methods for the determination of solution structures are increasingly important. However, the information obtainable from spectroscopic techniques is by no means as complete as that from single-crystal diffraction studies.

A valuable approach leading towards the determination of solution structures is the combination of molecular mechanics calculations with solution experimental data that can be related to the output parameters of force field calculations[164]. Examples of the combination of molecular mechanics calculations with spectroscopy will be discussed in Chapter 10. Here, we present two examples showing how experimentally determined isomer distributions may be used in combination with molecular mechanics calculations to determine structures of transition metal complexes in solution. The basis of this approach is that the quality of isomer ratios, computed as outlined above, is dependent on the force field and thus linked to the quality of the computed structures. That is, it is assumed that both coordinates on a computed potential energy surface, the energy and the nuclear coordinates, are reproduced with the same accuracy with empirical force field calculations.*

$[Co(trab)_2]^{3+}$ (trab = butane-1,2,4-triamine) has five isomers and a total of 16 conformers (all nondegenerate combinations of chair and skew-boat of the six-membered chelate rings; Table 8.3; for the structures see Fig. 8.3). Two of the

---

* As discussed in Section 3.3, this assumption depends on the type of force field used, i.e., whether there is any link between the parameterization and thermodynamic or spectroscopic, (i.e., physically meaningful) parameters.

**Table 8.3:** Calculated and experimentally determined isomer distribution of $[Co(trab)_2]^{3+}$ [199].

| Isomer [a] | Calculated [%] | Observed [b] [%] |
|---|---|---|
| A | 21 | 21 |
| B | 16 | 15 |
| C | 16 | 15 |
| D | 4 | 4 |
| E | 43 | 45 |

[a] For the assignment of the isomers see Fig. 8.3
[b] HPLC and $^{13}$C-NMR

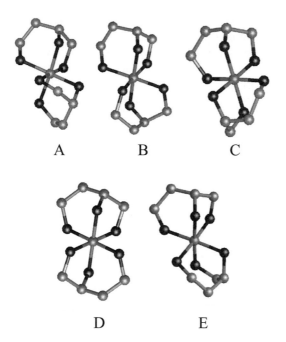

**Figure 8.3:** Calculated structures of the five isomers of $[Co(trab)_2)]^{3+}$ [199].

chromatographic bands of the three isomers A, B, C with optically pure trab were overlapping, i.e., only two peaks were resolvable using high-performance liquid chromatography (HPLC). The two additional species D, E, with one (*R*)- and one (*S*)-trab coordinated, were separated from the isomers A, B, C but again eluted as a single peak [199]. With $^{13}$C-NMR spectroscopy signals for all five isomers were resolved but for symmetry reasons only the signals due to isomer E could be attributed unambiguously. The calculated abundances of the five isomers agree well with the relative NMR intensities and the integrals of the chromatograms (sum of two peaks each for isomers A + (B or C), and D + E) and, therefore, allow an assignment of three of the five isomers. The chromatographic peaks and $^{13}$C-NMR shifts of the two isomers B and C (same abundances) are not unambiguously attributable (see Table 8.3). The two sets of experimental data together with the molecular mechanics calculations are self-consistent, indicating a high accuracy of the calculated structures of the five isomers of $[Co(trab)_2]^{3+}$. These are shown in Fig. 8.3.

Budotitane ($[Ti(bzac)_2(OEt)_2]$; bzac = 1-phenylbutane-1,3-dionate, OEt = ethoxide) is a promising anti-cancer drug but its experimental structure is unknown since no crystals were isolated from the isomeric mixture present in the pure complex[246]. There are five possible isomers (Table 8.4). $^1$H-NMR spectra revealed that only three isomers with abundances of 60%, 21% and 19% are stable in solution. Using symmetry based considerations they were deduced as all having *cis* configurations. However, an unambiguous attribution of the signals to the three *cis* isomers A, B and C was not possible[246]. The calculated isomer ratio based on minimized strain energies is in good agreement with the experimentally determined abundances and, therefore, allowed an assignment of the three sets of $^1$H-NMR transitions to the three stable isomers (see Table 8.4)[198]. The optimized

**Table 8.4:** Experimentally determined and calculated isomer distributions of bis($\beta$-diketonato) complexes of cobalt(III) and titanium(IV)[198].

| Compound[a] | Isomer[b] | Calculated [%] | Observed [%] |
|---|---|---|---|
| $[Ti(bzac)_2(OEt)_2]$ | A | 57 | 60 |
| (Budotitane) | B | 17 | 19 |
| | C | 26 | 21 |
| | D | 0 | 0 |
| | E | 0 | 0 |
| $[Ti(bbac)_2(OEt)_2]$ | *cis* | 100 | 100 |
| | *trans* | 0 | 0 |
| $[Co(acac)_2(NH_3)_2]^+$ | *cis* | 69 | 73 |
| | *trans* | 31 | 27 |
| $[Co(bzac)_2(en)]^+$ | A | 51 | 52 |
| | B | 22 ⎫ | 48[c] |
| | C | 26 ⎭ | |

[a] Structures

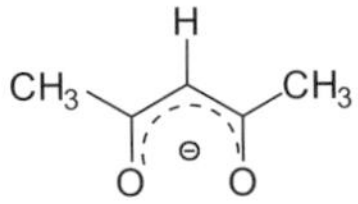

acac 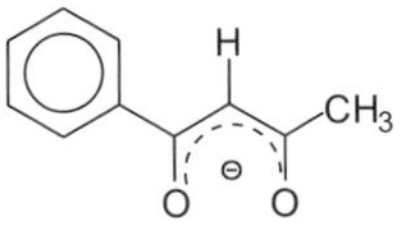          bzac 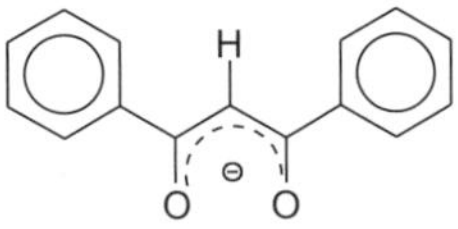          bbac

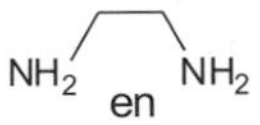

en

[b] For the assignment of the isomers see Fig. 8.4
[c] Not resolved

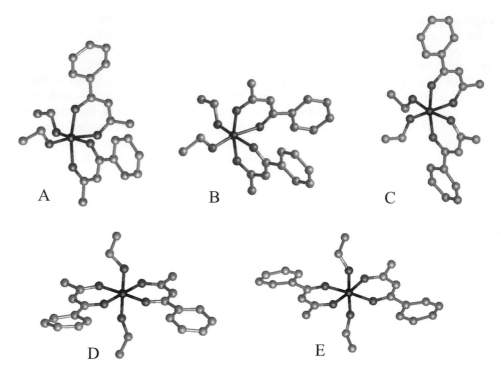

**Figure 8.4:** Calculated structures of the five isomers of Budotitane[198].

structures of all five isomers are shown in Fig. 8.4. The *trans* influence observed in a number of similar structures was modeled with different sets of bonding parameters (see also Section 12.2).

Simple molecular models of Budotitane suggested that the *trans* isomers are sterically preferred, and that the stabilization of the *cis* configurations is due to $p_\pi$-$d_\pi$ bonding[246,247]. However, the molecular mechanics analysis shows that the assumption that the *trans* isomers are less strained than the *cis* configurations is in error. Thus, Budotitane is an instructive example where the necessity of a full conformational analysis for the elucidation of steric strain is highlighted. Clearly, an additional stabilization of the *cis* isomers by $\pi$-bonding cannot be ruled out. However, based on the simple scheme usually used, its possible quantification is beyond the scope of empirical force field calculations. The fact that $p_\pi$-$d_\pi$ bonding is not relevant in Budotitane was further supported by the conformational analysis of a series of cobalt(III) complexes with similar ligand systems, using the same parameterization scheme. In contrast to titanium(IV), cobalt(III) has filled $d_\pi$ orbitals, and the experimentally determined preference for *cis* configurations is again predicted rather accurately (see Table 8.4)[198].

# 8.4 Mechanistic Information

The quantitative prediction of the stereochemistry of a chemical reaction by strain energies requires the knowledge of the reaction mechanism, i.e., the selective intermediates and/or transition states involved, and an accurate force field for the transient species. As discussed above, these are two demanding problems and, with a small number of exceptions involving combined quantum-mechanical/molecular-mechanical methods there are no reports of studies in this area that have used molecular mechanics for quantitative predictions at the same level of accuracy as for conformational analyses. Quantum-mechanical approaches are more appropriate in this area.

The application of empirical force field calculations to the design of asymmetric transformations is a worthy task, and some examples of studies in this area have been discussed above. On the basis of two examples we will now discuss some general aspects highlighting the limitations of qualitative considerations emerging from molecular mechanics calculations for the interpretation and support of assumed reaction pathways.

The hydroxide-promoted exchange of the α-proton of amino acids is catalyzed by metal ions[248]. If the amino acid is coordinated to a chiral complex fragment, the ratio of the two enantiomeric forms of the amino acid is controlled by a chiral intermediate[3]. One of the chiral fragments presented above in the discussion of the stereoselective ligand exchange reactions, {[Co((S),(S)-ppm)]} (see Table 8.2), was used to promote the epimerization of optically pure α-alanine[249]. The proposed mechanism is shown in Fig. 8.5.

The experimentally determined $(S)/(R)$-ratio of 18/82 was compared with the relative stabilities of the two diastereomeric products ([Co((S),(S)-ppm)((R)-ala)]$^{2+}$/ [Co((S),(S)-ppm)((S)-ala)]$^{2+}$), calculated by strain-energy minimization. The reported strain energies, based on a single conformer for each of the two diastereomeric products (identical to the crystal structure of the complex with coordinated $(R)$-alanine[250]), are in good agreement with the experimentally determined data (23/77 vs. 18/82). A full conformational analysis led to a ratio of 30/70 when only conformational flexibility is allowed or 33/67 when other isomers were also included in the analysis[222]. The assumption in the original paper was that the enantioselectivity is based on the relative energies of the diastereomeric forms of the cobalt(III) products[249]. Fortunately, a qualitatively similar result is expected if the stereoselectivity is controlled by the deprotonated intermediates. However, a quantitatively accurate prediction of the product ratio is not expected in this case.

The copper(II)-mediated condensation of primary amines with formaldehyde and nitroethane (Fig. 8.6) is highly stereoselective.

With [Cu((S),(R)-ala)$_2$] as starting material only one out of three possible isomeric products is formed (A in Fig. 8.6), and with [Cu(en)$_2$]$^{2+}$ one out of two isomers (B in Fig. 8.6) is obtained[251,252]. One possible intermediate is shown in Fig. 8.6[252]. Molecular mechanics calculations indicated that the intermediates leading to the observed species are less strained than any others. However, these calculations are only of limited value since the exact nature of the intermediate responsible for the selectivity is not known with certainty.

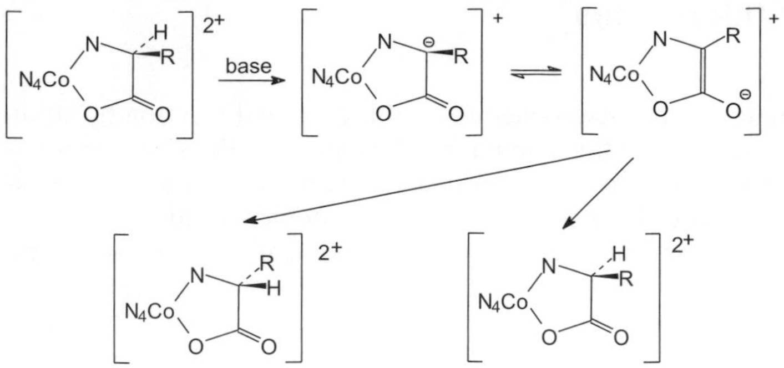

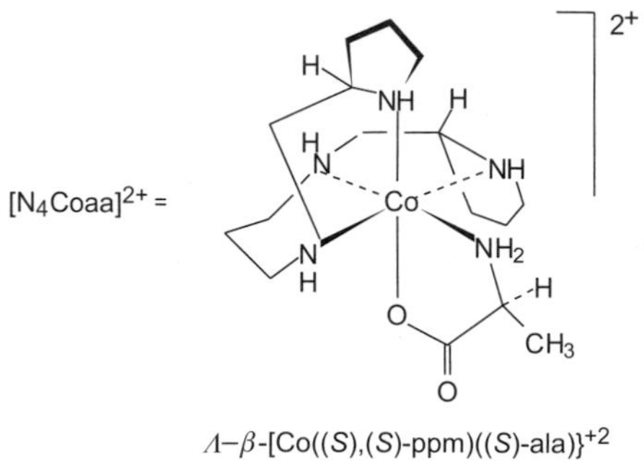

$[N_4Coaa]^{2+} =$

$\Lambda-\beta-[Co((S),(S)\text{-ppm})((S)\text{-ala})\}^{+2}$

**Figure 8.5:** The epimerization of α-amino acids promoted by a chiral cobalt(III) tetraamine cation.

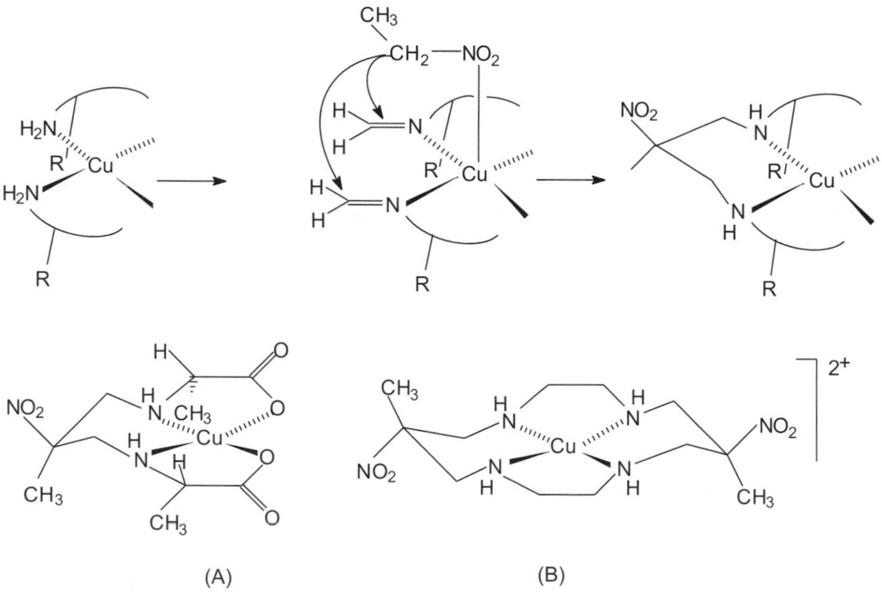

(A)                          (B)

**Figure 8.6:** The stereoselective template reaction of formaldehyde and nitromethane with primary amines coordinated in a *cis*-arrangement.

# 9  Metal Ion Selectivity

Ligands that selectively form complexes with certain metal ions are of importance in areas such as recycling and refinement of metals, the purification of solutions (or soluble solids), environmental remediation, treatment of metal poisoning as well as qualitative and quantitative metal ion analysis. The metal ion selectivity of a ligand is given by the difference in the stability constants of the complexes formed with all metal ions likely to be present, and these are related to the corresponding free energies of the complexation reactions. The factors influencing the stability of a coordination compound include the ease of desolvation of the metal ion and of the ligand, the strain imposed on the ligand by coordination to the metal ion, the strain imposed on the metal ion by coordination of the ligand, the metal-ligand bonding energy and solvation of the coordination compound. Note that the situation may be complicated by the fact that more than one solvent is present (e.g., in solvent-solvent extraction processes), that more than one ligand is involved (including simple anions and solvent molecules) and that coordination numbers and geometries may vary between the metal ions. The potential and limitations of molecular mechanics modeling in the area of metal ion selectivities have been reviewed recently[66].

Important, though not generally appreciated, problems in the application of molecular mechanics to the prediction of metal ion selectivities are the neglect of solvation and entropic effects (see Sections 3.7 and 3.8) and the variation of the metal-ligand bonding energy as a function of the metal ion. The difficulty in calculating relative stabilities on the basis of force field calculations, when different metal ions are involved, emerges directly from the fact that the estimate of relative stabilities via the comparison of total strain energies is in general limited to a series of conformers and isomers (see for instance Section 8.1 and relevant chapters in Part I and III). The determination by molecular mechanics calculations of the relative stabilities of a series of complexes with metal ions having differing preferences in terms of donor atoms is therefore a questionable approach.

The well known Irving-Williams series[253−256] and the HSAB principle[257] are established empirical concepts related to the stability of metal complexes, and the question of whether these are accounted for in a general molecular mechanics approach have not always been appreciated when metal ion selectivities have been modeled via empirical force field calculations. The assumption that the position of the minima and the steepnesses of the potential energy functions for the metal-ligand bonds for varying metal ions and constant donor sets accurately reflect variations in the bonding energies is an overinterpretation, at least in cases where electronic factors in the corresponding bonds differ considerably. Therefore, even with a well-tuned force field, which is based not only on structural but also on thermodynamic or spectroscopic data, the thermodynamics of metal-ligand bonding is rarely reproduced accurately enough. Exceptions are systems, where a series of metal ions has similar electronic properties (e. g., alkali metal ions, lanthanoids). With a given and constant coordination geometry, the strain enforced by the ligand and imposed on the ligand are in these cases the most important terms contributing to the complex stability.

There are examples, where molecular mechanics has been used with success in the area of metal ion selectivity, and these will be discussed in the next sections. Depending on the type of study, metal-dependent energy terms have to be subtracted from the strain energy of the metal complex. Also, the change of coordination number and geometry accompanying the ligand substitution process may have to be accounted for. The degree of preorganization of the ligand is dependent on the geometry of the product metal complex, and this is an important aspect to consider if stability differences between complexes with a given ligand and metal ions with differing geometric preferences are involved. For all the reasons outlined above, molecular-mechanics-based results in the area of metal ion selectivity have to be interpreted with care.


## 9.1   Chelate Ring Size

With most metal ions the complex stability decreases as the size of the chelate rings formed by open-chained polyamine ligands increases from five to six membered. Examples of this effect are presented in Table 9.1, where experimental data are compared with predictions based on strain energy differences between metal-free and coordinated ligands (Eq. 9.1).

$$\Delta U = U_{ML_n} - U_M - nU_L \tag{9.1}$$

The good agreement between the experimentally observed and calculated data indicates that the stability differences as a function of the chelate ring size are dominated by steric strain. Note that the effect studied in the example presented in Table 9.1 is the selection of a ligand type (five- or six-membered chelate) by a single metal ion (nickel(II)), rather than the selection of a metal ion by a ligand. Generally, this is an easier but, in terms of applications, less interesting task for molecular modeling[66]. Note also that the electron distribution in a ligand can be changed by coordination to a metal ion. It follows that the parameterization of the metal-free and the coordinated ligand should be different. This clearly is of importance in the field of metal ion selectivity (see Eq. 9.1) but it has not generally been appreciated. A recent analysis of this problem indicated that, for amine li-

**Table 9.1:** Experimentally determined and calculated stability constants of high spin nickel(II) amines with five- and six-membered chelates [120].

| Complex [a] | MM | | Observed | |
|---|---|---|---|---|
| | U [kJ mol$^{-1}$] | $-\Delta U$ [kJ mol$^{-1}$] [b] | $\Delta H$ [kJ mol$^{-1}$] | $-\Delta(\Delta H)$ [kJ mol$^{-1}$] |
| Ni(en) | 0.27 | | -2.15 | |
| Ni(tn) | 0.73 | 0.37 | -1.86 | 0.29 |
| Ni(en)$_2$ | 0.80 | | -4.37 | |
| Ni(tn)$_2$ | 1.71 | 0.73 | -3.59 | 0.79 |
| Ni(en)$_3$ | 1.09 | | -6.69 | |
| Ni(tn)$_3$ | 3.14 | 1.78 | -5.09 | 1.60 |
| Ni(dien) | 1.45 | | -2.84 | |
| Ni(dpt) | 1.98 | 0.35 | -2.53 | 0.31 |
| Ni(dien)$_2$ | 2.84 | | -6.05 | |
| Ni(dpt)$_2$ | 5.10 | 1.91 | -4.21 | 1.84 |
| Ni(2,2,2-tet) | 2.26 | | -3.35 | |
| Ni(2,3,2-tet) | 1.75 | -0.60 | -4.28 | -0.93 |

[a]  Structures

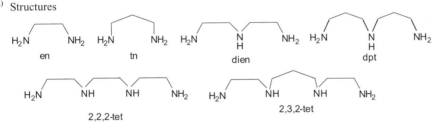

en     tn     dien     dpt

2,2,2-tet     2,3,2-tet

[b]  Corrected for strain energy differences of the free ligands

gands and up to trivalent transition metal ions, the effects are rather small, and this is one of the reasons for the accuracy observed in Table 9.1 [96].

An interesting observation is that the stability difference between metal complexes with ligands containing five- and six-membered rings is dependent on the metal ion size [120,158,258]. This effect has been studied for a relatively large number of ligands, including both macrocyclic and open-chain systems with various donor atom combinations. The two examples shown in Figs. 9.1 and 9.2 show the generally observed trend, indicating that an increase in chelate ring size from five- to six-membered increases the selectivity toward small metal ions.

This effect was attributed to the directionality of the lone pairs of the donor atoms (e. g., primary amines) of unstrained chelate ligands with a constant bite (donor atom – donor atom distance), leading to longer preferred metal-ligand distances for five-membered chelate rings (e.g., 2.5 Å for ethane-1,2-diamine (en) vs. 1.6 Å for propane-1,3-diamine (tn)) [120]. However, the N-M-N valence angles predicted by this model are 69° and 109.5° for en and tn, respectively, substantially different from the values preferred by most metal ions. Indeed, experimentally observed chelate angles of first row transition metal complexes involving diamines are usually about 85° for five-membered and 95° for six-membered

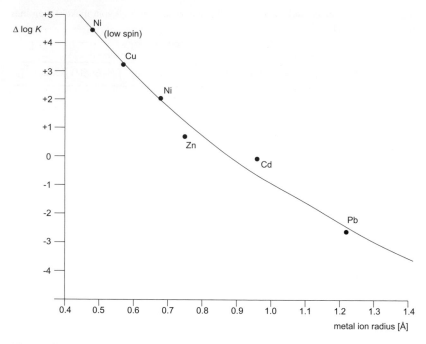

**Figure 9.1:** Stability differences for five- and six-membered chelate rings: 1,8-diamino-3,6-diazooctane (2,2,2-tet) vs. 1,9-diamino-3,7-diazanonane (2,2,2-tet); see footnote (a) to Table 9.1. Data from [120].

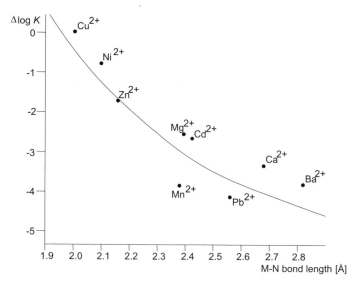

**Figure 9.2:** Stability differences for five- and six-membered chelate rings. $EDTA^{4-}$ vs. $TMDTA^{4-}$ (for ligand structures see Fig. 9.3). Data from [258].

rings. Thus, the strain imposed by the ligand preference (long bonds) is opposed to some extent by electronic preferences of the metal center (note, however, that bond stretching potentials generally are much steeper than angle bending potentials).

Nevertheless, the experimentally observed trend (Figs. 9.1 and 9.2) seems to contradict the expectations of hole size calculations (see below). Based on a simple geometric model, where the hole size is measured and the expected stability related to the hole size vs. metal ion size ratio, the order of selectivity of 12- to 16-membered tetraaza macrocycles differs from that predicted on the basis of chelate ring sizes and, more importantly, from the experimentally observed data [258]. Obviously, an approach involving molecular mechanics with a well tuned force field, where the total strain energy of the metal complexes, corrected for strain relaxation due to coordination of the metal-free ligand, is related to the complex stability, should take all important factors, including the orbital directionality of the ligand atoms in the chelate rings (M-L-Y angles), the electronic preferences of the metal ion (L-M-L' angles), and the metal ion size (M−L bonds), into account.

An important aspect of the prediction of stabilities by molecular mechanics calculations is that all possible isomers and conformers have to be considered. With macrocyclic ligands the preferred geometry can change from metal ion to metal ion, and this has to be accounted for when the relative stabilities are predicted. An example is shown in Fig. 9.5 below, where large, medium and small metal ions are predicted to enforce $\delta\lambda$, $\delta\delta$ and $\lambda\delta$ conformations, respectively, on *trans*-diammac[65,182]. Still greater variation is observed with complexes of EDTA type ligands, where four structural types are known (Fig. 9.3). In such a case, unless the coordination number and geometry of the complex can be predicted, there is no chance of predicting the metal ion selectivity. Based on this observation and the fact that the coordination geometries are strongly dependent on the chelate ring size (EDTA vs. TMDTA) and the electronic properties of the metal center, correlations between M−L bond length and complex stability (Fig. 9.2) may not always be the result of simple "size" effects.

## 9.2 Macrocycle Hole Size

By far the widest area of application of molecular mechanics and related methods toward the design of metal ion selective ligands is the calculation of the hole size of macrocyclic ligands. The simplest method for determining the hole size $R_A$ is to measure from crystal structural data the mean distance $R_H$ of the donor atom positions from their centroid and correct it with the covalent radius $R_C$ for the size of the donor atoms (Fig. 9.4). The metal ion selectivity is then a function of the relative stability constant $K' = f(R_A/R_P)$, where $R_P$ is the Pauling radius of the metal ion [263,264]. This method, however, is not valid for the design of new ligands since at least one experimental structure of a metal complex of the ligand to be studied has to be available for the calculations. Also, since the ligand geo-

six-coordinate

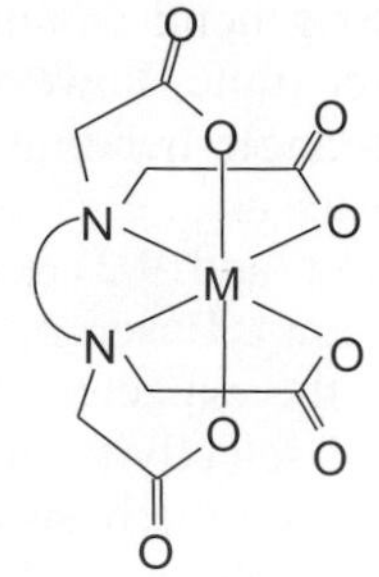

e.g. [Fe(EDTA)]$^-$ [259]

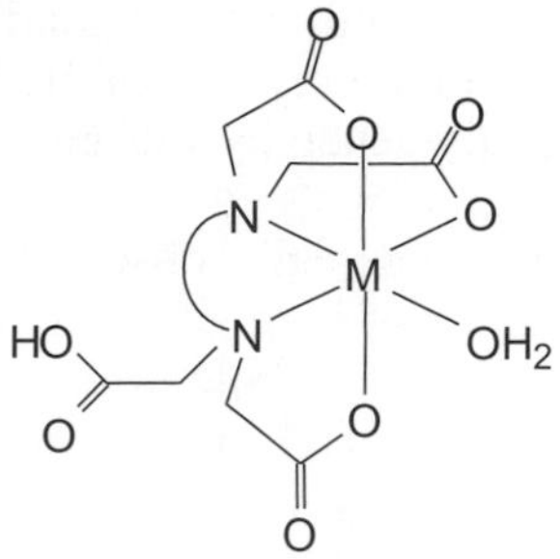

six-coordinate pentadentate

e.g. [Cr(HEDTA)OH$_2$] [260]

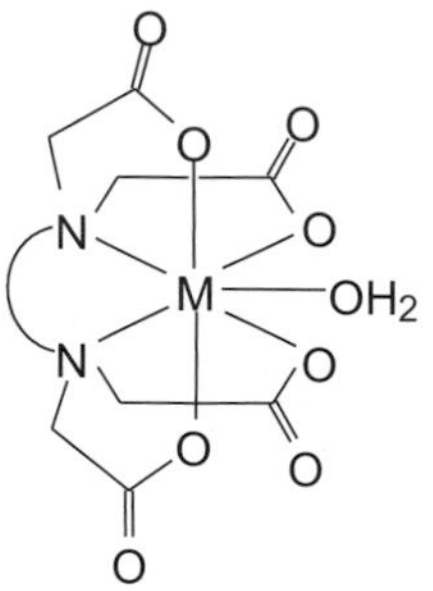

seven-coordinate
pentagonal bipyramidal
e.g. [Fe(EDTA)OH$_2$]$^-$ [261]

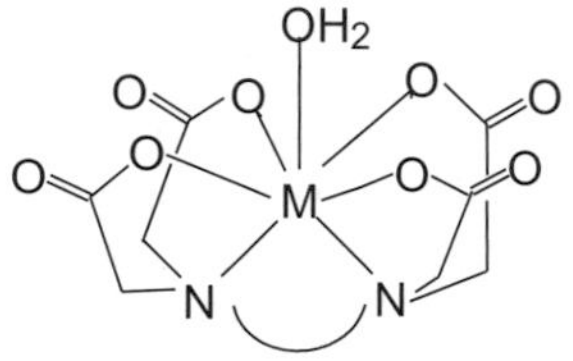

seven-coordinate
capped trigonal prismatic
e.g. [V(EDTA)OH$_2$]$^-$ [262]

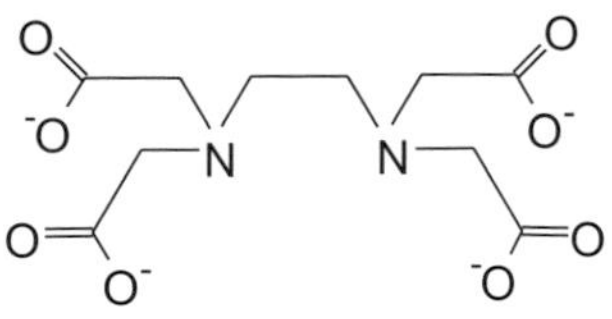

EDTA$^{4-}$

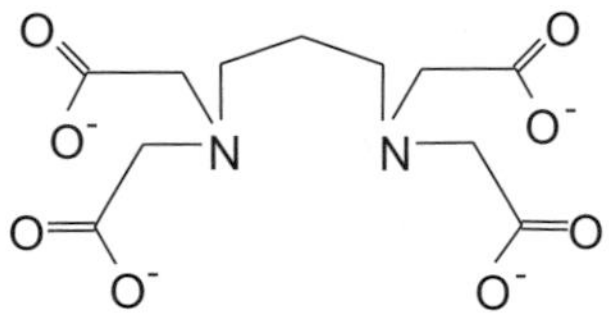

TMDTA$^{4-}$

**Figure 9.3:** Established geometries of metal complexes with EDTA-type ligands.

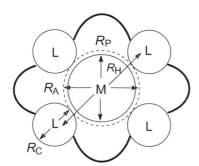

**Figure 9.4:** Geometrical determination of the macrocycle hole size [263].

metry is not variable in this model, it cannot adjust to the demand of the metal ion, and conformational flexibility is inhibited.

Neither of these problems occur with approaches based on molecular mechanics calculations. The best fit of a metal ion M to a macrocyclic ligand L is related to the minimum in the sum of all M-L bonding interaction terms (Eq. 9.2).

$$E_{ML} = \frac{1}{2} k_{ML} \left( r_{ML} - r_0 \right)^2 \tag{9.2}$$

This term vanishes by setting either $k_{ML} = 0$, or $r_{ML} = r_0$. While the first approach corresponds to a single-point calculation[265,266], with the latter method the total strain energy, mapped as a function of the metal-ligand distance, can be obtained[158]. Remember here that an accurate treatment of the metal ion selectivity should take into account not only the strain energy of the product complex but also its geometry and the strain of the metal-free ligands, as well as solvation and entropic terms. Although it is in principle possible to include all these terms in a molecular mechanics approach, this has not yet been achieved.

Three methods have been used to calculate total strain energies as a function of metal-ligand distances. In the first approach the strain energy is plotted as a function of $r_0$ with all other metal ion dependent force field parameters kept constant at "reasonable" values (for example those for nickel(II))[120,158]. The size-matching metal ion is then that with an equilibrium M−L distance ($r_0$) closest to the minimum of the resulting curve. This approach was challenged for two reasons: the interpretation of the resulting functions relies on an accurate knowledge of the strain-free metal-ligand bond distances $r_0$, which are unknown, and the curves calculated with this method depend on $k_{ML}$, which is not constant for varying metal centers, and not accurately known for any metal ion (the molecular mechanics force constants and the ideal metal-ligand distances are not physically meaningful parameters, see Section 3.5)[267]. In particular, it was demonstrated that there is a small but significant dependence of the position of the minima of the ideal strain energy vs. metal-ligand distance curves on the force constant $k_{ML}$[267]. The proposed alternative, involving restraints, (i.e., exceedingly large force constants) to fix the metal-ligand bond lengths, was queried because the resulting strain energies are overestimated and might be misleading[268]. A proposal that overcomes the problems of both these approaches involves using mathematically constrained metal-ligand bond distances and replacement of the valence angle terms around the metal ion by 1,3-interactions[65]. "Unreasonably large" metal-ligand stretching force constants are not necessary in this approach, and all metal-ligand stretching terms vanish so that the resulting energies are independent of $k_{ML}$ and the corresponding $r_0$ values (see Section 17.16).

One problem that has only rarely been appreciated is the fact that with unsymmetrical ligands the metal-ligand distances for each bond have to be varied within specific and different limits. An illustrative example is the hole size calculation of the hexaamine ligand *trans*-diammac which, with many first row transition metal ions, exhibits significantly different metal-ligand distances for the primary and secondary amines[65,182]. As shown in Fig. 9.5 there are significant differences in

a)

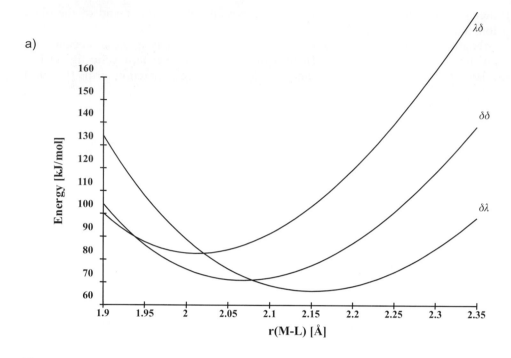

b)

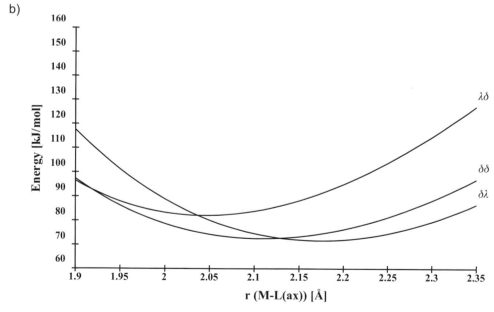

**Figure 9.5:** Hole size of the three conformers of *trans*-diammac (a) six identical metal-ligand distances and (b) different metal ligand distances for equatorial and axial bonds [65,182].

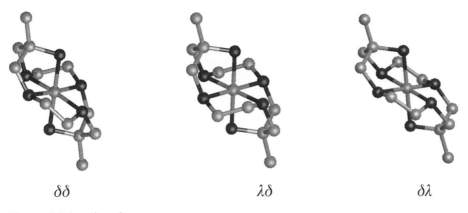

$$\delta\delta \qquad\qquad \lambda\delta \qquad\qquad \delta\lambda$$

**Figure 9.5** (continued)

the strain energy vs. metal-ligand distance curves when (a) all metal-ligand distances are varied simultaneously and (b) the constrained bond distances to axial and equatorial amines are different, as is observed experimentally*. The minimum for each isomer (hole size), the energy difference between the three minima (global energy minimum, conformational equilibrium) and the crossing points of the curves of different conformers (change of conformation) are different for the two types of calculation.

Hole sizes produced by another highly unsymmetrical ligand, a tetradentate bispidine derivative (see Fig. 9.6)[186] have been investigated using the two methods described above and using a new technique that allows the sum of all metal-donor distances to be constrained, while each individual distance can vary freely[186]. The three curves in Fig. 9.6(a) are nearly identical, i.e., the assumption of linear variations appears to be valid. However, from Fig. 9.6(b) it emerges that the variations in bond lengths are not linear. Therefore, with highly unsymmetrical ligands the errors in the hole-size calculations arising from the neglect of different variations of individual bonds might lead to appreciable errors. Note, that hole sizes and ligand hole plasicities cannot be validated experimentally since they are constructed to visualize and quantify the dependence of the shape and size of a ligand cavity on the metal ion size but without a metal ion present. Real metal ions will change the rela-

---

* The two sets of curves were calculated by constraining the metal-ligand bond distances, using MO-MEC[52] and a published force field[96]. For the two sets, the metal-ligand bond energies are set to zero and the curves are therefore independent of the metal center. In Fig. 9.5(b) the M–L distances for axial and equatorial ligands are different (the values for the axial ligands are plotted). While the distances for the axial bonds were varied between 1.90 Å and 2.35 Å as in Fig. 9.2(a), the values for the equatorial bonds were varied between 1.93 Å and 2.20 Å, 1.95 Å and 2.265 Å, 1.92 Å and 2.19 Å for the $\delta\delta$, $\delta\lambda$, $\lambda\delta$ conformers, respectively. The slopes for the bond distance variations for the axial and equatorial ligands for each of the three conformers were obtained by a linear regression of the corresponding bond distances, calculated with the same force field, for the cobalt(III), chromium(III), *low-spin* iron(III), cobalt(II) and nickel(II) complexes.

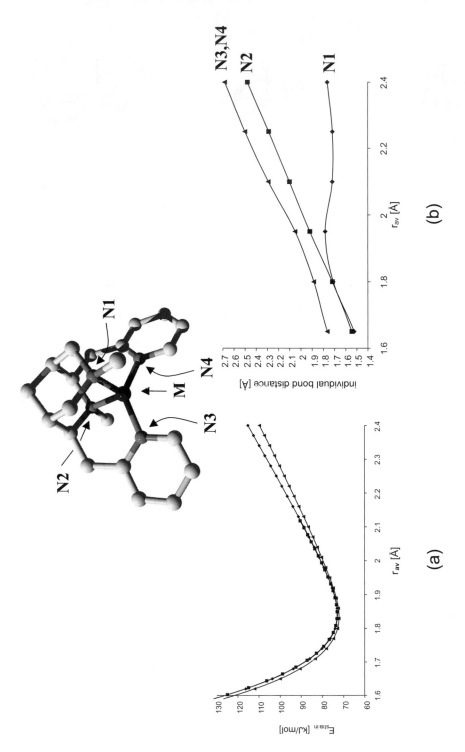

**Figure 9.6:** Shape and size of the cavity of a highly asymmetric bispidine ligand; reproduced with permission from reference[186].

tive bond distances and therefore the shape of the curves in plots such as that of Fig. 9.6(b). The role of hole size calculations is to quantify the ability of the ligand to impose a specific structure on the complex; in the area of metal ion selectivities, the enforcement of a structure on metal ions is of importance for the destabilization of complexes of some metal ions and the stabilization of others.

Apart from the different approaches to calculating hole sizes of macrocyclic ligands, there are also considerable differences in the force fields that have been used[57,58,116,120,231,265,267,269–273]. Unfortunately, no comparative study that systematically analyzes the various methods and force fields is available. Therefore, it is worth noting that the computation of the cavity size of 12- to 16-membered tetraaza macrocycles with two very different models and force fields led to remarkably similar results (Table 9.2).

**Table 9.2:** Cavity size for the *trans*-III conformers of tetraaza macrocyclic ligands, calculated by molecular mechanics.

| Macrocycle[a] | Cavity size [Å] | |
| | Ref. [274] | Ref. [275] |
| --- | --- | --- |
| 12-aneN$_4$ | 1.83 | 1.81 |
| 13-aneN$_4$ | 1.92 | 1.92 |
| 14-aneN$_4$ | 2.07 | 2.05 |
| 15-aneN$_4$ | 2.22 | – |
| 16-aneN$_4$ | 2.38 | – |

[a] Nomenclature:

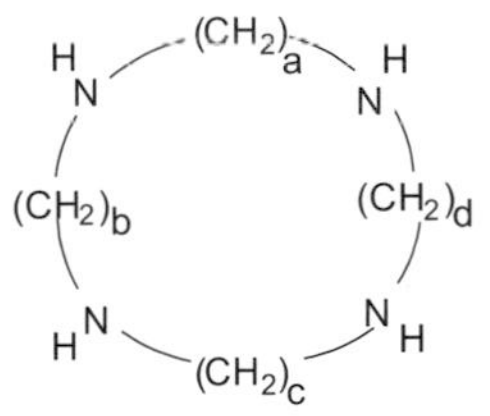

12-ane-N$_4$: a=b=c=d=2
13-ane-N$_4$: a=3;b=c=d=2
14-ane-N$_4$: a=c=3;b=d=2
15-ane-N$_4$: a=b=c=3;d=2
16-ane-N$_4$: a=b=c=d=3

# 9.3 Preorganization

Preorganization[276,277] as applied to macrocyclic ligands is an important concept, and the size of the cavity is only one aspect. The orientation of the ligating groups with respect to each other and the metal ion are also of importance. For instance, a macrocyclic ligand with four secondary amine donor groups might have two of the amine lone pairs disposed toward one side of the macrocycle plane and two to the other. Clearly, this ligand will need to reorganize conformationally in order to coordinate the metal ion and therefore it cannot be considered to be fully preorganized. A similar effect is that of the structural differences between tetraaza and tetrathia macrocyclic metal free ligands: the lone pairs in the latter are usually disposed toward the outside of the cavity while the former are often disposed toward the center and are, therefore, more preorganized[278]. This is one reason for the lower stability of complexes with tetrathia macrocyclic ligands. Preorganization of a ligand means the enforcement of a specific, often high energy conformation, and this is one important aspect related to the energization or entatic state principle[279–282].

Highly preorganized ligands are usually rigid molecules, and the rigidity is the result of hindered rotations around single bonds in the organic backbone, resulting from bulky substituents, bridging by small rings, multiple bonds and/or hydrogen bonding[281,282]. All of these effects are accessible by molecular mechanics.

How important the lack of preorganization is depends on the energy cost and barrier associated with the reorganization required on coordination. The energy cost can be estimated by computing the strain energy of the metal-free and the coordinated ligand, as described above in the discussion on cavity size. The energy barrier can be estimated by using constraints to drive torsion angles and so to move from the metal-free ligand conformation to that of the coordinated ligand. However, with $N$-3 independent torsion angles in an $N$-membered macrocycle there are usually too many possible pathways to be investigated by deterministic molecular mechanics methods. Therefore, the techniques of molecular dynamics were applied to establish the tendency of a macrocyclic ligand to adopt a conformation suitable for binding a metal ion[283,284]. The frequency of occurrence and longevity of suitable conformations in molecular dynamics runs is a qualitative estimate of the degree of preorganization. Using these methods, new, more highly preorganized ligands were designed[285].

Two methods for quantifying ligand preorganization have recently been developed. The structural reorganization of a ligand was described as a two step process which involves the conformational ligand reorganization term (energy difference between the most stable conformation of the metal-free ligand and that of the coordinated ligand) and the reorganization term which is due to the strain imposed by the metal ion (Eq. 9.3, Fig. 9.7). The first term ($\Delta U_{conf}$) can be related to the rigidity of a ligand and the second term ($\Delta U_{comp}$), the complementarity, can be related to the cavity size and shape[66,286].

$$\Delta U_{reorg} = \Delta U_{conf} + \Delta U_{comp} \qquad (9.3)$$

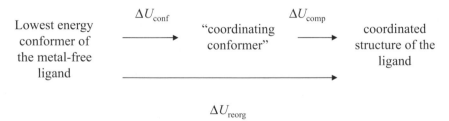

**Figure 9.7:** Definition of ligand complementarity.

Strain energy and structural reorganization parameters were used to quantify the preorganization of structurally reinforced tetraazamacrocyclic ligands[287]. Both parameters are based on the most stable ligand structure of the metal-free and the coordinated ligand. The strain energy ratio $E_L/E_C$ ($U_{strain}$ of the metal-free and the coordinated ligand, respectively) is a measure for the thermodynamics of the ligand reorganization (similar to $\Delta U_{reorg}$ in Eq. 9.3), and the sum of the absolute values of the differences of the intramolecular donor – donor distances $\sum_{n=1}^{6} |\Delta d_n|$ (see Fig. 9.8) is the corresponding structural parameter. A fully preorganized ligand would have an $E_L/E_C$ value of 1 and $\sum_{n=1}^{6} |\Delta d_n| = 0$.

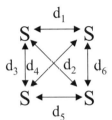

**Figure 9.8:** Definition of the structural reorganization parameter.

## 9.4   Quantitative Correlations Between Strain and Stability Differences

Linear and non-linear correlations of structural parameters and strain energies with various molecular properties have been used for the design of new compounds with specific properties and for the interpretation of structures, spectra and stabilities[66]. Quantitative structure-activity relationships (QSAR) have been used in drug design for over 30 years[288] and extensions that include information on electronic features as a third dimension (the electron topological approach, ET) have been developed and tested[48] (see Section 2.3.5). Correlations that are used in the areas of electron transfer, ligand field properties, IR, NMR and EPR spectroscopy are discussed in various other Chapters. Here, we will concentrate on quantitative structure-property relationships (QSPR) that involve complex stabilities[124,289–291].

The example discussed here involves the separation of lanthanoid(III) ions by solvent-solvent extraction with the three bis-alkylhydrogenphosphate ligands presented in Fig. 9.9 (see also Section 15.3)[124]. Similar QSPR's with lanthanoid(III) ions and/or these types of ligands have been reported[289,292]. The strain energy changes due to the complexation of lanthanoid(III) ions with the ligands HR of Fig. 9.9 (Eq. 9.4) is given in Eq. 9.5, where $U_M$ is the contribution of the Ln(III) aqua ion, $U_{HR}$ of the ligand (Fig. 9.9), $U_{Mcom}$ is that of the Ln(III) complex with six coordinated ligands (all coordinated as monodentates; three are deprotonated,

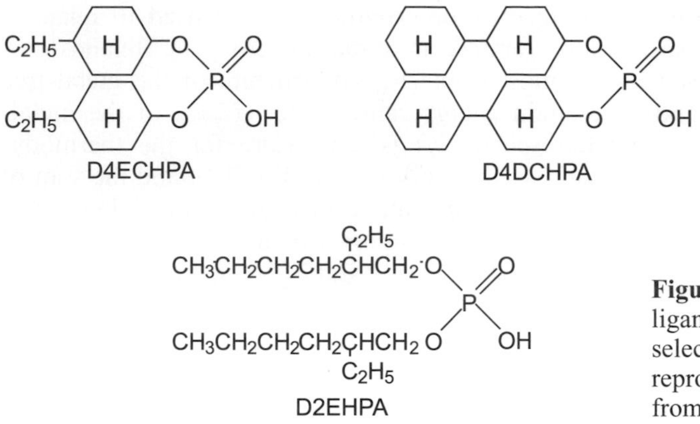

**Figure 9.9:** Organophosphate ligands used for metal-ion selective lanthanoid extraction; reproduced with permission from ref. [124].

leading to neutral complexes; three water molecules complete the coordination sphere[66,124]. $U_H$ is the contribution of the released protons and $U_{aq}$ that of the released water molecules.

$$M(OH_2)_9^{3+} + 6\,HR \rightleftharpoons \left[M\{(HR)(R)\}_3\,(OH_2)_3\right] + 3\,H^+ + 6\,H_2O \qquad (9.4)$$

$$\Delta U_C = U_{Mcom} + 3\,U_H + 6\,U_{aq} - U_M - 6\,U_{HR} \qquad (9.5)$$

In order to get relative strain energies and to simplify the analysis La(III) is defined as a reference point, and Eq. 9.6 is the corresponding strain energy equation. The correlation of these computed strain energy data sets with corresponding, experimentally available thermodynamic data leads to the QSPR defined in Eq. 9.7.

$$\Delta U_C^M - \Delta U_C^{La} = (U_{Mcom} - U_{Lacom}) - (U_M - U_{La}) \qquad (9.6)$$

$$\Delta U_C^M - \Delta U_C^{La} = \alpha \log\left(\frac{K_{ex}^M}{K_{ex}^{La}}\right) \qquad (9.7)$$

This is shown in Fig. 9.10. There is a small deviation of the apparent QSPR constant $\alpha$ from the theoretically expected value (1.26 vs. 1.08). Very similar values for $\alpha$ have been observed in other examples[124,289,292], and the deviation is assumed to be due to the neglect of entropy, ion-pairing and solvation[66]. QSPR's such as that shown in Fig. 9.10 are not necessarily of direct use for the design of new potent and selective extracting ligands, and this has been discussed in detail[66]. However, the accuracy of the correlation involving three different ligands indicates that the structures and the extraction mechanism are constant. High selectivity is indicated by large strain-energy differences and, correspondingly, large relative extractability differences. As expected, this is the result of increasing bulk of the ligand substituents, and this information can be used, together with molecular modeling, to design new extractants.

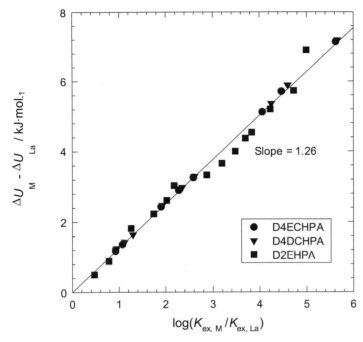

**Figure 9.10:** QSPR plot for selective lanthanoid complex formation by the organophosphate ligands shown in Fig. 9.9; reproduced with permission from ref. [124].

## 9.5 Conclusions

There are some major limitations to the prediction of metal ion selectivities by molecular mechanics calculations alone. This is due to a number of approximations inherent in the methods discussed above, and these restrict the reliability of the calculated stabilities. The main points are as follows.

- In an approach based on selectivity by hole size, geometric preferences of the metal ions (i.e., electronic effects) are largely excluded. However, different metal ions have not only different metal-ligand bond distance preferences (hole size) but also different preferences in terms of coordination number and angular geometry. For example, a ligand preferring trigonal prismatic coordination geometry will induce more strain on complexation with cobalt(III) than with cobalt(II), irrespective of the hole size.

- The same force field is used throughout for the calculation of curves of the type presented in Fig. 9.5. Therefore, there is an implicit but unstated assumption that the force field parameters defining M-L-Y valence angles, M-L-Y-X torsion angles and metal-centered nonbonded interactions, where these are included, do not vary with the metal ion. These are probably reasonable assumptions in general but there may well be cases where they lead to misleading results.

– Different conformations and configurations of the ligands have different cavity sizes (see Fig. 9.5). Therefore, all possible configurations and conformations need to be taken into account. That is, with relatively complex systems a full conformational analysis is required. This is particularly difficult to do in an exhaustive way when the metal ion can adopt different geometries [273].

– In order to predict the stability of specific metal ligand combinations, account needs to be taken of all of the thermodynamic contributions of which the strain induced in the ligand is only one. The desolvation of the ligand and the metal ion and the energy of the newly formed metal-ligand bonds are important factors, and the latter two of these certainly vary substantially. When comparing the binding of a single metal ion to series of chemically similar ligands there is little variation in the energy of the ligand or solvent bonds to the metal and the analysis is valid, but, when the metal ion is varied a more complete analysis might be required.

– The other important term is not the strain energy of the metal complex but the difference between the strain in the coordinated ligand and in the metal-free ligand. This corresponds to the energy cost associated with reorganizing the ligand to coordinate to the metal and depends on the preorganization of the ligand. Within the limitations described above this is not dependent on the metal but certainly varies substantially with the ligand and therefore needs to be taken into account when attempting to design a metal-ion-specific ligand.

The technique of predicting metal ion selectivity by molecular mechanics calculations for the computer assisted design of novel ligand systems seemed to hold great promise. So far, this is largely unfulfilled. Some reasons for this failure have been listed above. However, the problems are also associated with less predictable chemical variations, e.g., the formation of dinuclear species, when a mononuclear complex would or might have been expected, noncoordination of some donor groups and variable coordination numbers and/or geometries depending on the metal ion.

# 10   Spectroscopy

There are two applications driving the development of the computational prediction of spectra: the design of new materials with given spectra (i.e., design and synthesis of spectroscopic model compounds or new materials) and the determination of solution structures by the correlation of computed with experimentally determined spectroscopic properties. Both applications are of particular relevance in the field of bioinorganic chemistry since crystal structural information is often difficult to obtain for metalloproteins while solution spectra are generally more accessible. The determination of structures based on solution spectra in combination with molecular mechanics may provide some important information on the structure of the active species in solution not accessible by other methods. Also, the study of the reactivity of low-molecular-weight model compounds having the same spectroscopic characteristics as biomolecules may lead to information on enzyme mechanisms.

There are two potential limitations to the application of empirical force field calculations to predicting solution properties. First, environmental effects (crystal lattice or solvent sheath) are usually neglected and second, the determination of the global energy minimum is not unequivocal. While in theory there are methods for dealing with the first problem, there is no totally reliable solution to the latter (Section 3.7 and Chapter 5). Coupling the optimization of structures by empirical force field calculations with the simulation of experimentally available spectroscopic data is an appealing method for solving both problems. If the spectroscopic simulation, based on an energy-minimized structure, is similar to the experimental data, then the calculated structure is likely to represent the global energy minimum of the compound in the environment where the spectroscopic data was collected [164] *.

The fact that, in the case of solution spectra, the spectroscopic data are obtained from the solvated species while the corresponding molecular mechanics structure usually represents the "naked" species is somewhat unsatisfactory. This is one reason for some observed, albeit small, differences between structural parameters in solution and in the solid state. Nevertheless, the accuracy obtained often is surprisingly high, and this might be explained by the fact that the parameteriza-

---

* In cases with an equilibrium distribution of various conformations all the relevant geometries should to be modeled (structure and spectra simulation), and the spectra averaged on the basis of the Boltzmann distribution should be compared to the solution properties.

tion of the force field is generally based on crystal structural data, and the calculated structures therefore represent the molecule in an averaged solid state environment (see also Section 3.7)[293].

In this chapter, applications of the combination of empirical force field calculations with infrared, UV-Vis-NIR, EPR and NMR spectroscopy are discussed. Also, some specific electronic effects of transition metal-ligand bonding, interpreted on the basis of combined molecular mechanics-spectroscopic studies are presented. Vibrational frequencies are directly available from molecular mechanics calculations if a second-derivative energy-minimizing routine is used (full-matrix Newton-Raphson, see Section 4.2.4). The calculation of energy levels via ligand field or angular overlap model (AOM) calculations, based on computed structures (molecular mechanics), can be used to provide information on the d-d transitions and on EPR spectra. All relevant NMR parameters, i.e., chemical shifts (including paramagnetic shifts), coupling constants, and nuclear Overhauser (NOE) effects, are dependent on the molecular geometry and can therefore be used for structural correlations.

## 10.1   Vibrational Spectroscopy

Although vibrational frequencies have been calculated and compared with experimental data since the early days of molecular mechanics refinements using full-matrix second-derivative procedures[13,73], there are few reports on the application of this method in the field of transition metal chemistry. One of the examples is a study on linear metallocenes[108]. Here, the molecular mechanics force constants were obtained by adjusting starting values by fitting the calculated vibrations to thoroughly analyzed experimental spectra. The average difference (rms) between experimental and calculated vibrations was of the order of ca. $30 \, \text{cm}^{-1}$.*
Table 10.1 shows the observed and calculated frequencies associated with the skeletal modes of ferrocene (see Fig. 10.1 for the representation of these modes).

**Table 10.1:** Observed and calculated vibrational frequencies of the skeletal modes of ferrocene [108].

|          | Observed [$\text{cm}^{-1}$] | Calculated [$\text{cm}^{-1}$] |
|----------|-----------|------------|
| $A_{1u}$ | 44        | 42         |
| $E_{1u}$ | 179       | 177        |
| $A_{1g}$ | 309       | 282        |
| $E_{1g}$ | 389       | 363        |
| $A_{2u}$ | 478       | 493        |
| $E_{1u}$ | 492       | 498        |

---

* A similar accuracy was obtained in calculated vibrations of organic molecules[73].

**Table 10.2:** Skeletal force constants for linear metallocenes [108].

| Term [a] | Force constant (equilibrium value) | | | | | | | | |
|---|---|---|---|---|---|---|---|---|---|
| | $[VCp_2]$ | $[CrCp_2]$ | $[FeCp_2]^+$ | $[FeCp_2]$ | $[CoCp_2]^+$ | $[RuCp_2]$ | $[OsCp_2]$ | $[CoCp_2]$ | $[NiCp_2]$ |
| M-D ($a$) [kcal mol$^{-1}$ Å$^{-2}$] | 135(1.928) | 150(1.798) | 205(1.677) | 205(1.649) | 205(1.682) | 205(1.816) | 206(1.855) | 185(1.726) | 140(1.817) |
| M-D-C ($\gamma$) [kcal mol$^{-1}$ rad$^{-2}$] | 85(90) | 93(90) | 117(90) | 100(90) | 100(90) | 125(90) | 130(90) | 45(90) | 28(90) |
| D-M-D ($\beta$) [kcal mol$^{-1}$ rad$^{-2}$] | 50(180) | 40(180) | 40(180) | 40(180) | 40(180) | 52(180) | 56(180) | 40(180) | 40(180) |
| 1,5-dihedral ($\alpha$) [kcal mol$^{-1}$] | – | – | – | 0.36(180) | – | 3.4(180) | 3.4(180) | – | – |

a) D is a dummy atom at the centroid of the Cp ring
1 kcal = 4.184 kJ

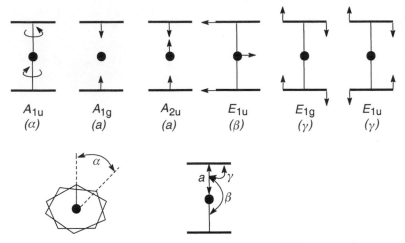

$$
\begin{array}{cccccc}
A_{1u} & A_{1g} & A_{2u} & E_{1u} & E_{1g} & E_{1u} \\
(\alpha) & (a) & (a) & (\beta) & (\gamma) & (\gamma)
\end{array}
$$

**Figure 10.1:**  Skeletal modes in linear metallocenes[108].

A comparison between experimental and calculated spectroscopic parameters for various cyclopentadienyl (Cp) compounds indicated that the internal Cp modes are transferable while the skeletal modes are dependent on the nature of the metal-Cp bonds. The skeletal force constants for a number of metallocenes are given in Table 10.2.

The force field derived from spectroscopic data was used to optimize structures of a series of [M(Cp)$_2$] complexes (M = V(II), Cr(II), Fe(II), Fe(III), Ru(II), Os(II), Co(II), Co(III), Ni(II)), including strapped ferrocenes, and gave good agreement with experimental structures (accuracy in bond lengths of the order of 0.01 Å and in angles (valence and torsional) of about 3–4°)[108]. The calculated barriers to rotation of the cyclopentadienyl rings were also in reasonable agreement with experimentally determined values.

## 10.2   Electronic Spectroscopy

The d-orbital energy levels in a metal complex are dependent on the electron configuration of the metal center, the number and type of ligand atoms involved, and their geometric arrangement. For a given stoichiometry and structure (experimentally determined or calculated, e. g., via empirical force field calculations), the energy levels and thus the ligand field spectra can be calculated. Before discussing applications of this approach, for the design of coordination compounds with pre-specified spectroscopic properties or for the determination of structures, the interesting related question of the relationship between bond length and d-d transition energy, that has been extensively debated in the literature will be discussed now.

The ligand field strength of transition metal complexes with a given donor set (e. g., $[M(NR_3)_6]^{n+}$) is, as expected, strongly dependent on the M-N bond length. This effect was first noted in cobalt(III) complexes of the type *trans*-$[Co(L)X_2]^+$ (L = *n*-aneN$_4$ (*n* = 12,13,14,15,16); X = Cl$^-$, CN$^-$, N$_3^-$)[274] (Table 10.3; the ideal M-N bond distances of the five macrocyclic ligands, calculated by molecular mechanics, are also given in Table 10.3). The 13-membered ring is the smallest with a *trans* configuration, and the corresponding cobalt(III) complex has the highest ligand field. This was interpreted as being the result of compression enforced by the macrocyclic ligand. In a number of subsequent publications[275,294-296] it was shown that it is not always the macrocyclic tetraaza ligand with the smallest hole size that leads to the largest ligand field of transition metal complexes (see Table 10.3).

**Table 10.3:** In-plane ligand field splitting ($Dq^{xy}$) or transition energies (cm$^{-1}$) for *trans*-octahedral transition metal complexes with tetraaza mocrocyclic ligands[a]. Data as compiled in [120].

| Ligand | Ideal M-N length [Å] | Co(III) | Fe(II) | Ni(II) (S = 0) | Ni(II) (S = 1) | Cu(II) | Cr(III) | Rh(III) |
|---|---|---|---|---|---|---|---|---|
| 12-ane-N$_4$ | 1.82 | – | – | 2115 | – | 16.810 | – | – |
| 13-ane-N$_4$ | 1.92 | 2750 | 2208 | 2140 | – | 18.310 | – | – |
| 14-ane-N$_4$ | 2.07 | 2562 | 2029 | 2043 | 1460 | 19.900 | 2450 | 25.000 |
| 15-ane-N$_4$ | 2.22 | 2362 | 1842 | 1955 | 1240 | 17.610 | 2123 | 23.800 |
| 16-ane-N$_4$ | 2.38 | 2295 | | | 1100 | | | 22.730 |
| en$_2$ | | 2530 | | 1961 | 1150 | 18.180 | 2200 | 24.630 |

[a] Nomenclature

12-ane-N$_4$: a=b=c=d=2
13-ane-N$_4$: a=3; b=c=d=2
14-ane-N$_4$: a=c=3;b=d=2
15-ane-N$_4$: a=b=c=3;d=2
16-ane-N$_4$: a=b=c=d=3

en = ethane-1,2-diamine

One of the conclusions that emerged directly from the data was that the highest ligand field strength is not produced by the ligand leading to the shortest metal-ligand atom distance but by the one which best fits the metal ion demand. This interpretation seemed to be at variance with the widely accepted fact that the ligand field splitting is inversely proportional to approximately the fifth or sixth power of the metal-ligand distance $r$[166,207,297-300].

A factor which was not considered in detail in the qualitative analysis of the data presented in Table 10.3 is the dependence of the ligand field splitting on the angular geometry. The ligand field strength decreases with increasing deviation of the ligand atoms from the direction of the e$_g$ set ($\sigma$-bonding) of the d orbitals

(90° angles). This was demonstrated with copper(II) complexes of a series of 13- to 16-membered tetraaza-macrocyclic ligands similar to the ones shown in Table 10.3. The highest ligand field was produced with the 14-membered ring, while the 13-membered ring, with copper-nitrogen bond lengths similar to those of the 14-membered ring complex, leads to a considerably lower ligand field splitting. This is a consequence of decreasing overlap, caused by the square pyramidal geometry of the chromophore[166] (see also Table 10.7, below). In addition, a further decrease of the ligand field splitting is expected and observed when the direction of the ligand bonding orbitals do not coincide with the metal-ligand atom direction, i.e., when misdirected valencies (bent bonds) are present[166].

Another important point is that the ligand field splitting increases (increasing overlap) with increasing alkyl substitution on the amine ligands because of inductive effects[120,158,296,301]. However, increasing substitution of the ligand backbone often leads to steric crowding that can be relaxed by an elongation of the metal-ligand bonds. Since the two mechanisms, inductive effect and elongation, are opposing in terms of the ligand field splitting, the net effect is unpredictable (at least without thorough calculations), and the importance of inductive effects is difficult to quantify.

In recent studies, involving the prediction of electronic and EPR spectra of hexaamine complexes of chromium(III), low spin iron(II), cobalt(III), nickel(II) and copper(II)[90,166,197,302] with a combination of molecular mechanics and AOM (angular overlap model)* calculations, the two effects could be separated (Table 10.4; see also text below) because the structural factors (steric crowding; molecular mechanics) and the electronic factors (inductive effects; AOM) are parameterized separately.

Note that the increasing donor strength in the series ammonia < primary amine < secondary amine < tertiary amine should lead to increasing force constants and/ or decreasing equilibrium distances in the force field parameterization. A single set of parameters for an M-N interaction generally leads to good agreement between calculated and experimental structures, probably because the differences are small and the coordination geometry is dominated by the ligand structure.

---

* The angular overlap model is based on a simplified MO treatment of the metal-ligand interaction and has been developed for the analysis of electronic spectra[303-308]. Similarly to ligand field theory, the AOM is based on the molecular symmetry of the complex, and the AOM parameters ($e_\sigma$, $e_\pi$, $e_\delta$) may be derived from the corresponding ligand field parameters ($Dq$, $Ds$, $Dt$). However, the AOM parameters do, in contrast to the ligand field parameters, have chemical significance ($\sigma$, $\pi$, $\delta$ – donor or acceptor strength). The basis of the AOM is that a perturbation of the metal d-orbital occurs by weak covalent interactions with ligand orbitals of suitable symmetry. The destabilization of the d orbitals is proportional to the two atom overlap integral. The energy change of each d-orbital is given by $\Delta E(d) = [F(d,\sigma)]^2 e_\sigma + [F(d,\pi_x)]^2 e_{\pi_x} + ...$, and the total change in energy of each d-orbital is the sum of the effects of all the ligand atoms on that orbital. The $e$-parameters are dependent on the metal ion, the ligand atom and the metal-ligand bond length, while $F$ is dependent on the angular geometry. For example, the d-orbital energies in an octahedral complex are $E(d_{z^2}) = 3 e_\sigma$ ($1 e_\sigma$ each from the ligands along the z-axis, $1/4 e_\sigma$ each from the ligands in the $xy$ plane), $E(d_{x^2-y^2}) = 3 e_\sigma$ ($3/4 e_\sigma$ each from the four ligands in the $xy$ plane), $E(d_{xz}) = E(d_{yz}) = E(d_{xy}) = 4 e_\pi$.

**Table 10.4:** Quantification of inductive effects via normalized $e_\sigma$ parameters obtained by the MM-AOM method[90,166].

| | $e_\sigma$ (norm) $[cm^{-1}]$[a) | | | |
|---|---|---|---|---|
| | $r(Cr^{III}) = 2.08$ Å | $r(Co^{III}) = 1.98$ Å | $r(Ni^{II}) = 2.13$ Å | $r(Cu^{II}) = 2.03$ Å |
| $NH_3$ | 7200 | 7245 | 3582 | – |
| $RNH_2$ | 7400 | 7433 | 3857 | 6400 |
| $R_2NH$ | 8000 | 7715 | 4133 | 6700 |
| $R_3N$ | 8700 | 8186 | 4592 | – |

a) $e_\sigma = \dfrac{C}{r^6}$

The data in Table 10.3 were taken as an indication that the ligand which best fits the metal ion demand, in terms of the hole size, induces the highest ligand field splitting, and not the ligand leading to the smallest metal-ligand distance[158]*. This interpretation involves, in a qualitative way, the angular distortions discussed above. Based on this interpretation, the question was asked whether metal-ligand bond compression exists at all. To achieve short metal-ligand bonds the conformational freedom and the flexibility of the ligand have to be reduced to prevent either a tetrahedral twist or the metal ion from moving out of the ligand plane. This was successfully done with a number of structurally reinforced ligand systems[120,158,301].

An interesting question related to this debate is whether any metal-ligand bond has a "normal" length or whether it is elongated or compressed. To answer this question one needs to know the strain-free bond distance, a parameter which is not directly accessible[65]. An instructive example in this respect is shown in Fig. 10.2: *trans*-diammac is a pendent arm macrocyclic hexaamine (two primary and four secondary amines), which is known to enforce short metal-ligand bond lengths for most of the first row transition metal ions. In many examples this leads to the strongest ligand field splitting observed for hexaamine complexes, stronger than for hexaamine cage ligands with six secondary amines (see also Table 10.6 below)[90]. Shown in Fig. 10.2 is a plot of the strain energy as a function of the metal-ligand bond length for all three possible conformers of $[M(\textit{trans}\text{-diammac})]^{n+}$. Also shown in Fig. 10.2 are the calculated bond distances (which are in good agreement with experiment) for a number of metal complexes of *trans*-diammac, indicating that for all metal ions studied, except for zinc(II), the observed bond distances are smaller than the best fit to the ligand[65,182]. It emerges that, although the metal-ligand distances are very short, these are actually elongated and not compressed by the ligand.

The MM-AOM approach – AOM calculations based on molecular mechanics refined structures – was used to predict d-electron transitions of chromium(III),

---

* The underscored $Dq^{xy}$ values are those where the enforced M–N distance (hole size, first column in Table 10.3) is closest to the ideal M–N distance ($r_0$ value) of the force field used in that study.

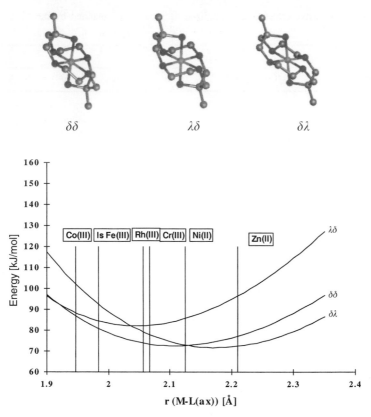

**Figure 10.2:** Hole size of the three nondegenerate conformers of [M(*trans*-diammac]$^{n+}$ (calculated as in Chapter 9, Fig. 9.5 (b)).

low spin iron(III), low spin iron(II), cobalt(III) and nickel(II) hexaamines and copper(II) tetraamines[90,166,197,281,302,308–310]. AOM calculations allow one to compute d-electron energy levels based on the geometry of the chromophore and the bonding parameters ($e_\sigma$ and $e_\pi$) for all ligand atoms[306,307]. Usually, the AOM is used to *interpret* electronic properties (UV-vis-NIR, EPR spectra, magnetic moments). For the *prediction* of spectrocopic data based on an established structure (experimental or molecular mechanics), a known and transferable set of electronic parameters must be used. This is not, *a priori,* a given property of the AOM approach and therefore is problematic[90,166,207,311]. The successful application of the MM-AOM approach with constant parameter sets for chromium(III), cobalt(III), low spin iron(III), low spin iron(II) and nickel(II) hexaamines and copper(II) tetraamines[90,166,197,281,302,309,310] does not imply that the electronic parameters are strictly transferable but that, for the systems studied so far, the errors introduced by assuming transferability are acceptable.

The electronic parameters used for the MM-AOM calculations are given in Table 10.5 (see footnote on p. 129). Observed and calculated d-d transitions for chromium(III), cobalt(III) and nickel(II) hexaamines are listed in Table 10.6. The error limits are of the order of 800 cm$^{-1}$ or less, while the observed transitions cover a range of 2000–5000 cm$^{-1}$, dependent on the metal ion.

**Table 10.5:** Electronic parameters for MM-AOM calculations [62,91,166].

| Parameter [cm$^{-1}$] | Cr(III) | Is Fe(III) | Co(III) | Ni(II) | Cu(II) [d) |
|---|---|---|---|---|---|
| $F_2$ [a) | 1211 | 786 | 1046 | 1243 | – |
| $F_4$ [a) | 101 | 57 | 85 | 79 | – |
| $e_\sigma(NH_3)$ [b,c) | 7200 | – | 7245 | 3582 | – |
| $e_\sigma(RNH_2)$ | 7400 | 7500 | 7433 | 3857 | 6400 |
| $e_\sigma(R_2NH)$ | 8000 | 7500 | 7715 | 4133 | 6700 |
| $e_\sigma(R_3N)$ | 8700 | – | 8186 | 4592 | – |
| $e_\sigma$(pyridine) | – | 7900 | – | – | – |
| $e_\pi$(pyridine) | – | –500 | – | – | – |

a) Condon-Shortley interelectronic repulsion parameters.
b) for low spin iron(III) a single $e_\sigma$ for the various amines was used.
c) normalized for Cr-N = 2.080, Fe-N = 1.985, Fe-N$_{py}$ = 1.968,
Co-N = 1.980, Ni-N = 2.130, Cu-N = 2.027, Cu-O = 2.440 Å;
for the caculations the values are adjusted with 1/r$^6$.
d) $k = 0.7$, $\xi = -580$, $K = 0.43$, $P = 0.036$, $\alpha^2 = 0.74$.

**Table 10.6:** Observed and calculated spin allowed d-d transitions of chromium(III), cobalt(III) and nickel(II) hexaamines [90].

| Compound[a) | Observed [cm$^{-1}$] | | Caculated [b) [cm$^{-1}$] | |
|---|---|---|---|---|
| [Cr(dtne)]$^{3+}$ | 20790; | 27780 | 20250; | 27680 |
| *mer*-[Cr(dpt)$_2$]$^{3+}$ | 21100; | 27700 | 20050; | 27340 |
| *cis*-[Cr(cyclam)(NH$_3$)$_2$]$^{3+}$ | 21370; | 28170 | 21610; | 29000 |
| [Cr(teatacn)]$^{3+}$ | 21410; | 27930 | 21990; | 29440 |
| [Cr(tn)$_3$]$^{3+}$ | 21570; | 28190 | 21190; | 28540 |
| *mer*-[Cr(dien)$_2$]$^{3+}$ | 21700; | 27850 | 21360; | 28900 |
| [Cr(NH$_3$)$_6$]$^{3+}$ | 21640; | 28940 | 21420; | 28790 |
| [Cr(en)$_3$]$^{3+}$ | 21800; | 28490 | 21780; | 29170 |
| [Cr(cis-diammac)]$^{3+}$ | 22120; | 28570 | 21790; | 29303 |
| [Cr(sar)]$^{3+}$ | 21860; 22500 | 28820 | 22180; 22850 | 29860 |
| *s-fac*-[Cr(dien)$_2$]$^{3+}$ | 22170; | 29070 | 22110; | 29550 |
| [Cr(tacn)$_2$]$^{3+}$ | 22780; | 29400 | 23120; | 30550 |
| *trans*-[Cr(cyclam)(NH$_3$)$_2$]$^{3+}$ | 22470; 23810 | 29940 | 22060; 24030 | 30320 |
| [Cr(trans-diammac)]$^{3+}$ | 23420; | 30120 | 23525; | 31430 |
| *mer*-[Co(dpt)$_2$]$^{3+}$ | 19530; | 27700 | 19410; | 27670 |
| [Co(dtne)]$^{3+}$ | 20200; | 28900 | 19790; | 28590 |
| [Co(tn)$_3$]$^{3+}$ | 20600; | 28500 | 20420; | 28740 |
| [Co(teatacn)]$^{3+}$ | 20830; | 28820 | 21180; | 29570 |
| [Co(NH$_3$)$_6$]$^{3+}$ | 21050; | 29500 | 21180; | 29540 |

**Table 10.6** (continued)

| Compound[a] | Observed [cm⁻¹] | | | | Calculated[b] [cm⁻¹] | | | |
|---|---|---|---|---|---|---|---|---|
| [Co(sar)]³⁺ | 21230; | 29150 | | | 21170; | 29530 | | |
| [Co(en)₃]³⁺ | 21370; | 29410 | | | 21510; | 29960 | | |
| *u-fac*-[Co(dien)₂]³⁺ | 21370; | 29500 | | | 21250; | 29670 | | |
| *mer*-[Co(dien)₂]³⁺ | 21460; | 29200 | | | 21270; | 29680 | | |
| [Co(*cis*-diammac)]³⁺ | 21790; | 29940 | | | 21960; | 30390 | | |
| [Co(tacn)₂]³⁺ | 21830; | 30030 | | | 21720; | 30100 | | |
| [Co(*trans*-diammac)]³⁺ | 22370; | 30490 | | | 22450; | 30910 | | |
| [Ni(NH₃)₆]²⁺ | 10750; | 17500; | 28200 | | 10720; | 17130; | 27760 | |
| [Ni(tn)₃]²⁺ | 10900; | 17800; | 28200 | | 10900; | 17370; | 28050 | |
| [Ni(dtne)]²⁺ | 10910; | 19380; | 27550 | | 10840; | 18600; | 27280 | |
| *mer*-[Ni(dpt)₂]²⁺ | 10970; | 17570; | 28070 | | 10550; | 17110; | 27480 | |
| *mer*-[Ni(dien)₂]²⁺ | 11500; | 18700; | 29100 | | 11350; | 18350; | 28910 | |
| [Ni(tach)₂]²⁺ | 11550; | 18850; | 29600 | | 11630; | 18330; | 29270 | |
| [Ni(en)₃]²⁺ | 11700; | 18350; | 29000 | | 11600; | 18300; | 29220 | |
| [Ni(*cis*-diammac)]²⁺ | 12380; | 18700; | 20410 | 30960 | 12180; | 18110; | 19630 | 30260 |
| [Ni(sar)]²⁺ | 12390; 12820 | 19760; | 20410 | 30300 | 12480; 12630 | 19270; | 19740 | 30900 |
| [Ni(tacn)₂]²⁺ | 12500; | 19800; | | 30800 | 12490; | 19480; | | 30740 |
| [Ni(*trans*-diammac)]²⁺ | 12530; 13160 | 20500; | | 31970 | 11950; | 19320; | | 30450 |

[a] Structures

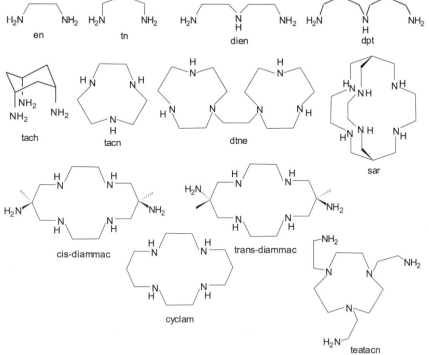

[b] averages where appropriate.

Semi-empirical and ab-initio studies of similar systems have led to less accurate predictions[312-314]. Observed and calculated d-electron transition energies together with spin Hamiltonian parameters for copper(II) tetraamines are listed in Table 10.7*. The accuracy of these predictions is comparable with that obtained for the hexaamines in Table 10.6, and the trends expected for bond length and angular distortions are well reproduced (see above).

The studies discussed here*, have been used to establish the scope and accuracy of the MM-AOM method. The results obtained indicate that combined MM-AOM calculations are potentially a powerful tool for the design of novel compounds with particular spectroscopic properties and for the determination of solution structures. The study involving EPR spectroscopy and MM-AOM calculations, discussed in the next section, was the first example of this approach. Others have been published since and confirm the general applicability of the method. Another, particularly instructive example, which also involves the computation of redox potentials is given in Chapter 11.

An important parameter of ligand field spectra that has not often been modeled is the intensity of electronic transitions. Admixing of odd-parity functions to the d-orbitals due to static distortions can lead to a relief of the parity selection rule. Therefore, structural parameters which describe the amount of distortion of a pseudo-centrosymmetric chromophore can be correlated to intensities. This is shown in Fig. 10.3, where the tetrahedral twist angles $\theta$ and the extinction coefficents (as well as the transition energies) of the d-d transitions of a series of copper(II) compounds of macrocyclic ligands are plotted against the size of the macrocyclic ligand (see Table 10.7 for the structures of the ligands and the full set of experimentally determined and computed data; similar plots of other molecular properties, including EPR parameters and redox potentials have also been published[197,281]). An interesting and more general way to describe the symmetry of a coordination compound is the continuous symmetry measures approach [41,42,315]. Plots very similar to that shown in Fig. 10.3 can be produced when the symmetry measure $S$ is correlated with the intensities of the electronic spectra [316].

The colour of a compound depends on the electronic transition energies, the relative intensities of the transitions, if more then one is present and the band profiles. There have been attempts to compute real traces of d-d spectra (transition energies, linewidths and intensities)[317,318]. The approach is based on known or assumed structures and, therefore, it would be well suited for a combination with structural optimization based on molecular mechanics. The transition energies are computed with the angular overlap model (AOM), just as described for the MM-AOM approach, described above. The intensity distributions (relative intensities) are computed, based on a parameterization of the electric dipole transition moments (static model)[319] and an additional vibronic coupling model, based on parity mixing due to normal bonding models, derived by normal coordinate analysis of appropriate spectra[320,321]. The computation of band profiles is based on gaus-

---

* The data discussed here are those which were used for the validation of the MM-AOM methods and published before 1995. More recent examples, published between 1995 and 2000 appear in the references. These are all of similar accuracy.

**Table 10.7:** Observed and calculated d-d transitions and spin hamiltonian parameters of copper(II) tetraamine cations [166].

| Compound[a] | | (xy) [cm$^{-1}$] | (xz) [cm$^{-1}$] | (yz) [cm$^{-1}$] | (z$^2$) [cm$^{-1}$] | $g_1$ | $g_2$ | $g_3$ | $A_1$ [10$^4$ cm$^{-1}$] | $A_2$ [10$^4$ cm$^{-1}$] | $A_3$ [10$^4$ cm$^{-1}$] |
|---|---|---|---|---|---|---|---|---|---|---|---|
| [Cu(en)$_2$X$_2$]$^{n+}$ | obs. | 19700 | 19700 | 17900 | 14100 | 2.06 | 2.06 | 2.17 | – | – | – |
|  | MM-AOM | 20430 | 19080 | 19870 | 13310 | 2.04 | 2.04 | 2.16 | 26 | 29 | 205 |
| [Cu(1)(2)]$^{2+}$ | obs. | 22210 | 22210 | 22210 | 19880 | 2.04 | 2.06 | 2.14 | (11) | (40) | (207) |
|  | MM-AOM | 20160 | 18730 | 19620 | 18680 | 2.04 | 2.04 | 2.17 | 25 | 30 | 202 |
| [Cu(1)$_2$X]$^{n+}$ | obs. | 19300 | 19300 | 19300 | 16430 | 2.05 | 2.05 | 2.16 | 31 | 31 | 200 |
|  | MM-AOM | 19890 | 18390 | 19210 | 13900 | 2.04 | 2.04 | 2.16 | 23 | 32 | 203 |
| [Cu(3)X]$^{n+}$ | obs. | 18020 | 18020 | 18020 | – | – | – | – | – | – | – |
|  | MM-AOM | 18640 | 13540 | 17760 | 11260 | 2.03 | 2.07 | 2.17 | 7 | 66 | 190 |
| [Cu(3)X$_2$]$^{n+}$ | obs. | 20410 | 20410 | 20410 | – | – | – | – | – | – | – |
|  | MM-AOM | 23250 | 22100 | 23020 | 17200 | 2.04 | 2.04 | 2.14 | 27 | 32 | 212 |
| [Cu(4)X$_2$]$^{n+}$ | obs. | 20830 | 20400 | 20800 | – | 2.05 | 2.05 | 2.15 | 35 | 35 | 199 |
|  | MM-AOM | 20400 | 19200 | 20020 | 13600 | 2.04 | 2.04 | 2.16 | 26 | 30 | 205 |
| [Cu(5)X]$^{n+}$ | obs. | 18180 | 18180 | 18180 | – | – | – | – | – | – | – |
|  | MM-AOM | 18340 | 16590 | 17530 | 11170 | 2.04 | 2.05 | 2.17 | 13 | 41 | 197 |
| [Cu(5)X$_2$]$^{n+}$ | obs. | 17180 | 15390 | 16950 | – | 2.06 | 2.06 | 2.21 | 26 | 26 | 193 |
|  | MM-AOM | 17320 | 16010 | 16820 | 11130 | 2.05 | 2.05 | 2.19 | 24 | 28 | 194 |
| [Cu(6)X]$^{n+}$ | obs. | 17240 | 16130 | 16130 | – | 2.06 | 2.06 | 2.21 | 23 | 23 | 184 |
|  | MM-AOM | 17850 | 14150 | 15800 | 13180 | 2.04 | 2.06 | 2.18 | 12 | 40 | 195 |

**Table 10.7** (continued)

a) $OH_2$ was used as an axial ligand for all MM calculations, except for $[Cu(5)X]^{n+}$ and $[Cu(6)X]^{n+}$, where the pendent nitro group is coordinated.

Structures

$[Cu(en)_2(OH_2)_2]^{2+}$

$[Cu(3)(OH_2)]^{2+}$

$[Cu(5)(OH_2)]^{2+}$

$[Cu(1)(2)]^{2+}$

$[Cu(3)(OH_2)_2]^{2+}$

$[Cu(5)(OH_2)_2]^{2+}$

$[Cu(1)_2(OH_2)]^{2+}$

$[Cu(4)(OH_2)_2]^{2+}$

$[Cu(6)(OH_2)]^{2+}$

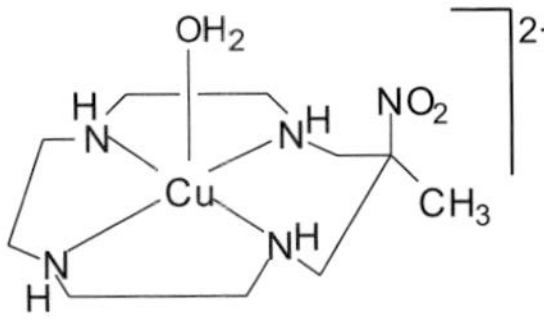

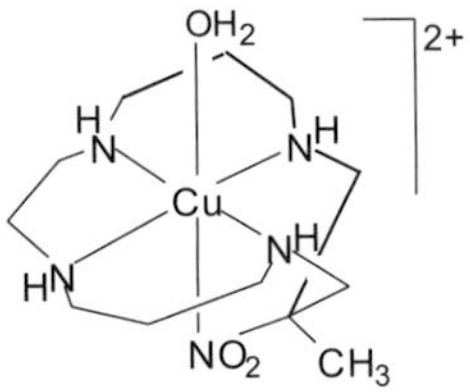

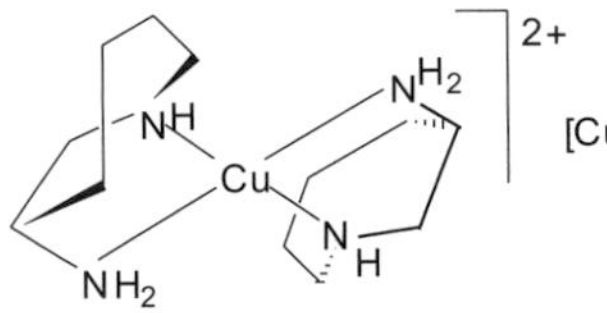

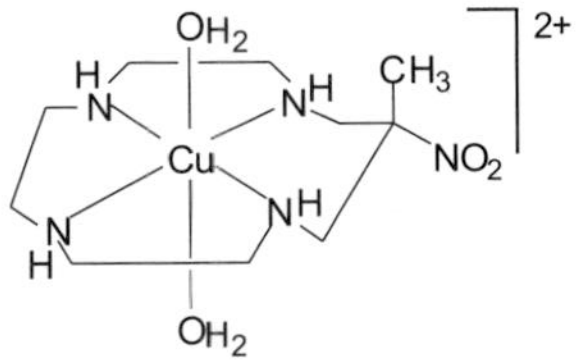

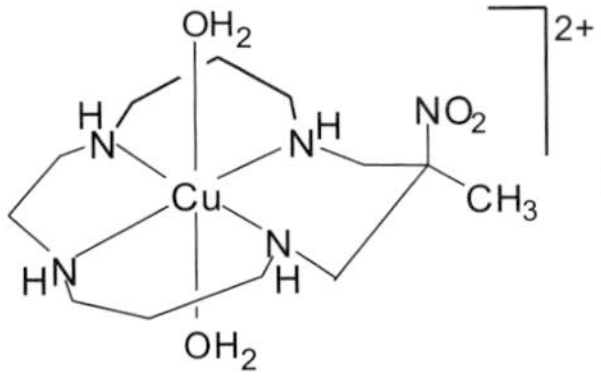

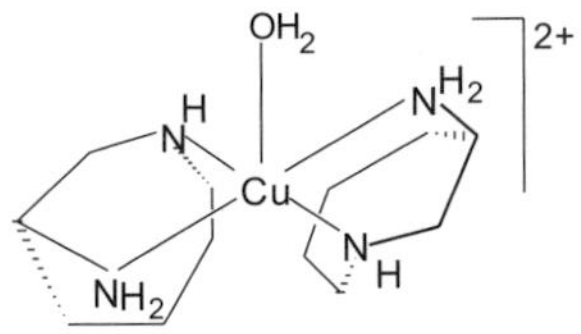

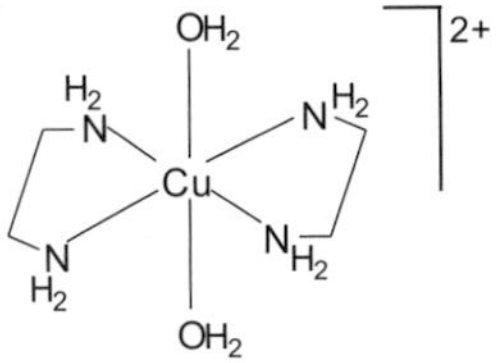

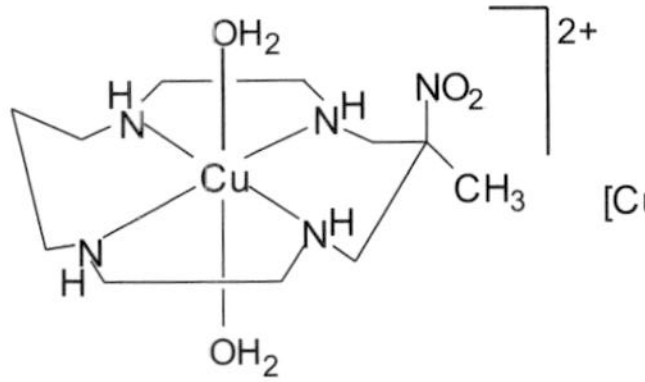

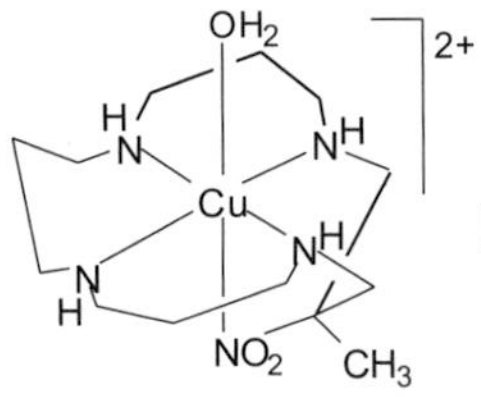

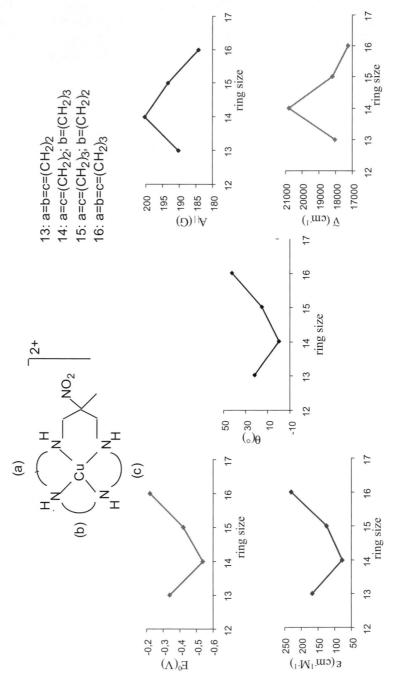

**Figure 10.3:** Correlations between tetrahedral twist angle $\theta$ and properties of tetraaminecopper(II) complexes [197,281].

sian curves with relative widths, governed by the slopes of their transition energies with respect to changes in all energy parameters[317]. A number of published examples indicate that this combination of models leads to qualitatively well reproduced ligand field spectra[311,317,320–322].

## 10.3 EPR Spectroscopy

The high sensitivity of the g-values of low-spin iron(III) to structural variations and their large anisotropy imply that the prediction of the EPR spectra must be based on highly accurate structures[207]. The MM-AOM method for low-spin iron(III) complexes was tested on a number of examples involving bi-, tri- and hexadentate ligands with amine and pyridyl donor sets (Table 10.8).

**Table 10.8:** Observed and calculated g-values for low spin iron(III) complexes[207].

| Compound | $g_1$ | | $g_2$ | | $g_3$ | |
|---|---|---|---|---|---|---|
| | obs. | calc. | obs. | calc. | obs. | calc. |
| $[Fe(trans\text{-diammac})]^{3+ a)}$ | 1.63 | 1.45 ($\delta\delta$) | 2.46 | 2.54 | 2.84 | 2.95 |
| | | 0.95 ($\lambda\delta$) | | 1.96 | | 3.29 |
| | | 1.89 ($\delta\lambda$) | | 2.05 | | 2.46 |
| $[Fe(en)_3]^{3+ a)}$ | – | 1.45 (lel$_3$) | | 1.45 | | 3.34 |
| | – | 1.35 (lel$_2$ob) | | 2.05 | | 2.97 |
| | – | 1.60 (lelob$_2$) | | 2.42 | | 2.48 |
| | – | 1.35 (ob$_3$) | 2.69 | 2.55 | 2.69 | 2.55 |
| $[Fe(phen)_3]^{3+ b)}$ | 1.51 | 1.61 | 2.69 | 2.62 | 2.69 | 2.62 |
| $[Fe(bpy)_3]^{3+ b)}$ | 1.63 | 1.74 | 2.64 | 2.53 | 2.64 | 2.53 |
| $[Fe(terpy)_2]^{3+ b)}$ | 1.76 | 1.73 | 2.54 | 2.21 | 2.54 | 2.73 |

a) Structures

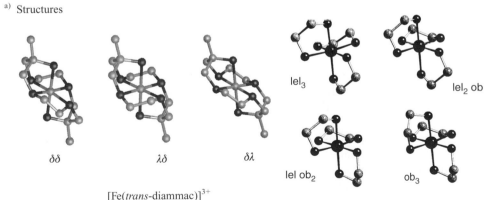

$[Fe(trans\text{-diammac})]^{3+}$

b) Nomenclature: phen = 1,10-phenanthroline; bpy = 2,2'-bipyridine; terpy = 2,2':6',2"-terpyridine.

Generally, there is good agreement between the experimentally observed and calculated  EPR spectra. Areas, where considerable errors resulted from a large variation of the g-values as a function of a specific distortion mode, were identified by model AOM calculations of the g-values as a function of the corresponding distortion mode. Two structural problems were addressed with the MM-AOM method applied to low spin iron(III):

(1)  The three reported structures of [Fe(*trans*-diammac)]$^{3+}$ all have disordered five-membered chelate rings, making an assignment of the structures (solid state or solution) to one of the three possible conformers (see Chart in Table 10.8) virtually impossible[217]. The main difference between the three chromophores is the magnitude of the tilt angle $\theta$ between the axis of the two primary amines and the macrocyclic plane. The observed and calculated g-parameters are strongly dependent on this distortion[207,323]. Based on the quality of agreement between experimentally determined and calculated EPR parameters of a series of low-spin iron(III) complexes and on the fact that the $\delta\lambda$ conformer has an exceedingly large strain energy (112.2 vs. 91.5 and 90.7 kJ mol$^{-1}$) it was concluded that the four observed experimental spectra (three solid samples with slightly different structures and one solution; Table 10.8 gives the data for the solution spectrum only) are due to the $\delta\delta$-conformer, and that the structural disorder is between $\delta\delta$ and $\lambda\lambda$ conformations. A minor contribution by the $\lambda\delta$ conformer could not be excluded[207].

(2)  The simulation of the EPR spectra of "[Fe(terpy)$_2$]$^{3+}$" was not possible with a symmetrical distortion mode involving both tridentate ligands[207]. Based on the generally good agreement between experimentally observed and calculated spectra a coordination polyhedron with one of the two terpy ligands acting as a bidentate was proposed. This interpretation is in agreement with all published data of the complex. Also, there is precedent for terpy coordination in this mode. However, other modes of distortion cannot be excluded, and further results have to be awaited in this area to solve this structural and spectroscopic problem with certainty. The problem involving "[Fe(terpy)$_2$]$^{3+}$" is therefore a good example for demonstrating the potential and also the limitations of the MM-AOM approach.

EPR spectra of weakly coupled dinuclear systems with each nucleus having $S = 1/2$ (dipole-dipole interactions) can be simulated with the spin Hamiltonian parameters (g- and A-values of each metal center) and the four geometric parameters $r$, $\tau$, $\eta$ and $\xi$ that define the orientation and distance of the two chromophores (see Fig. 10.4)[324–326]. The number of parameters involved in the simulation of the spectrum varies from 8 to 16, and this is often too large for an unambiguous structural assignment, even for well resolved spectra. Also, the structural information only relates to the relative orientation of the two g-tensors and therefore at most gives some information concerning the orientation of the two chromophores and no direct data related to the ligand conformation and overall shape of the molecule. Molecular mechanics alone also does not necessarily lead to an

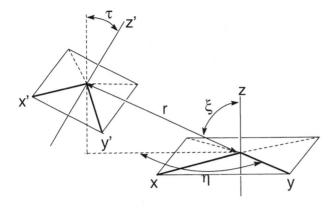

**Figure 10.4:** Structural parameters used for the simulation of EPR spectra involving weakly coupled $S=1/2/S=1/2$ systems.

unambiguous definition of the structural properties. This is due to the often high flexibility of dinucleating ligands and the associated number of possible conformers.

A combination of the two techniques was shown to be a useful method for the determination of solution structures of weakly coupled dicopper(II) complexes (Fig. 10.5)[165]. The MM-EPR approach involves a conformational analysis of the dimeric structure, the simulation of the EPR spectrum with the geometric parameters resulting from the calculated structures and spin Hamiltonian parameters

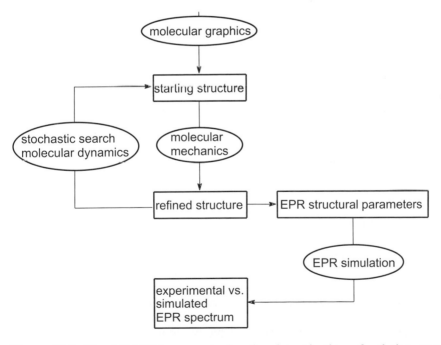

**Figure 10.5:** The MM-EPR approach for the determination of solution structures of weakly coupled $S=1/2/S=1/2$ systems.

derived from similar complexes, and the refinement of the structure by successive molecular mechanics calculation and EPR simulation cycles. This method was successfully tested with two dinuclear complexes with known X-ray structures and applied to the determination of a copper(II) dimer with unknown structure (Table 10.9 and Fig. 10.6) [165].

**Table 10.9:** Structural parameters for weakly coupled dinuclear copper(II) complexes [165].

| Compound | Method | r [Å] | ξ [°] | τ [°] | η [°] |
|----------|--------|-------|-------|-------|-------|
| Cu$_2$L1 | MM-EPR | 6.7 | 67 | 0 | 0 |
|          | X-ray | 6.9 | 70 | 0 | 0 |
| Cu$_2$L2 | MM-EPR | 8.0 | 70 | 75 | 29 |
|          | X-ray | 8.0 | 58 | 80 | 33 |
| {CuL3}$_2$ | MM-EPR | 6.7 | 73 | 5 | 6 |

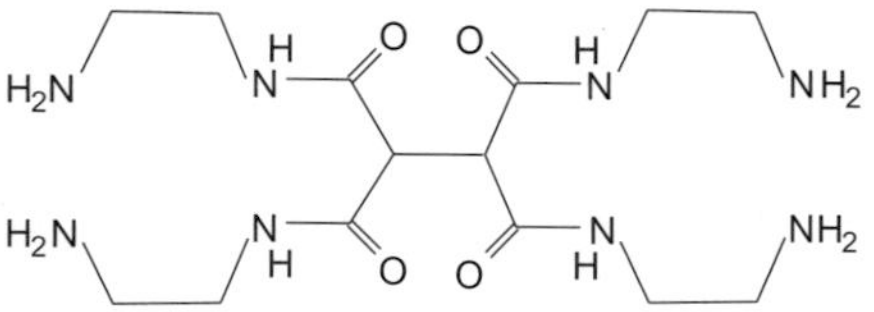

(L1)

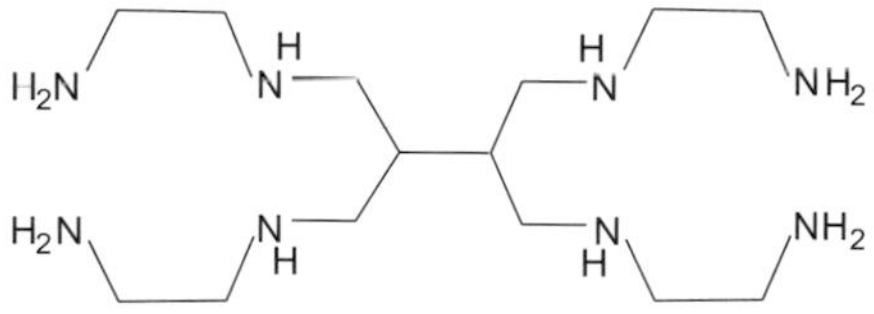

(L2)

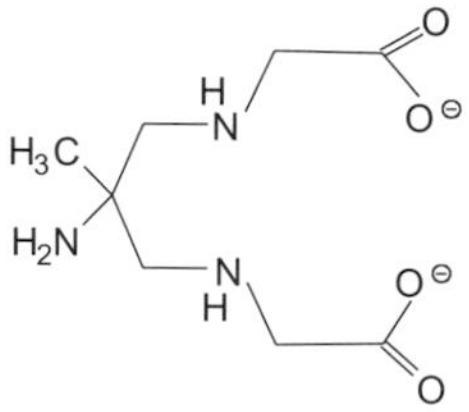

(L3)

The calculated EPR spectrum (MM-AOM, see above) of the mononuclear copper(II) complex shown in Fig. 10.6 is in good agreement with the experimental data, and the MM-EPR structure of the dinuclear compound is supported by thermodynamic data: for all ligands of the type of L3 (see Table 10.9 and Fig. 10.6) except the two with $R_1 = R_2 = H$ and $R_1 = H$, $R_2 = CH_3$ there is only a monomeric copper(II) complex observable. This observation is paralleled by the fact that the computed strain energies of the dinuclear compounds, corrected by a stoichiometric factor of 2, are larger than those of the monomeric compounds with the more highly substituted ligands, while the dimeric compound is more stable for the two least substituted ligands [165].

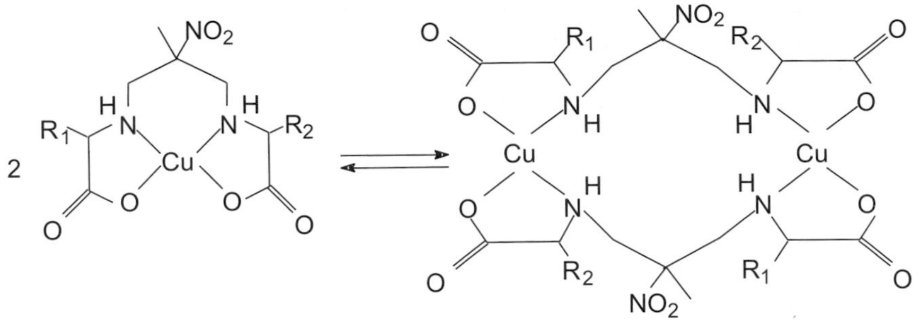

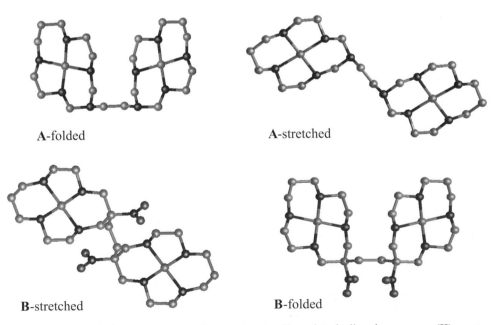

**Figure 10.6:** The equilibrium involving monomeric (X-ray structure) and dimeric (MM-EPR) [CuL3]$_n$ (R$_1$ = R$_2$ = H; n = 1,2)[165].

The validity of the MM-EPR method is further demonstrated with the two structurally related dicopper(II) complexes A and B whose calculated structures (MM-EPR) are presented in Fig. 10.7. For both bis-macrocyclic ligands two identical cyclam-type 14-membered tetraaza macrocyclic ligands (A, 1,3,6,10,13-pentaazacyclotetradecane; B, 13-nitro-1,4,8,11-tetraazacyclotetradecane) are coupled by an ethane bridge. While there might be some repulsion of the nitro groups in ligand B, the bridgehead atoms (nitrogen in the dinuclear compound A, and carbon in complex B) are not expected to have a pronounced influence on the structure of the dicopper(II) complexes. Complex B has a frozen solution EPR spec-

**A-folded**                    **A-stretched**

**B-stretched**                 **B-folded**

**Figure 10.7:** Solution structures of two structurally related dinuclear copper(II) complexes[204].

trum that is typical of a mononuclear copper(II) complex, indicating that there is no coupling between the two copper(II) sites, probably due to the large separation of the two metal centers in a stretched conformer. In contrast, the less substituted compound A exhibits an EPR spectrum that is typical of a weakly coupled dicopper(II) system, and, therefore, indicating a folded geometry[204,327]. This structural difference might have been expected due to the steric demand of the nitro substituents (see Fig. 10.7).

However, the strain energies of the two conformers each of A and B that are shown in Fig. 10.7 (34 kJ mol$^{-1}$, 44 kJ mol$^{-1}$, 52 kJ mol$^{-1}$, 58 kJ mol$^{-1}$, for A-folded, A-stretched, B-folded, B-stretched, respectively; other conformations have higher energy) indicate that for both complexes the folded conformation is more stable than the stretched geometry, although the energy increase to the less stable conformer is larger in the case of the dicopper(II) complex A (10 kJ mol$^{-1}$ vs. 6 kJ mol$^{-1}$), for which the folded geometry was observed.

As is often the case (see Sections 3.2.6 and 3.7), the molecular mechanics analysis above does not include any electrostatic interaction energies. To include these, the charge distribution and the charge compensation by ion pairing to counter ions (perchlorate) need to be known. Model calculations indicate that an effective charge of around +1.6 per copper center, a value that is expected from thermodynamic considerations, leads to electrostatic repulsion energies of ca. 17 kJ mol$^{-1}$ and 10 kJ mol$^{-1}$, respectively, for the folded and stretched conformers. In agreement with the experiment (EPR spectra) this qualitative analysis indicates a preference for the folded structure of A, and for the stretched structure of B[204].

Because of the approximations involved in this analysis the thermodynamic results have to be considered with caution. This is not only due to a rather crude analysis of the electrostatic effects but also, and this is a general problem, to the neglect of solvation in the molecular mechanics refinement. However, the structures presented in Fig. 10.7 are valuable because they are based not only on the structure optimization by molecular mechanics but also on spectroscopic data. This example is therefore instructive for two reasons: first, it demonstrates that, depending on the study, the often-neglected electrostatic effects may be of considerable importance. Second, not only can experimental observables help to refine solution structures, they can also prevent a wrong conclusion. As in this example, the combination of experimental data with molecular mechanics calculations is often the only way to get reliable structural information.

The MM-EPR approach has been used successfully in a number of recent studies[205,206,328,329]. The most novel is that of the solution structure refinement of a dicopper(II) compound of a cyclic octapeptide[205] which is only the second structure of a dicopper(II) compound of this type of biologically important ligand and the first of a metal compound of an artificial cyclic octapeptide. An important development in this area is a new method for the simulation of EPR spectra (SOPHE)[325,326], which allows the simulation of coupled EPR spectra of polynuclear species with more than two metal centers with any electron spin $\neq$ 0, based on sets of parameters similar to those discussed above.

## 10.4 NMR Spectroscopy

Molecular mechanics has been used in combination with NMR spectroscopy to solve structural problems. MM-NMR techniques have been extensively used to solve protein structures [220,330]. The main NMR information used in the modeling process is based on Karplus relations and NOE effects. A number of studies involving platinum anticancer drugs bound to DNA fragments have led to important information with respect to the mode of action of this drug (see Section 17.19). Recent applications also involve the simulation of paramagnetic shifts in proteins with metal centers such as cobalt(II) [331]. In such systems, the fact that protons close to metal centers have short relaxation times ($T_1$ and $T_2$) can be used to establish connectivity patterns [332–335]. Applications in the area of simple coordination compounds are quite rare [118,200–203] although these can be of importance as models for metalloproteins, especially for the determination of solution structures, where the modeling of the metal center can be one of the more serious problems (see also Chapter 13).

The structure determination of biopolymers using NMR spectroscopy usually involves interactions of protons [216,330]. Typically, interactions of protons (nuclear Overhauser effect, NOE) that are close in space but separated by several subunits of the biopolymer are used to establish the folding of the backbone. Distance restraints are then used to compute a structure which is checked by back-calculation of the NOE spectra and comparison with experimental results [336]. For large and highly flexible systems molecular dynamics is invaluable for scanning the conformational space.

MM-NMR techniques were recently applied successfully to a relatively small nickel(II) compound, a derivative of the coenzyme F430, where no long range coupling is observed [202]. Other studies using a combination of NMR techniques and molecular mechanics for the determination of solution structures of low molecular weight transition metal compounds include combined MM-NOE studies of palladium(II) phosphine complexes [118,200], the conformational analysis of a series of molybdenum complexes with tripodal phophine and carbonyl ligands, based on NOE spectra and force field calculations, and involving a Boltzmann-weighted average over all conformations, and structures of cobalt(III) complexes refined with empirical force field calculations based on the Karplus relation and observed dihedral angles [201]. The molecular mechanics models used in some of these studies are rather crude. However, the structural data are generally in good agreement with the experimental data, and the studies published so far indicate the large potential in combining NMR data with empirical force field calculations for the refinement of structures in solution.

# 11 Electron Transfer

Both the thermodynamics and kinetics of electron transfer reactions (redox potentials and electron transfer rates) have steric contributions, and molecular mechanics calculations have been used to identify them. A large amount of data has been assembled on $Co^{3+}/Co^{2+}$ couples, and the majority of the molecular mechanics calculations reported so far have dealt with hexaaminecobalt(III/II) complexes.

The basis for the application of molecular mechanics calculations to the thermodynamics and kinetics of electron transfer reactions is shown in Fig. 11.1. The redox potential is a function of the energy difference between the ground states of the oxidized and reduced forms of a complex, and to some extent this is related to the difference between the strain energies of the oxidized and reduced forms ($E^o = f(\Delta H_s^o)$). The rate for the self-exchange process can also be related to strain energies. Various methods have been used to compute the transition state structure and its energy[211,231,337,338], and there is some debate on the scientific value and the accuracy that can be expected from these approaches. The situation is complicated by the fact that, due to the neglect of solvation, entropic terms and electronic coupling, incomplete conformational analyses and inconsistencies in the force fields, there are large discrepancies between computed and experimental data[211,231] and, therefore, the quality of the approaches cannot be fully analyzed. Special care has to be taken in this area not to over-interpret computed data, even if they are fortuitously in agreement with the experiment. The requirements in an electron self-exchange reaction are that the total energy be conserved and therefore, the two reactants reorganize to the same nuclear configuration prior to electron transfer. There is an infinite set of such identical configurations, and the structure identified as the transition state must be that with the smallest reorganization energy. The two primary approaches to computing this structure and energy (see Fig. 11.1) are: (i) to determine the crossing points of the strain energy vs. bond distance curves of the oxidized and reduced forms; (ii) to compute the inner-sphere reorganization energy $\Delta G^*$ as $\lambda_{in,v}/4$, where $\lambda_{in,v}$ is the vertical internal reorganization energy which is computed by mutating the optimized structure of the oxidized form to the reduced form and computing its energy without refinement and vice versa.

The curves in Fig. 11.1b demonstrate that the situation is more complex when more than one isomer or conformer is present in the system. With the three conformers of [Co(*trans*-diammac)]$^{3+/2+}$ three distinct redox potentials are expected,

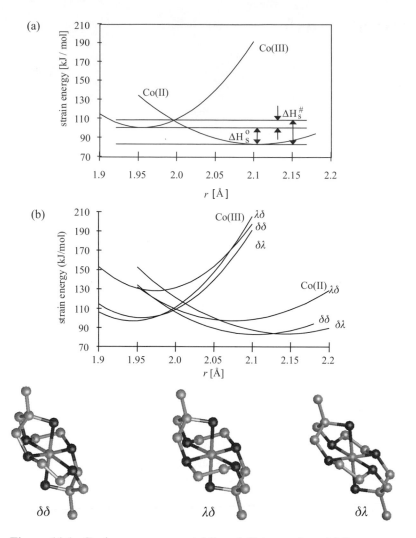

**Figure 11.1:** Strain energy vs. metal-ligand distance plots. (a) Parameters used in the calculations. (b) Calculated curves for the three conforms of [Co(*trans*-diammac)]$^{3+/2+}$ [231].

and the effective electron transfer rate (lowest $\Delta G^*$) does not necessarily involve the same conformer in the oxidized and reduced forms.

The determination of the structure of the encounter complex (relative orientation of the two reactants) and the ensuing information on the stereoselectivity of the electron transfer is a further possible application of molecular mechanics in this field, but this has not yet been evaluated.

## 11.1 Redox Potentials

Reduction potentials of hexaaminecobalt(III) complexes span a range of more than 1.4 V, with the lowest potential (−0.63 V) exhibited by [Co(*trans*-diam-mac)]$^{3+/2+}$ and the highest potential (+0.28 V) found for [Co(tmen)$_3$]$^{3+/2+}$ (for li-gand structures see Table 11.1; recently, a Co(III)/Co(II) couple with a potential of +0.84 V has been reported[310], see below). *trans*-diammac leads to relatively short metal-ligand bonds and therefore stabilizes the cobalt(III) state. The hexa-amine cage ligand discussed below (Fig. 11.3) leads in the lel$_3$ conformation to re-latively long cobalt-amine bonds, destabilizing the cobalt(III) state. Thus, steric effects can make a substantial contribution to the variation in reduction potentials.

The redox potentials and the strain energies at the cobalt(III) and cobalt(II) oxi-dation states of the most stable conformers of a number of hexaaminecobalt(III/II) complexes are listed in Table 11.1 (selected data from[231]). The strain energy

**Table 11.1:** Observed and calculated redox potentials of cobalt(II/III) hexaamine couples[132,231].

| Compound[a] | Total strain energy [kJ mol$^{-1}$][b] | | $\Delta H_s^o$ [kJ mol$^{-1}$][c] | $E_{1/2}^{obs}$ [V][d] | $E_{1/2}^{calc}$ [V] |
|---|---|---|---|---|---|
| | Co$^{III}$ | Co$^{II}$ | | | |
| [Co(en)$_3$]$^{2+/3+}$ | 45.12 | 4.48 | 40.64 | −0.18 | −0.10 |
| [Co(tmen)$_3$]$^{2+/3+}$ | 154.33 | 86.22 | 68.11 | +0.28 | +0.28 |
| [Co(*trans*-diammac)]$^{2+/3+}$ | 98.56 | 82.00 | 3.01 | −0.63 | −0.63 |
| [Co(sar)]$^{2+/3+}$ | 127.73 | 79.73 | 25.52 | −0.45 | −0.32 |
| [Co(sep)]$^{2+/3+}$ | 115.58 | 60.95 | 33.63 | −0.30 | −0.20 |
| [Co(tacn)$_2$]$^{2+/3+}$ | 109.57 | 82.23 | 8.45 | −0.41 | −0.55 |
| [Co(NH$_3$)$_6$]$^{2+/3+}$ | 29.38 | −8.78 | 38.16 | +0.06 | −0.14 |

[a] Structures

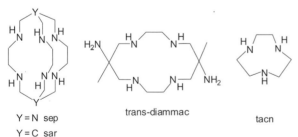

Y = N  sep
Y = C  sar

trans-diammac

tacn

en = ethane-1,2-diamine
tmen = 1,2-dimethylbutane-1,2-diamine

[b] Strain energy contribution to $\Delta G^o$ in Fig. 11.1 (a). Lowest strain energy difference between identical conformers.
[c] Lowest energy conformers; force field of[58].
[d] vs. SHE.

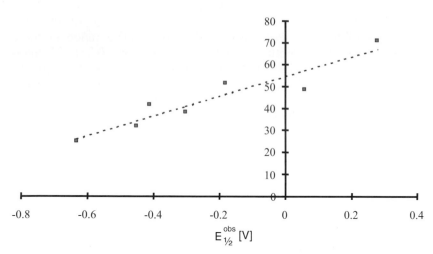

**Figure 11.2:** Experimentally determined redox potentials of the hexaaminecobalt(III/II) complexes from Table 11.1 as a function of the strain energy difference between the oxidized and reduced forms[231].

difference between the two oxidation states was found to correlate with the experimentally determined reduction potentials[132]. Fig. 11.2 is a plot of the redox potentials of hexaaminecobalt(III/II) complexes with primary amine donors as a function of the strain energy differences between the oxidized and reduced forms (data from[231]).

The experimentally determined redox potentials are given as solid points while the line corresponds to the calculated potentials. Based on Eq. 11.1, where $F$ is the Faraday constant ($F = 96.5$ kJ mol$^{-1}$) and $n = 1$, the slope of the line should be 96.5 kJ mol$^{-1}$ V$^{-1}$, if differences in $\Delta S$ are neglected and strain relaxation is the only contribution to the variation in redox potential.

$$\Delta G^\circ = -n\,F\,E^\circ \tag{11.1}$$

The slope of 50 kJ mol$^{-1}$ V$^{-1}$ indicates that the variation of redox potentials of over 1.0 V is due to a large extent to strain relaxation. Other possible contributions to redox potentials are electronic effects[132], specific hydrogen bonding[339], ion-pairing[132,151,339], solvation[339] and hydrophobicity[340]. The fact that different slopes are obtained for correlations with primary and secondary amines (50 vs. 65 kJ mol$^{-1}$ V$^{-1}$)[231] was thought to be due to a combination of variations in the nucleophilicity and rigidity of the ligands[231]. Eq. 11.2 combines these separate correlations ($a$ = number of alkyl groups at the nitrogen donors).

$$E^\circ = \frac{\Delta H_{\text{strain}}}{40.378} - 1.354 - a\left(\frac{\Delta H_{\text{strain}}}{1285.1} - 0.0525\right) \tag{11.2}$$

The mean accuracy of the computed potentials is $\pm 0.08$ V, and the reasonable linearity and the fact that the intercepts are identical within the error indicate that the neglected terms (solvation, entropy) are approximately dependent of the strain energies, as one might expect on simple qualitative considerations[231].

Results of similar accuracy have also been obtained for copper(II/I) couples[341]. This was unexpected since these redox processes are usually accompanied by changes in the coordination number. The justification for using the strain energy approach to estimate redox potentials of copper(II/I) couples was that the axial ligands in copper(II) compounds are at long distances and, therefore, do not contribute much to the strain energies, and that the bond energy differences for the axial ligands in five- and six-coordinate complexes might be similar due to the generally observed differences in bond distances (approx. 2.3 Å and 2.5 Å for $4+1$ and $4+2$ chromophores, respectively)[341]. A recent report indicates that the stability constants of copper(I) compounds do not vary much ($\log K_{Cu^I L} \approx 14 \pm 2$), while those of copper(II) span a large range ($1 \leq \log K_{Cu^{III} L} \leq 20$)[342]. Therefore, the variation of the reduction potentials ($-0.66\,V \leq E^0 \leq 0.89\,V$) is almost entirely due to changes in the stability (and strain) of the copper(II) form.

From Fig. 11.1b it emerges that the simple electrochemical model (Eq. 11.3) cannot be correct, and the model of Eq. 11.4 has been proposed[343,344].

$$A^n + e^- \overset{E^o}{\rightleftharpoons} A^{(n-1)+} \tag{11.3}$$

$$
\begin{array}{ccc}
A_1^{n+} + e^- & \xrightarrow{\;E_1^o\;} & A_1^{(n-1)} \\
\Big\updownarrow K_1 & & \Big\updownarrow K_1' \\
A_2^{n+} + e^- & \xrightarrow{\;E_2^o\;} & A_2^{(n-1)+} \\
\Big\updownarrow K_2 & & \Big\updownarrow K_2' \\
\vdots\; K_i & & \vdots\; K_i' \\
A_i^{n+} + e^- & \xrightarrow{\;E_i^o\;} & A_i^{(n-1)+}
\end{array}
\tag{11.4}
$$

Here, the $i$ conformers each of the oxidized and the reduced forms are related by the $2(i-1)$ equilibrium constants $K_i$ and $K_i'$, respectively, and by the $i$ redox potentials $E_i^o$. A quantitative analysis of the redox potential in the square scheme of Eq. 11.4 requires a knowledge of all equilibrium constants. For labile systems this is only possible when theoretical methods can be applied. Molecular mechanics has been used in this context to calculate the conformational equilibria and then to predict the electrochemical behavior of $[Co(sep)]^{3+/2+}$[151], $[Co(dien)_2]^{3+/2+}$[151] and $[Co\{(S)\text{-pn}\}_3]^{3+/2+}$[345] (sep is defined in Table 11.1, dien in Table 8.1, pn in Table 8.2).

The three isomers of $[Co(dien)_2]^{3+/2+}$ have, as predicted by molecular mechanics calculations (see also Table 8.1 in Section 8.1 for calculated and observed isomer ratios of $[Co(dien)_2]^{3+}$), measurably different redox potentials[228]. However, the strain energy differences between various conformers of each isomer were calculated to be too small for a measurable difference of the redox potentials, and the order of stability in both oxidation states was the same[151]. A similar problem occurred with $[Co(sep)]^{3+/2+}$[151]. For $[Co\{(S)\text{-}pn\}_3]^{3+/2+}$ the four redox potentials between isostructural pairs lie within the predicted range of 20 mV. However, due to lack of resolution, a quantitative analysis was not possible.

Very different redox potentials were recently reported for the two stable conformers of a hexaaminecobalt(III) complex with a cage ligand (see Fig. 11.3). Two isomers with very different properties were isolated: one is yellow (first d-d transition at 480 nm) with a potential of $E^0 = 0.0\,V$, the other is blue (600 nm) with $E^0 = 0.84\,V$. The crystal structure of the yellow compound was solved (Fig. 11.3(a))[346], for the blue compound no diffracting crystals were isolated. That structure was solved by a combination of molecular mechanics, the simulation of the ligand field spectra (MM-AOM) and the redox potential (MM-Redox), see Fig. 11.3(b)[310].

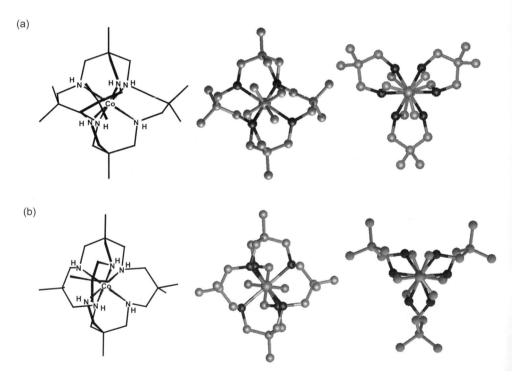

(a)

(b)

**Figure 11.3:** Computed structures of two conformers (a, lel$_3$; b, ob$_3$) of a cobalt(III) complex with the large ring hexaamine cage ligand (reproduced with permission from reference[310]).

## 11.2 Electron Transfer Rates

The Marcus theory provides an appropriate formalism for calculating the rate constant of an outer-sphere redox reaction from a set of non-kinetic parameters [347–350]. The simplest possible process is a self-exchange reaction, where $\Delta G = 0$. In an outer-sphere electron self-exchange reaction the electron is transferred within the precursor complex (Eq. 11.5).

$$[Co^*]^{3+} + [Co]^{2+} \overset{K}{\rightleftharpoons} [Co^*]^{3+} \cdots [Co]^{2+} \overset{k_{et}}{\to} [Co^*]^{2+} \cdots [Co]^{3+}$$
$$\rightleftharpoons [Co^*]^{2+} + [Co]^{3+} \tag{11.5}$$

$$k = \kappa A r^2 \exp(-\Delta G^*/RT) \tag{11.6}$$

The electron exchange rate $k$ (Eq. 11.6) is a function of the transmission coefficient $\kappa$ (approximately 1 for for reactions with substantial electronic coupling (>4 kJ), i.e., for adiabatic reactions), the effective collision frequency in solution ($Z \sim 10^{11}$ $M^{-1}$ $s^{-1} \sim Ar^2$) and the free energy term $\Delta G^*$.

In this simple form of the Marcus theory, two terms, the reorganization of the inner coordination shell ($\Delta G^*_{in}$) and that of the solvent sheath ($\Delta G^*_{out}$), both describing the degenerate transition state geometry, contribute to $\Delta G^*$. While molecular mechanics might also be used to model the encounter complex, force field calculations have mainly been applied to the estimation of the inner coordination shell reorganization term ($\Delta G^*_{in}$) [211,336–338,351–353].

By leaving all other terms constant one cannot expect accurate predictions of the self-exchange rates, and the magnitude of variations in the rates due to various terms (electronic coupling, encounter complex formation etc.) has been estimated [354]. Therefore, the calculated rates presented in Table 11.2 are surprisingly

**Table 11.2:** Observed and calculated electron self-exchange rates of cobalt(III/II) hexaamine complexes [231].

| Compound [a] | Calc [b] | Obs |
|---|---|---|
| $[Co(en)_3]^{3+/2+}$ | $2.55 \cdot 10^{-4}$ | $3.4 \cdot 10^{-5}$ |
| $[Co(tmen)_3]^{3+/2+}$ | $1.66 \cdot 10^{-8}$ | $8.5 \cdot 10^{-8}$ |
| $[Co(sar)]^{3+/2+}$ | 0.72 | 2.1 |
| $[Co(sep)]^{3+/2+}$ | 0.052 | 5.1 |
| $[Co(trans\text{-}diammac)]^{3+/2+}$ | 11 | (900) |
| $[Co(tacn)_2]^{3+/2+}$ | 0.013 | 0.19 |
| $[Co(NH_3)_6]^{3+/2+}$ | $4.3 \cdot 10^{-5}$ | $1 \cdot 10^{-7}$ |

[a] For structures, see Table 11.1.
[b] Force field of [58]; strain energy contribution to $\Delta G^{\#}$ (see Fig. 11.1a):
   $\Delta H^{\#} = 2\,H_s - H_s^{Co(III)} - H_s^{Co(II)}$

accurate. These have been obtained by the computation of strain energy vs. M-L curves for the oxidized and reduced forms of all conformers and computing the lowest energy transition state. Since the force field used did not include any angular constraints, it is likely that the corresponding crossing point is the common nuclear configuration with lowest energy.

# 12   Electronic Effects

The structure and stability of a metal complex are both intimately related to the distribution of electrons within the complex. Thus, the two types of information directly available from empirical force field calculations are subject to any variability in the electron distribution. The problem of how such "electronic effects" can be accounted for in a general parameterization scheme is a many-faceted theme. Central to the molecular mechanics concept is that, in contrast to quantum mechanics, the electron distribution in a molecule is regarded as localized and invariant. While the influence of substituents can be accounted for by the assignment of different atom types, modeling of highly and variably delocalized systems is less trivial. Conjugated $\pi$-systems can be treated with $\pi$-electron MO-calculations coupled to molecular mechanics[355]. With compounds involving metal centers such calculations remain a considerable problem. However, the principal challenge in the molecular modeling of transition metal compounds is that of describing the influence exerted by the d-electrons.

Compounds with partly filled d-orbitals have two peculiarities:

- a number of electronic ground states may be accessible; and
- when there is an unsymmetrical occupation of the d-orbitals, not all bonds to the metal ion are equivalent.

The electronic ground state that a particular metal center adopts is a function of the chromophore. In many cases the ground state can be derived from chemical knowledge (e.g. octahedral cobalt(III) ($^1A_{1g}$) or tetrahedral Ni(II) ($^3T_1$) complexes). However, based on the molecular mechanics formalism alone, this problem cannot be solved in a general way. Let us consider coordination compounds that are close to the spin-crossover limit (for example hexacoordinate iron(II) ($^1A_{1g}/^5T_{2g}$)). In these cases it is not possible to assign the atom type of the metal center without further information (experimental or theoretical). Therefore, molecular mechanics alone is not always able to predict the structural properties.

Even with an assumed or experimentally determined knowledge of the electronic ground state (e.g., by ligand field spectroscopy) modeling of specific electronic effects due to partly filled d-orbitals such as Jahn-Teller effects, *trans* influences and $\pi$-backbonding is not trivial. However, if molecular mechanics is used as a technique for the approximate calculation of energy surfaces with a set of functions and corresponding parameters that have been derived from experimental data (see Chapter 3) new functions and parameters that account for d-orbital effects can be added to the set of potential energy functions used in the classical molecular mechanics method. Some examples are described in this Chapter.

Three approaches have been used to directly include electronic effects in molecular mechanics-based modeling of transition metal compounds, and these have been described in Sections 3.2, 3.2.2 and 3.6: Hybrid QM/MM methods, valence bond concepts used as a basis for a new molecular mechanics model and a combination of molecular mechanics with angular overlap model terms. The most far reaching method is that of embedding ab-initio quantum mechanics into molecular mechanics models, e.g., the IMOMM methods, where parts of a molecular structure (typically the chromophore) are computed with quantum-mechanical methods whereas the rest is optimized by molecular mechanics[356]. The VALBOND model is an elegant rule-based force field, which computes gross hybridizations from the topology of the starting structure and refines them by Bent's rule-type algorithms[79]. An elegant approach with respect to electronic effects based on open d-shells is the CLF/MM method, which uses a cellular ligand field (CLF, i.e., an angular overlap model (AOM)) – based term in the set of potential energy functions[77,357]. This approach has been used to accurately compute Jahn-Teller distorted copper(II) compounds, high- and low-spin nickel(II) species with a single set of force field parameters and *trans* influences in planar $d^8$ complexes[77].

## 12.1   d-Orbital Directionality

In principle, the valence angles around metal ions can be described in much the same way as valence angles in organic molecules, i.e., with a harmonic function. Solutions to the problem of modeling complexes with higher coordination numbers that lead to regular polyhedra with two or more different types of angles (for instance 90° and 180° for octahedral, and 90°, 120° and 180° for trigonal bipyramidal structures) have been described[76,123] (see also Sections 3.2.2 and 3.6). A more serious problem is that, with a given coordination number, various geometries may be accessible, e.g., square pyramidal or trigonal bipyramidal for pentacoordination and pentagonal bipyramidal or monocapped trigonal prismatic for heptacoordination. One possibility for dealing with this ambiguity is to decide which is the appropriate type of coordination polyhedron before minimization and then use a geometry specific parameter set. However, this leads to a loss of the predictive power of molecular mechanics calculations. Three approaches to modeling the d-electron directionality have been reported: The VALBOND module that uses a hybrid orbital-based term for the molecular shape (see above), the CLF/MM model that uses a cellular ligand field stabilization energy term for the coordination geometry (see above), and an approach based on a combination of nonbonded interactions and a harmonic sine function for the optimization of the angular geometry, which will now be discussed in detail.

The points-on-a-sphere (POS) approach represents a general way of modeling coordination polyhedra and is based on the remarkably successful prediction of the coordination geometry based on ligand-ligand repulsion alone[86]. Non-bonded interactions between donor atoms have been included in a force field parameterization in which they replace the ligand-metal-ligand angle functions[59]. This method has been used successfully to calculate structures and strain energies of a wide range of coordination compounds with a variety of co-ordination numbers and geometries[57,65,89]. In essence, the inclusion of 1,3-non-bonded interactions around the metal center does not explicitly define any d-orbital directionality (electronic effects). However, the ligand-ligand interactions, which in this approach determine the coordination geometry, are strongly dependent on the metal-ligand distances, which in turn are a function of the ligand field strengths.

Nevertheless, from more recent studies it emerges, not unexpectedly, that the model based on 1,3-interactions alone sometimes leads to structural predictions where the angular geometry is not modeled satisfactorily[90]. From this and the fact that the approach of using exclusively ligand-ligand interactions to define the geometry of transition metal chromophores was successful in many examples, it follows that repulsion is an important, but not the sole, factor in determining the ligand arrangement around a metal center. That is, the angular geometry about the metal center is best modeled by including ligand-ligand interactions *and* an additional term describing the electronic metal preferences.

For octahedral and square planar geometries the ligands are located along the cartesian axes. The simplest function with minima at $90°$ and $180°$ is a harmonic sine function. Since the σ-bonding $e_g$ set of d-orbitals is directed along the cartesian coordinate axes, it is possible that this type of function can be used in general cases (at least when σ-bonds are predominant), i.e., not only for square planar, square pyramidal and octahedral geometries. This clearly would enhance the utility of the method since no assumptions in terms of the coordination geometry would then need to be made prior to structure optimization. However, the enforcement of planarity by the harmonic sine function could require large force constants and therefore lead to some distortion within the plane. A plane twist function has therefore been developed with a harmonic potential to enforce planarity without enforcing any direction of the metal-donor vectors within the plane[109].

The amount of the electronic contribution to the total strain (force constant $k_{LML'}$, Eqs. 12.1, 12.2;

$$E_{LML'} = \frac{1}{8} k_{LML'} \sin^2 2\theta \qquad (12.1)$$

$$k_{LML'} = c \cdot F \cdot \frac{k_{ML} + k_{ML'}}{2} \qquad (12.2)$$

the parameters for the 1,3-interactions are the same as in the pure interligand repulsion approach, and independent of the metal ion) is a function of the ligand field strength[58,231] and is determined as follows. The angle bending force constant $k_{LML'}$ is first fitted to a series of cobalt(III) hexaamines with widely varied chelate and trigonal twist angles. The resulting constant $c$ is then adjusted for effects due to the d-orbital occupancy of the metal center ($F$; see Table 12.1; spin-pairing energies are neglected) and the ligand field properties (spectrochemical series) which are assumed to be a function of the metal-ligand bonding force constant of the two relevant ligand atoms, $k_{ML}$ and $k_{ML'}$.

In principle, the resulting angle bending force constants are generic parameters (Table 12.2). However, in the light of the discussions on the relationship between force field parameters and physical reality presented in Section 3.5, the generic

**Table 12.1:** Crystal field effects in octahedral fields.

| $d^n$ | Configuration | Stabilization energy in Dq and P[a) |
|---|---|---|
| high spin | | |
| $d^0$ | | 0 |
| $d^1$ | $t_{2g}^1$ | $-4\,Dq$ |
| $d^2$ | $t_{2g}^2$ | $-8\,Dq$ |
| $d^3$ | $t_{2g}^3$ | $-12\,Dq$ |
| $d^4$ | $t_{2g}^3\,e_g^1$ | $-6\,Dq$ |
| $d^5$ | $t_{2g}^3\,e_g^2$ | 0 |
| $d^6$ | $t_{2g}^4\,e_g^2$ | $-4\,Dq + P$ |
| $d^7$ | $t_{2g}^5\,e_g^2$ | $-8\,Dq + 2\,P$ |
| $d^8$ | $t_{2g}^6\,e_g^2$ | $-12\,Dq + 3\,P$ |
| $d^9$ | $t_{2g}^6\,e_g^3$ | $-6\,Dq + 4\,P$ |
| $d^{10}$ | $e_g$ | $0\,Dq + 5\,P$ |
| low spin | | |
| $d^4$ | $t_{2g}^4$ | $-16\,Dq + P$ |
| $d^5$ | $t_{2g}^5$ | $-20\,Dq + 2\,P$ |
| $d^6$ | $t_{2g}^6$ | $24\,Dq + 3\,P$ |
| $d^7$ | $t_{2g}^6\,e_g^1$ | $18\,Dq + 3\,P$ |

[a) crystal field stabilization and spin pairing.

**Table 12.2:** Force constants for the harmonic sine function for hexacoordinate transition metal hexamines[58].

| Metal ion | $k_{LML'}$ [mdyn Å] |
|---|---|
| $Cr^{III}$ | 0.025 |
| $Co^{II}$ | 0.017 |
| $Co^{III}$ | 0.050 |
| $Ni^{II}$ | 0.025 |
| $Cu^{II}$ | 0.013 |

angle bending force constants are, instead, good starting points, which can be improved (where necessary) through fitting to experimental data.

Some examples (different electronic ground states) of experimentally determined structures and geometries obtained by strain energy minimization with the two models, i.e., with and without the harmonic sine term for the metal-ligand-metal interaction, are presented in Table 12.3.

**Table 12.3:** Some examples showing the improvements in structural predictions with the harmonic sine function [58,231].

| Complex [a] | Parameter [b] [°] | X-ray | Urey–Bradley [c] | Urey–Bradley and electronic term [d] |
|---|---|---|---|---|
| $[Co(tmen)_3]^{3+}$ | $\varnothing$ | 44 [358] | 41 | 44 |
| $[Ni(tn)_3]^{2+}$ | N-Ni-N | 174–177 [359] | 165–167 | 170–174 |
| $[Cu((R)\text{-}ahaz)((S)\text{-}ahaz)]^{2+}$ | $\theta$ | 0 [360] | 8 | 0 |

[a] Structures

$[Co(tmen)_3]^{3+}$     $[Ni(tn)_3]^{2+}$     $[Cu((R)\text{-}ahaz)((S)\text{-}ahaz)]^{2+}$

[b] $\varnothing$: trigonal twist angle; octahedron: $\varnothing = 60°$; trigonal prism: $\varnothing = 0°$;
  $\theta$: tetrahedral twist angle; square plane: $\theta = 0°$, tetrahedron: $\theta = 90°$.
[c] 1,3-interaction alone [57].
[d] 1,3-interaction and harmonic since function [58,231].

While bond distances were generally well reproduced with the earlier, simpler force field [57,65], the addition of the harmonic sine term has led to a considerable improvement in terms of the reproduction of the angular distortions, which were previously overestimated in highly strained complexes.

## 12.2 The *trans* Influence

The *trans* influence is defined as the extent to which a coordinated group lengthens the bond *trans* to itself. The origin of the *trans* influence is thought to be electronic: ligands that are *trans* to each other can compete for a stabilizing interaction with an empty metal $\sigma$-orbital and/or a filled metal d-orbital through $\pi$-back-bonding. The ligand that is able to form stronger bonds with the metal causes the bond to the ligand in the *trans* position to weaken and lengthen. It is well established that the relative extent of the *trans* influence is dependent on both the donor group and metal center[361].

Thus far, *trans* influences have usually been modeled by using separate sets of force field parameters for the bonds that are *trans* to each other[198]. In this way molecular mechanics loses some of its predictive power. In a series of metal ions and donor groups with known electronic properties it should be possible to establish an algorithm that generates corrections to the bonding parameters with regard to the type and disposition of the ligand atom. An interesting development in this respect is the concept of half-integer bond orders used to model $\pi$-backbonding in transition metal complexes[127]. More exciting in this respect is the CLF/MM model which has been used to model *trans* influences in square planar $d^8$ complexes with a single set of parameters[77]. Since d-orbitals are centrosymmetric individual perturbations from centrosymmetrically related donors cannot be separated. This problem was overcome and good quality predictions were obtained by the definition of artificial ligand-ligand stretching potentials[77].

## 12.3   Jahn-Teller Distortions

In a non-linear molecule with an electronically degenerate state, a structural distortion (static or dynamic) must occur that lowers the symmetry and, therefore, removes the degeneracy and lowers the total electronic energy[362,363]. This situation applies to a number of metal complexes and coordination geometries, hexacoordinate $d^9$ complexes being the best known example. Copper(II) complexes are both structurally and spectroscopically well characterized, and many biologically relevant compounds have been identified and studied[364−367]. Therefore, it is not surprising that a number of studies reporting molecular mechanics calculations of copper(II) species have appeared in the literature[57,65,120,123,163,165,166,204,273,368−372]. The non-trivial problem of accounting for Jahn-Teller distortions in the molecular mechanics formalism has been tackled with four different strategies:

− The axial ligands (often anions in the solid or solvent molecules in solution) are fixed at a given distance[120,368,369]. This method has been used to model the geometry of the in-plane ligands which depends on nonbonded interactions to the axial donors. Accurate results are only expected if the position of the axial ligands is known, and that is only possible if an experimentally determined structure is available. Consequently, the range of applications of molecular mechanics calculations is restricted and the predictive capability is lost.

− Two different parameter sets for axial and in-plane ligands are used[57,65,165,166,204,273]. While the exact structure of the chromophore does not need to be known for this approach, the direction of the elongation (or compression) has to be specified beforehand. Therefore, this method too is not generally applicable but it covers a wide range of axially distorted copper(II) compounds with structurally predetermined Jahn-Teller influences. For example, copper(II) complexes with four amine donors, bis (amino acid) compounds and complexes with tetraaza, tetrathia and mixed-donor macrocyclic ligands generally have weakly bound axial donors [57,65].

- The CLF/MM method (see Chapter 3 and above) uses a cellular ligand field stabilization term and arrives at relatively accurate predictions of Jahn-Teller distortions with a single set of parameters (see Table 12.4) [77].

- A general approach for predicting Jahn-Teller distortions of copper(II) hexaamines has been published, and it has the potential to be applied to donor atoms other than nitrogen, metal centers other than copper(II), and various types of coordination polyhedra [372]. The method is based on a harmonic first-order model [373] where the Jahn-Teller stabilization energy is the result of the $Q_\theta$ distortion mode (Eqs. 12.3, 12.4, Fig. 12.1).

$$Q_\theta = \left(\frac{1}{12}\right)^{\frac{1}{2}} \cdot (\delta x_1 + \delta x_2 + \delta y_3 + \delta y_4 - 2\,\delta z_5 - 2\,\delta z_6) \tag{12.3}$$

$$Q_\varepsilon = \frac{1}{2} \cdot (\delta x_1 + \delta x_2 - \delta y_3 - \delta y_4) \tag{12.4}$$

This distortion leads to a gain of electronic energy, i.e., the Jahn-Teller stabilization energy $E_{JT}$ as defined by Eqs. 12.5, 12.6,

$$E_{JT}^{xy} = \left\{(-0.01198)(\delta x)(\Delta)\right\}/r_0 \quad [\text{kJ mol}^{-1}] \tag{12.5}$$

$$E_{JT}^{z} = \left\{(-0.02396)(\delta z)(\Delta)\right\}/r_0 \quad [\text{kJ mol}^{-1}] \tag{12.6}$$

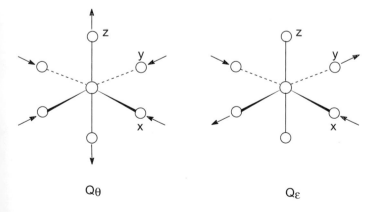

$$Q_\theta \qquad\qquad\qquad Q_\varepsilon$$

**Figure 12.1:** The two compounds of the $\varepsilon_g$ Jahn–Teller active mode.

**Table 12.4:** Optimized structures of Jahn–Teller distorted $CuN_6$ chromophores [372].

| Compound [b] | In-plane | | Axial | |
|---|---|---|---|---|
| | X-ray | Calcd. | X-ray | Calcd. |
| $[Cu(tach)_2]^{2+}$ [374] | 2.07 | 2.07 | 2.35 | 2.34 |
| $[Cu(bpy)_3]^{2+}$ [375] | 2.03 | 2.04 | 2.34 | 2.34 |
| $[Cu(phen)_3]^{2+}$ [376] | 2.04 | 2.04 | 2.33 | 2.35 |
| $[Cu(terpy)_2]^{2+}$ [377] | 2.04 | 2.04 | 2.29 | 2.34 |
| $[Cu(en)_3]^{2+}$ [378] | 2.08 | 2.06 | 2.34 | 2.37 |
| $[Cu(9\,ane3)_2]^{2+}$ [379] | 2.06 | 2.06 | 2.32 | 2.35 |
| $[Cu(pzpy)_2]^{2+}$ [380] | 2.01 | 2.03 | 2.39 | 2.36 |
| $[Cu(tptz)(pca)]^{2+}$ [381] | 2.03 | 2.06 | 2.38 | 2.37 |
| $[Cu(im)_6]^{2+}$ [382] | 2.03 | 2.01 | 2.59 | 2.39 |

[a] The direction of the elongation was correctly predicted in each case.
[b] References for the x-ray data; structures:

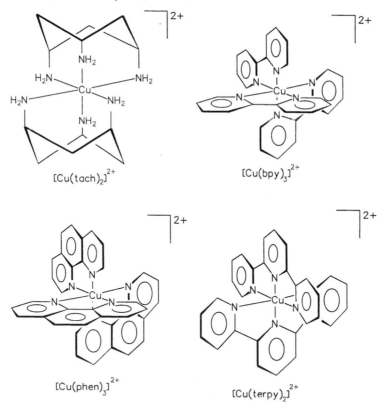

**Table 12.4** (continued)

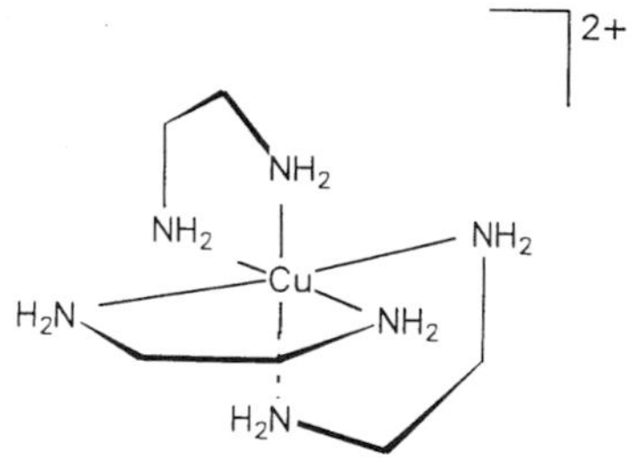

$[Cu(en)_3]^{2+}$

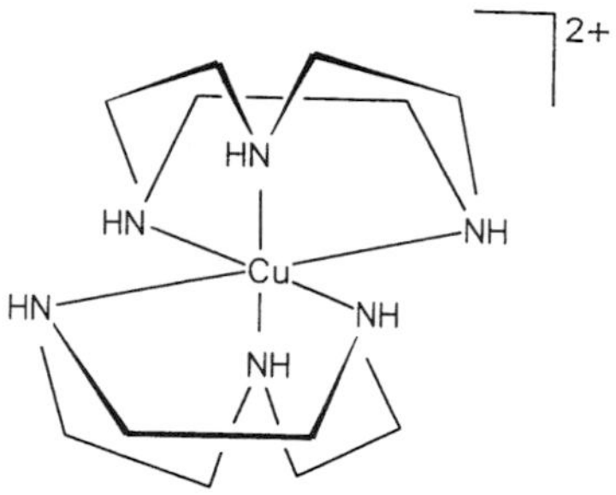

$[Cu(9ane3)_2]^{2+}$

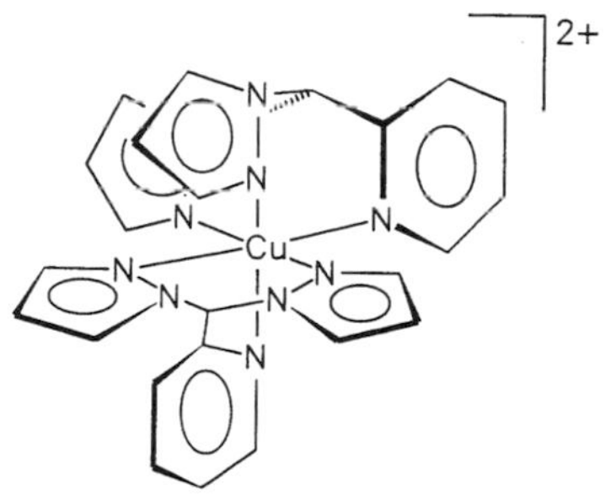

$[Cu(pzpy)_2]^{2+}$

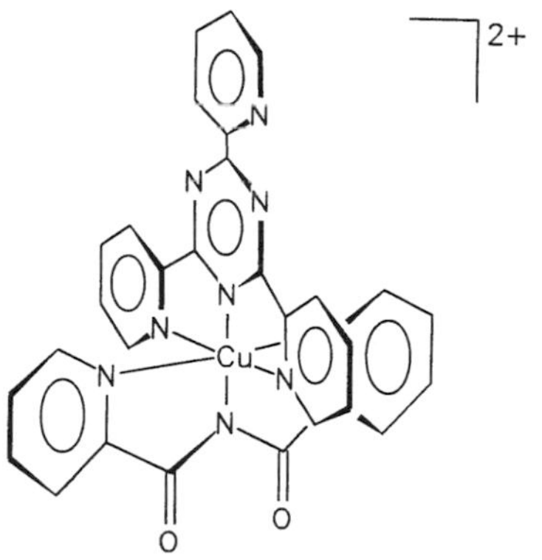

$[Cu(tpz)(pcc)]^{2+}$

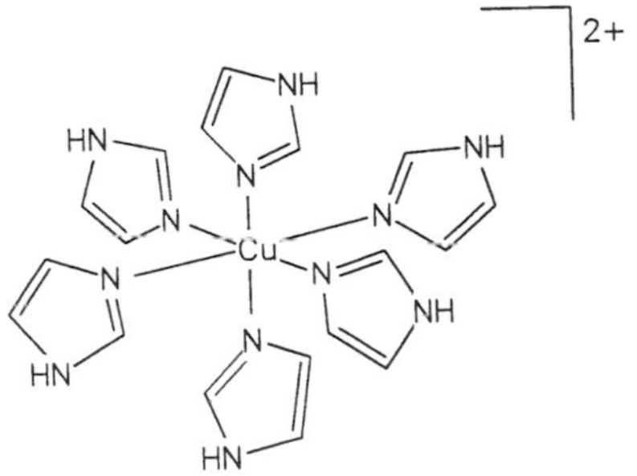

$[Cu(im)_6]^{2+}$

where $\Delta$ is the ligand field stabilization energy. The decrease in energy due to the electronic stabilization (there is a linear dependency on the total nuclear displacement) is counterbalanced by an increase in strain in the molecule, associated with bond elongation and compression. This situation is illustrated in Fig. 12.2, which shows a cross-section through the two potentials (strain and electronic) along one $Q_\theta$ direction. Strain energy minimization with the usual set of potential energy

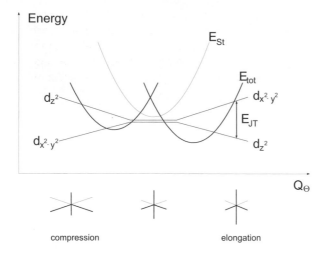

**Figure 12.2:** Total energy as a function of the strain energy $E_{ST}$ and the Jahn–Teller stabilisation energy $E_{JT}$.

functions and including the electronic term ($E_{JT}$, see Eqs. 12.5, 12.6) along the three possible axes for elongation leads to the prediction of the direction of the elongation (relative total energy in the three directions), and to an accurate prediction of the corresponding structures (see Table 12.4; included in the table are data produced with the CLF/MM approach[77]). The only case, with a large discrepancy between the calculated and observed structure is the hexakis(imidazole) complex. One possible reason for the failure of the simple model in this case is that the monodentate ligands are not restricted by intraligand strain from large elongations. Therefore, a harmonic representation of the bonding function might be a poor approximation in this case.

An interesting case of a Jahn–Teller distorted hexacoordinate copper(II) complex is shown in Fig. 12.3. The elongation along the O-Cu-O axis leads to a loss of delocalization within the hfacac ligand skeleton. Clearly, the simple

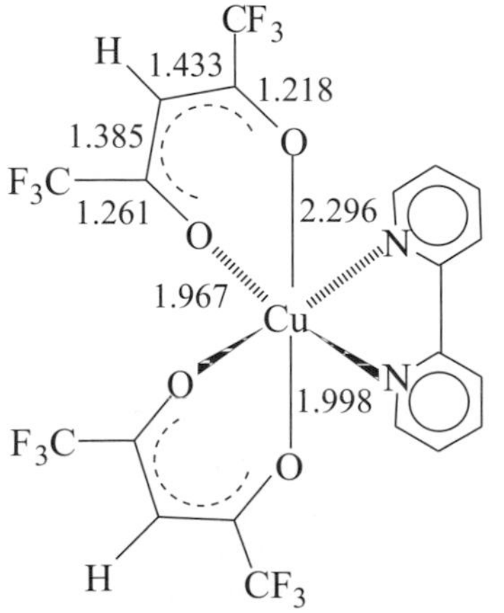

**Figure 12.3:** Structure of $[Cu(hfacac)_2(bpy)]$[383] (bond distances in Å).

"electronically doped" molecular mechanics model will not be able to predict this additional distortion within the organic part of the molecule. The question is whether any method can be found for solving a problem involving the interplay of two electronic factors while at the same time retaining the advantages associated with the simplicity and general applicability of molecular mechanics calculations.

# 13 Bioinorganic Chemistry

During the last decade there has been a rapid growth in the application of molecular mechanics and dynamics to biological macromolecules and to the study of their interactions with small molecule substrates. Structure-based drug design has been a powerful driving force behind much of this expansion and most of the more recent studies have made use of commercially available packages combining software and graphics. That there has not been a comparable growth in the study of bioinorganic systems is probably due in part to the fact that there are relatively few metal based drugs or drugs that act by binding to metal centers. However, a second deterrent has undoubtably been the difficulties associated with modeling the metal-centered sites of interest. Significant advances have been made but none of the commercially available packages is yet able to model routinely and accurately the full range of such sites. The importance of metalloprotein systems will drive further developments in coming years[384].

In this chapter we survey the molecular mechanics and dynamics studies of bioinorganic systems. We also address the problems involved in carrying out such studies and point to possible strategies for dealing with them.

## 13.1 Complexes of Amino Acids and Peptides

The detailed structural study of metalloproteins was preceded by the study of small molecule metal complexes of amino acids and peptides[385]. The development of force fields for modeling metalloproteins might, logically, also begin with molecular mechanics modeling of amino acid and peptide complexes that have metal-ligand interactions of the type seen in the metalloprotein of interest. In this way, force field parameters can be developed that accurately reproduce the details of precisely determined structures. Surprisingly then, there have been very few molecular mechanics studies of metal complexes of amino-acids or small peptides and most of these have been of cobalt(III) systems. For example, the three isomers of $[Co(L\text{-methionine})_2]^{3+}$ (Fig. 13.1) have been modeled[386]. This is a system of no direct biological relevance but is one which was chosen to begin the task of extending, in a step-wise fashion, an existing force field to enable the

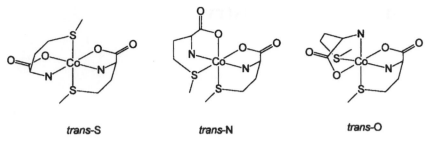

trans-S                    trans-N                    trans-O

**Figure 13.1:** The three isomers of bis(L)-methioninecobalt(III).

study of metal-polypeptide interactions. Some mixed ligand complexes of cobalt (III) with amino acids have also been investigated[249]. There have been a number of molecular mechanics investigations on copper(II) complexes of amino acids[154] and amino acid derivatives[163,371,387]. A force field for metalloporphyrins has also been developed by modeling small-molecule complexes[45].

The paucity of molecular mechanics modeling of small molecule analogues of active sites of metalloproteins is certainly due in some cases to the lack of suitable analogues. Highly constrained and unusual metal environments are common in metalloproteins. Examples are the geometrically unusual type I copper sites in blue copper proteins[388] and the three-coordinate iron sites in nitrogenase[389]. In such cases development of force field parameters can only proceed by logical extension from those developed for well-characterized systems. As yet, there are only few examples of small-molecule analogues that mimic the structures and properties of these and other sites[282].

Cyclic peptides can be viewed as a step on the way from the modeling of unconstrained peptides to folded proteins. The copper(II) complex of a cyclic octapeptide has been investigated by molecular mechanics and EPR spectroscopy and the structure was found to be in accord with those of closely related complexes[205]. Similar combined approaches are also applicable to metalloproteins.

## 13.2   Metalloproteins

One motivation behind the modeling of metalloproteins has been the need to visualize the structures of proteins that cannot be crystallized. In order to do this in a reliable way it is necessary to know the structure of a closely related protein. For example, a model of the type I copper-containing protein stellacyanin was developed using the crystal structure of cucumber basic protein (CBP) as a starting point[390]. The geometry about the copper center in the CBP structure (Fig. 13.2) was assumed to be the ideal geometry, and soft bond length and bond angle force constants were used in order to allow for some variation in the geometry. Improper torsion angles were used to constrain three of the donor atoms and the copper

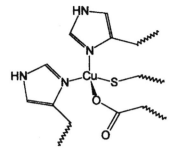

**Figure 13.2:** Proposed geometry of the blue-copper site in stellacyanin.

ions to be close to coplanar since this accorded with the crystal structures of many blue-copper proteins[390]. Such techniques for modeling the metal center have little or no predictive value but in this case the aim of the study was to develop a model of the secondary structure of the protein stellacyanin rather than one of its metal center.

The opposite approach of keeping the bulk of the protein geometry fixed to that observed crystallographically while optimizing the geometry of the active site and its immediate surrounds has also been investigated. In a study of plastocyanin and amicyanin, the geometry about the copper(I) centers was well reproduced[391].

An alternative approach to representing the metal center has been developed for zinc(II) centers and applied to the modeling of the interaction of natural substrates and inhibitors of the enzyme human carbonic anhydrase[105,392,393]. Structurally characterized four-, five- and six-coordinate small-molecule complexes of zinc(II) were analyzed to determine the distribution of bond lengths and angles about the zinc ion. A function was developed that was able to reproduce these structural features and was added to the program YETI[394], developed for modeling small molecule-metalloprotein interactions.

A limitation of this approach is that it is necessary to assign the coordination number, and, when modeling five-coordinate sites, it is necessary to assign the geometry as either square pyramidal or trigonal bipyramidal prior to energy minimization. This allows the relative energies of the two five-coordinate geometries to be investigated but limits the predictive value of the method. Also, it does not readily allow for the many intermediate coordination numbers and geometries to be reproduced. However, the method did yield useful models of the interactions of human carbonic anhydrase with either the natural substrate, bicarbonate, or a series of sulfonamide inhibitors. A revised model which does allow for flexibility in coordination number and geometry has been developed more recently and has been used to study both native and cobalt(II) substituted versions of carbonic anhydrase I[395].

The solution structures of a number of metalloproteins with paramagnetic metal centers were determined with molecular mechanics and dynamics in combination with NMR spectroscopy (see also Chapter 10)[332–335]. Due to the complexity of the molecules, for metalloproteins a crystal structure of the compound or a derivative is often needed for the definition of the starting geometry. Molecular dynamics is then used to find low-energy conformers. The dynamics calculations also allow the visualization of areas of large flexibility, and this can lead to some understanding of the enzyme mechanism.

Metalloenzymes pose a particular problem to both experimentalists and modelers. Crystal structures of metalloenzymes typically reveal only one state of the active site and the state obtained frequently depends on the crystallization conditions. In some cases, states probably not relevant to any aspect of the mechanism have been obtained, and in many cases it may not be possible to obtain states of interest, simply because they are too reactive. This is where molecular modeling can make a unique contribution and a recent study of urease provides a good example of what can be achieved[191]. A molecular mechanics study of urease as crystallized revealed that a water molecule was probably missing from the refined crystal structure. A conformational search of the active site geometry with the natural substrate, urea, bound led to the determination of a consensus binding model[191]. Clearly, the urea complex cannot be crystallized because of the rate at which the urea is broken down to ammonia and, therefore, modeling approaches such as this represent a real contribution to the study of metalloenzymes.

An important precursor of such studies is the investigation of model compounds that are fully characterized. For instance, the end-on mode of dioxygen binding in dicopper complexes has been investigated by molecular mechanics[396]. The successful reproduction of the structures of such model compounds provides a sound basis for the modeling of similar sites in metalloenzymes[397].

## 13.3   Metalloporphyrins

There have been a number of studies of the important class of proteins that have metalloporphyrin active sites. Models for the metalloporphyrin and corphinoid sites have been derived by modeling small molecules and extending the AMBER[45] and MM2 force fields[62,196,398]. A slightly modified version of the AMBER force field has been used in a molecular dynamics simulation of hydrogen peroxide binding to the heme iron in cytochrome *c* peroxidase[399]. The use of modeling in this case enabled the investigation of a putative 'inner-sphere' complex, proposed as a precursor to the activated form of the enzyme. The CHARMM force field was used in a study of the temperature dependence of both the structure and the internal dynamics of (carbonmonoxy)myoglobin[400]. Force constants for the iron-heme and iron-CO interactions were taken from vibrational data of model compounds and from iron carbonyl complexes. The geometry about the iron was assumed to be rigorously octahedral[400]. A detailed analysis was made of the movements of the iron center with respect to the heme group and it was found that the largest-amplitude motions were perpendicular to the heme plane[400].

The geometry of metalloporphyrins and other tetrapyrroles have been studied in detail by molecular mechanics. The effect of (i) the size of the metal ion, (ii) axial ligation by planar ligands, such as imidazoles, (iii) the phenyl group orientation in tetraphenyl porphinato complexes, and (iv) the flexibility of the porphyrin macrocycle, were analyzed using a modified version of MM2 [401]. It was found that as one goes from small ions to large ions the ideal porphyrin conformations go from ruffled or saddle-shaped conformers to planar forms and to domed structures for very large ions such as lead(II) (Fig. 13.3). The role of steric hindrance in discriminating between carbon monoxide and dioxygen in some haem model compounds has been studied with AMBER [45] and MM2 [398], and the analysis of structure-sensitive Raman lines in nickel and copper porphyrins have been complemented by modified DREIDING force field calculations [402–404].

The last step of methanogenesis by archaebacteria is catalyzed by coenzyme F430, which contains nickel coordinated to the most reduced tetrapyrrolic macrocycle found in nature to date. An empirical force field analysis has shown that the macrocycle is so flexible that it can coordinate coenzyme F430 in both a planar or a trigonal bipyramidal mode [196]. This study involved a modified MM2 force field in combination with an extensive conformational search using the random kick method, molecular dynamics and a Monte Carlo dihedral search. The same force field and search methods were used in a study following the crystal structure determination of a coenzyme F430 derivative, 12,13-diepi-F430M. The force field reproduced the new structure well and led to further important insights into the conformational changes accompanying the epimerizations of coenzyme F430 that occur upon its isolation [405]. The molecular mechanics analysis of the consequences of porphinoid reduction on the core size and flexibility of the macrocycle

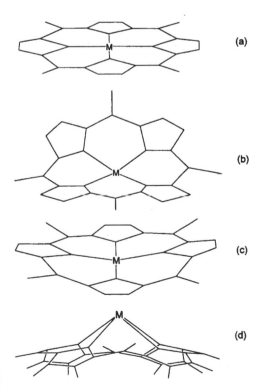

(a)

(b)

(c)

(d)

**Figure 13.3:** (a) Planar, (b) ruffled, (c) saddle, and (d) dome conformations of metalloporphyrins.

showed that the core size increases and the flexibility remains constant when the reduction occurs at the β-pyrrole position (chlorin, isobaceriochlorin, pyrrocorphin). In contrast, the core size decreases and the flexibility increases when reduction occurs at the methine positions[406].

Metal complexes of corrin ligands have also attracted interest. The MM2 force field has been extended to allow modeling of corrin systems including adenosylcobalamin (coenzyme $B_{12}$)[407]. The conformational flexibility of the 5'-deoxyadenosyl ligand was investigated and molecular dynamics calculations with NMR distance derived restraints were used to investigate the motions of the atoms and the corrin fold angles.

## 13.4    Metal-Nucleotide and Metal-DNA Interactions

Molecular mechanics and dynamics studies of metal-nucleotide and metal-DNA interactions to date have been limited almost exclusively to modeling the interactions involving platinum-based anticancer drugs, an area that has been reviewed extensively. As with metal-amino-acid complexes, there have been surprisingly few molecular mechanics studies of simple metal-nucleotide complexes that provide a means of deriving reliable force-field parameters. A study of bis(purine)-diamine-platinum(II) complexes successfully reproduced the structures of such complexes and demonstrated how steric factors influenced the barriers to rotation about the platinum(II)-nitrogen(purine) coordinate bonds associated with interconversion of the head-to-head (HTH) to head-to-tail (HTT) isomers (Fig. 13.4)[135].

In the process, force field parameters for the platinum(II)-nucleotide interactions were developed. In two more recent studies the force fields for purine-platinum(II) complexes have been reassessed and the influence of repulsions involving the metal ion have been investigated[98,99]. The van der Waals radii derived for the platinum(II) ion varied from 1.7 to 2.44 Å. A promising approach involving the use of ab-initio calculations to derive force constants has been applied to the interaction between platinum(II) and adenine[408].

There have been numerous studies of the interaction of the highly effective anticancer drug cisplatin, *cis*-[$PtCl_2(NH_3)_2$], with DNA. The majority of these stu-

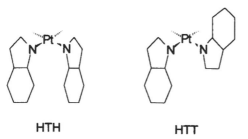

**HTH**                    **HTT**

**Figure 13.4:** The head-to-head and head-to-tail conformers of bis(purine)-platinum(II) units.

dies have concentrated on the adduct formed between the platinum(II) ion and two adjacent guanine bases (G) on one strand of DNA: this is the adduct formed most frequently in the interaction between platinum and DNA. For instance, such adducts in a number of DNA sequences have been modeled with the emphasis being on analyzing the effect of adduct formation on the local and global conformation of the DNA [409-411]. This analysis yielded values for the unwinding and bending which have since been used in the interpretation of experimental determinations of such structural parameters [136]. A number of similar models have been produced in which the details of the hydrogen bonding networks associated with the adduct vary [412,413]. All of these studies have revealed direct or indirect hydrogen bonds between the ammine ligands and the DNA molecule, which accords with the observation that replacement of all amine protons with methyl or methylene groups leads to a loss in anticancer activity.

The observation that cisplatin forms adducts with the GpG and ApG (G = guanine, A = adenine, p = phosphate) sequences of DNA but not with the GpA sequence has also been probed by molecular mechanics [140,414]. In this case it was found that the nature of the interactions of one of the ammine ligands depended on the base on the 3′ side (the second in the sequence). When this base is guanine the interaction is a strong hydrogen bond but when it is adenine the interaction is a repulsive interaction between the same ammine ligand and the exocyclic $-NH_2$ group of the adenine. This is consistent with formation of the adducts with GpG and ApG and the nonformation of the adduct with GpA.

Models of adducts that link one strand of DNA to the other (interstrand) have also been produced [141,415]. These too, reveal hydrogen bonding interactions consistent with established structure-activity relationships. The models have been used to aid in the design of new platinum(II) complexes that should form the interstrand adducts in preference to intrastrand adducts [416].

Stereo- and enantio-selectivity arising from steric interactions between DNA and bulky platinum(II) complexes have been investigated by molecular mechanics. Good correlations between the extent of binding and steric interactions were obtained for the *R* and *S* enantiomers of [PtCl_2(ahaz)] (ahaz = 3-aminohexahydroazepine) [417].

The interaction of the inactive *trans* analogue of cisplatin with DNA has also been modeled using molecular dynamics [418]. In this case the modeling was carried out as part of a broad study of the distortions of the DNA structure caused by *trans*-[PtCl_2(NH_3)_2]. The distortions were measured experimentally and the modeling was used as an adjunct to the interpretation of these experiments, an approach that is being used increasingly in studies of bioinorganic systems. The 1,3-adduct formed by *trans*-[PtCl_2(NH_3)_2] has also been studied using a novel internal coordinate molecular mechanics method [419].

There have been few modeling studies of analogues of cisplatin binding to DNA. The effect of increasing steric bulk on one of the amine groups has been investigated using molecular mechanics [412]. It was concluded that the increasing binding energy that followed from increasing steric bulk, correlates with decreasing anticancer activity. Also, the differences in anticancer activity of the *cis*, *trans*-(*S,S*), and *trans*-(*R,R*) isomers of cyclohexane-1,2-diamine have been investigated[420]. Small differences in the hydrogen bonding interactions referred to above were observed but it could not be determined whether these were sufficient to account for the differences in activity.

The effects of steric bulk of both platinum complexes and the nucleobases have been investigated using a novel "ligand repulsive energy" (LRE) methodology[421–423]. The employed methodolgy differs from other molecular mechanics studies but the analyses allow for the comparison of many ligands and provide an understanding of the role of substituents in slowing binding and of the role of shape features such as the "flatness".

New classes of platinum(II) anticancer drugs such as bisplatinum and *trans* complexes interact with DNA in novel ways and, therefore, it is not surprising that these too have been the subject of molecular mechanics studies. For instance, the kinetics of binding of $\{[trans\text{-}PtCl(NH_3)_2]_2(\mu\text{-}NH_2(CH_2)_4NH_2)\}^{2+}$ (1,1-t,t) to the dinucleotide dGpG and the structure of the product have been investigated using molecular mechanics guided by NMR results[424]. The trinuclear complex, BBR3464, is able to link widely separated bases on DNA, and molecular modeling has been used to determine the relative stabilities of 1,4-interstrand and 1,5-intrastrand adducts[425]. A modification of the AMBER force field has been used to study the binding of *trans* complexes of the type $[PtCl_2(NH_3)L]$ where L = quinoline or $NH_3$, to nucleobases and to DNA. In the models of DNA binding a stacking of the quinoline and the nucleobases was observed[122,426].

The use of NMR to study platinum(II)/oligonucleotide adducts has developed rapidly in recent years. All such studies require molecular mechanics or related modeling methods to supplement the 2D NMR data and the combination has been used very successfully. For example, isomeric forms of the [Pt(GpG)hpip] (hpip = homopiperazine) complex have been identified using NOESY crosspeaks between the hpip and GpG ligands[427]. The structures of oligonucleotides with cisplatin bound intrastrand to GpG and ApG sites have been determined[428–431] as have structures with interstrand and 1,3-GpTpG intrastrand complexes[432,433].

Molecular modeling of the noncovalent interaction between trisphenanthroline metal complexes and DNA has been used to aid in the interpretation of the NMR spectra of these systems[434]. The AMBER force field was used to model the DNA, and the metal complex/DNA systems were energy-minimized. Distances taken from the energy minimized models were used to calculate relaxation rates and compared with the experimentally determined data[434].

## 13.5   Other Systems

The naturally occurring ligand, enterobactin (Fig. 13.5) is one of the most efficient binders of iron(III). It forms complexes which have strictly right-handed chirality and the reasons for this have been analyzed using molecular mechanics methods, as have the reasons for the superiority of enterobactin as an iron-binding ligand[435].

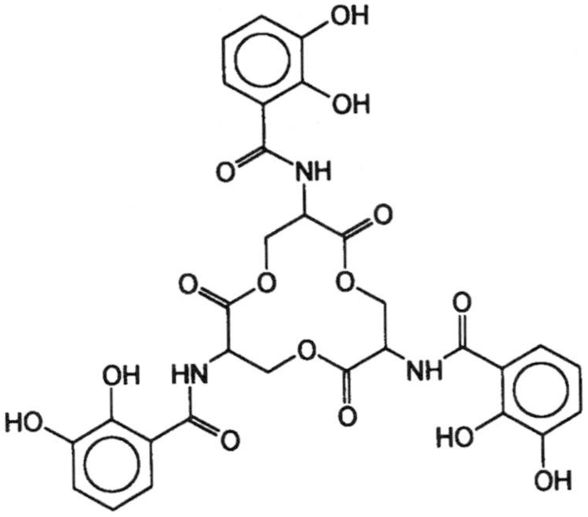

**Figure 13.5:** Enterobactin.

The bleomycins (Fig. 13.6) are a family of glycopeptide-derived antibiotics which are used in the treatment of various tumors. They bind iron in the blood and form an "active" hypervalent oxo-iron species. The two-dimensional structure is well known but no crystal structures of bleomycin or its metal complexes have been reported. The MM2 force field was modified and extended by modeling of the crystal structures of the cobalt complexes of two bleomycin analogues in order to develop a force field for metal-bleomycin complexes [62]. The only way in which the global minimum structure could be obtained was to conduct an exten-sive conformational search using molecular dynamics and Monte Carlo searching. The conformers and isomers of the cobalt complex of bleomycin itself were stu-died and it was concluded that both of the isomers proposed in the literature were sterically feasible [63]. However, for one of the coordination modes no conforma-tion, within 10 kJ mol$^{-1}$ of the lowest-energy conformer, was found that had the

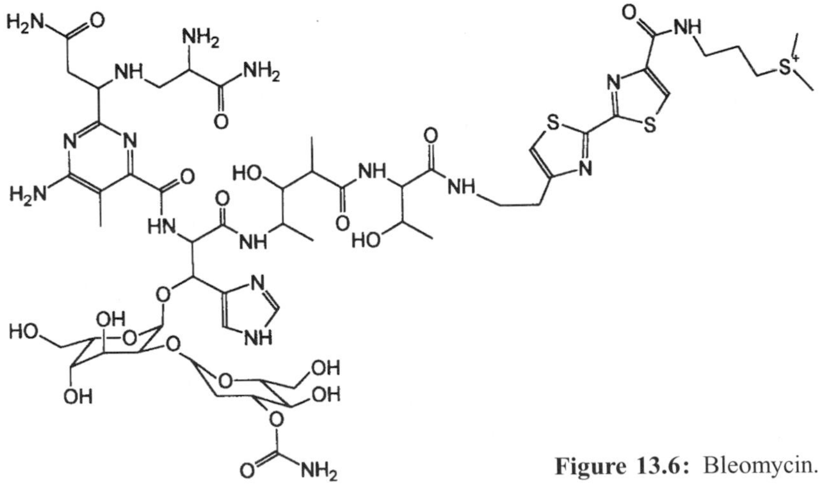

**Figure 13.6:** Bleomycin.

correct dimensions to fit into the minor groove of DNA. The low-energy conformers of the other coordination mode were able to bind to the minor groove of DNA, as predicted by experiment[63]. Molecular mechanics studies of bleomycin model compounds have shown that hydrogen bonds between bleomycin and $O_2$ or $H_2O_2$ may be of importance in the active form of bleomycin[436].

Transferrin model compounds and 44 related iron(III) crystal structures were used to modify the AMBER force field for subsequent studies of ferric transferrin. Energy minimization was conducted both in vacuo and, more interestingly, with the generalized Born/surface area (GB/SA) continuum treatment described in Section 3.7[437].

## 13.6   Conclusions

The modeling of large biomolecules and their interactions with metal ions is fraught with difficulties. A major problem arises from the flexibility of proteins and DNA and the resulting manifold of conformational geometries they can adopt. This leads to the caveat that any model is but one possible representation of the system under study, and any interpretations and predictions based on the model must be made in this light. A second difficulty is that the level of accuracy of molecular mechanics models of biomolecules is still substantially lower than that of small molecules. This is partly because of the inherent difficulty associated with developing models with large numbers of independent parameters when there are only a relatively small number of experimental structures and these are determined at a relatively low level of precision. The number of approximations that need to be made with respect to electrostatic and solvent interactions further limits the accuracy of the models.

Whilst these difficulties do not invalidate application of molecular mechanics methods to such systems, they do mean that the interpretation of the results must be different to what is appropriate for small-molecule systems. For these reasons, the real value of molecular modeling of macromolecule systems emerges when the models are used to make predictions that can be tested experimentally or when the modeling is used as an adjunct to the interpretation of experiments. Alternatively, the relatively crude molecular mechanics models, while not of quantitative value, are an excellent aid to the visualization of problems not readily accessible in any other way. Molecular dynamics is needed, especially for large molecules, to scan the energy surface and find low-energy minima. The combination of computational studies with experimental data can help in the assignment of the structure.

# 14   Organometallics

The last few years have seen an enormous growth in applications of molecular mechanics to organometallic systems. There is no fundamental reason why molecular mechanics should not be applied in this area. However, there are two significant problems:

– defining the atom connectivity, and
– unambiguously defining the type of bonding between the atoms.

Examples are the assignment of the hapticity of π-bonded systems coordinated to metal centers, the recognition of bonds such as metal-metal bonds, and the modeling of π-bonding, *trans*- and other strong electronic effects. These aspects are also of importance, but to a lesser extent, in the modeling of organic and classical coordination compounds. Therefore, it is not surprising that the application of molecular mechanics to the field of organometallic chemistry has developed slowly, and that quantum-mechanical modeling was and, in many cases, still is the method of choice for computing structures and properties of this class of compounds[29,438].

In order to model organometallic systems it is necessary to be able to define the type of bonding, knowing only the metal center and the type of ligands. It is difficult to imagine how a method that is based exclusively on empirical force field calculations can be used to predict the hapticity of π-ligands such as cyclopentadienyl or allyl systems and the coordination mode of carbonyl groups (Fig. 14.1). The additional information needed may be available through electron counting, and therefore be relatively easy to implement in modeling systems, but a more thorough understanding of the electronic properties may also be required. The fact that many systems are fluctional indicates that the equilibrium between various bonding modes is often a subtle balance between electronic and steric factors.

Thus, it is not surprising that, with few exceptions, force field calculations of organometallic systems start with a predefined bonding scheme. This is not unreasonable since the type of bonding can usually be determined from spectroscopic results, and it is often more or less constant within a class of similar compounds. Force field calculations can then be used to get a more detailed picture of the structural and dynamic properties of a molecule with a given connectivity. In spite of these restrictions to modeling of organometallics, the results obtainable are potentially useful, especially for catalytic reactions (see also Sections 8.2 and 8.4).

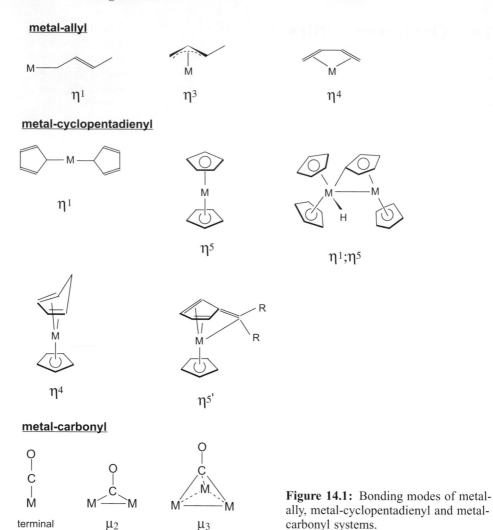

**Figure 14.1:** Bonding modes of metal-ally, metal-cyclopentadienyl and metal-carbonyl systems.

## 14.1   Metallocenes

A number of molecular mechanics studies of metal cyclopentadienyl complexes have been reported recently. The systems studied include linear metallocenes (in particular ferrocene), ferrocene derivatives (such as complexes with substituted cyclopentadienyl ligands, bis(fulvalene)diiron complexes, ferrocenophanes and mixed ligand complexes with carbonyls and phosphines), and non-linear cyclopentadienyl complexes with various metal centers[108,127,155,439–452].

A number of different approaches have been used to model metal-cyclopentadienyl fragments. In some of the reported studies a rigid body approach was used for the ligand[440,443,446,449]. While this might be reasonable for most metal-cyclopentadienyl complexes, small distortions, especially with substituted cyclopentadienyl ligands and bridged systems, might not be modeled accurately. An out-of-plane distortion of cyclopentadienyl substituents will inevitably have some influence on the barrier to rotation of the aromatic ligand. Thus, the rigid body approach might lead to the misinterpretation of dynamic effects. Also, the rigid geometry approach is not applicable to many other organometallic systems involving π-bonded ligands and thus lacks the appeal of generality.

The methods used to describe the interaction between the metal and the cyclopentadienyl ligand include an approach based on electrostatic and van der Waals interactions alone[441,446], bonding to the centroid of the cyclopendadienyl ligand[108,127,155,440,443,446,449,450–452], and bonding to three[442] or all five[447,448] carbon atoms of the cyclopentadienyl ring. The description of the bonding situation based solely on nonbonded interactions (electrostatic and van der Waals terms) is reasonable for alkali, alkaline earth and rare earth complexes, but transition metal metallocenes should probably be modeled with an approach involving some covalent interactions. Based on the premise that any thoroughly parameterized set of functions can be used in an empirical force field calculation (see Sections 3.1 and 3.2), this differentiation may not seem to be necessary. However, difficulties can arise because approaches using a purely nonbonded model or σ-bonds to three out of five carbon atoms are difficult to parameterize, especially if a wide range of metallocenes are to be modeled with a single set of parameters.

Three topological models for covalent bonding in metallocenes have been defined[108] (Fig. 14.2(b,c); also shown in Figure 14.2(a) is the purely electrostatic model):

- The metal is bound to each of the five carbon atoms of the cyclopentadienyl ligand, leading to a coordination number of 10 for a linear metallocene (Fig. 14.2(b)). This model reflects the bonding situation in a metallocene to some extent but, if donor-metal-donor angle functions are involved, the rotation of the cyclopentadienyl ligands around the $C_5$ axis might be difficult to parameterize and the vibrational modes are not easily defined since they involve interdependent bond length and valence angle changes. These problems vanish when donor-metal-donor angle bending is replaced by 1,3-nonbonded interactions (points-on-a-sphere model, see Section 3.2.2)[155]. Note, that there are also 1,3-nonbonded interactions involving the metal center and two carbon atoms of one ring, and these have to be excluded[155]. Computed structural, thermodynamic and spectroscopic data are, as expected, of good quality and this approach is easy to implement in molecular mechanics programs and easy to parameterize.

- The metal is bound to a dummy atom at the centroid of the five-membered rings (Fig. 14.2(c)). The rotation of the ligands around the ligand-metal-ligand axis is easily achieved, and this topology is a reasonable model for the bond-

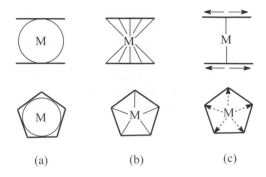

**Figure 14.2:** Topologies for linear metallocenes.

(a)                    (b)                    (c)

ing situation. The problem with this approach is not so much the bonding between the dummy atom and the metal but the artificial bonds that are needed
to bind the dummy atom to the carbon atoms of the cyclopentadienyl ligand.
These need to be strong and, therefore, might lead to some distortion of the ligand. Thus, the parameters describing the bonding within the ligand skeleton
are artificially high. Also, nonexistent vibrational modes are introduced, and
this might be an unwanted artefact if force field calculations are used to calculate vibrational spectra.

– The metal is bound to an interactionless dummy atom at the centroid of the
  five-membered rings (Fig. 14.2(c)). This approach is similar to the one above,
  viz., parameters for centroid-metal bonding have to be defined, but the forces
  on the dummy atoms are distributed among the cyclopentadienyl carbon
  atoms. Thus, the result in terms of the bonding situation is similar to the first
  model and the artefacts of the second diminish, while it still is comparatively
  easy to control ring rotation. The procedure for defining the centroid and re-
  distributing the forces might be as time consuming as the evaluation of the
  ring rotation in the $\eta^5$-model but the parameterization is more straightfor-
  ward.

The principle aim of the reported studies was to model structures, conformational equilibria and fluxionality. Parameters for the model involving interactionless dummy atoms were fitted to infrared spectra and allowed for the structures of
metallocenes (M = V(II), Cr(II), Fe(II), Fe(III), Ru(II), Os(II), Co(II), Co(III),
Ni(II)) and analogues with substituted cyclopentadienyl rings (Fig. 14.3) to be accurately reproduced[108]. The preferred conformation and the calculated barrier
for cyclopentadienyl ring rotation in ferrocene were also found to agree well with
the experimentally determined data (Table 14.1). This is not surprising since the
relevant experimental data were used in the parameterization procedure. However,
the parameters were shown to be self-consistent and transferable (except for the
torsional parameters which are dependent on the metal center). An important conclusion was that the preference for an eclipsed conformation of metallocenes is
the result of electronic effects. van der Waals and electrostatic terms were similar
for the eclipsed and staggered conformation, and the van der Waals interactions

**Table 14.1:** Experimentally determined and calculated structural and dynamic features of metallocenes [a]

| Metallocene [b] | $d$ [Å] | | $\alpha$ [°] | | $\beta$ [°] | | $\gamma$ [°] | | $\delta$ [°] | | $\Delta E$ [kJ mol$^{-1}$] [c] | |
|---|---|---|---|---|---|---|---|---|---|---|---|---|
| | calc. | obs. | calc. | obs. | calc. | obs. | calc. | obs. | calc. | obs. | calc. | obs. |
| FeCp$_2$ | | | 0.0 | 0.0 | | | | | +1.6 | −0.24 | 3.02 | 3.8 |
| Fe((CH$_3$)$_5$Cp)$_2$ | | | | | | | | | −0.04 | −0.06 | 18 | 36 |
| RuCp$_2$ | | | | | | | | | −0.14 | | 28.5 | 33.9 |
| OsCp$_2$ | | | | | | | | | +0.29 | | | |
| (FeCp)$_2$ (1) | 1.628 | 1.635 | 10.6 | 9 | 167.5 | 163.4 | 19.7 | 23.0 | | | | |
| (FeCp)$_2$ (2) | 1.631 | 1.630 | 0.02 | 1.1 | 174.3 | 170.0 | 10.9 | 13.1 | | | | |
| (FeCp)$_2$ (3) | 1.659 | 1.650 | | | 165.9 | 167.3 | 20.2 | 19.2 | | | | |

a) data from [108] and references therein.
b) for structures see Figure 14.3.
c) $\Delta E$: barrier to rotation

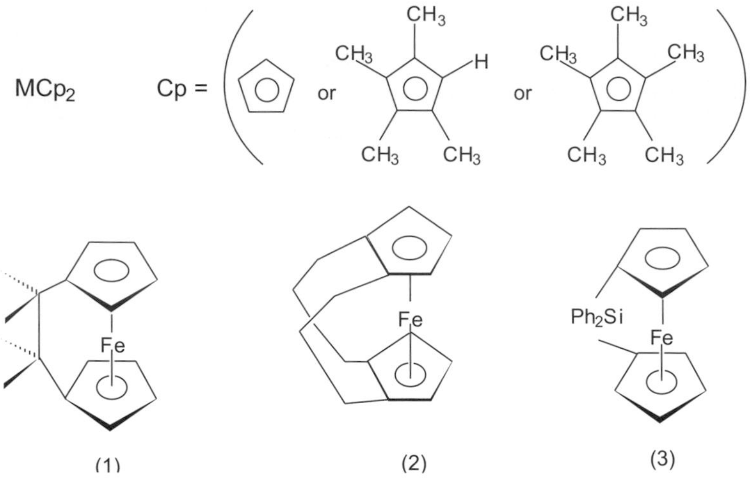

**Figure 14.3:** Metallocenes investigated by force field calculations (see Table 14.1).

were attractive[108]. It is important to note, however, that these conclusions are to some extent dependent on the parameterization scheme and particularly on the parameters used for the nonbonded interactions.

## 14.2   Transition Metal-Allyl Systems

($\eta^3$-allyl)metal systems are important intermediates in a number of regio- and en-antio-selective synthetic reactions. Two earlier reports describe molecular me-chanics calculations of ($\eta^3$-allyl)palladium(0) and ($\eta^3$-allyl)nickel(0) systems [244,245]. In both studies a version of the MM2 program[50] which could not deal with coordination numbers greater than four was used, and dummy atom tech-niques were required to describe the coordination sphere. From the possible topolo-gical models, shown in Fig. 14.4, the dummy[245] and the $\eta^{2\,[244]}$ model were em-ployed. Although these approaches should be improved in terms of general applic-ability (e.g., fluxional systems, application to other metal-$\pi$-systems) and in terms of vibrational analyses (see metallocenes, above), the results obtained are impress-ive. The calculated structures agree reasonably well with experimental data and the predicted stereoselectivities reproduce the experimentally observed trends (see also Section 8.2.2).

A combination of molecular mechanics, based on the dummy atom technique described above, with a QSAR-type treatment and a linear free energy relationship (LFER)-based model was used successfully to analyze the relative importance of steric influences on regioselectivity and stereoselectivity in the palladium-cata-lyzed allylation reaction[453].

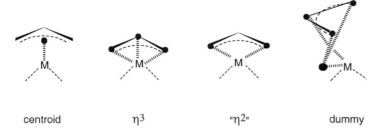

| centroid | $\eta^3$ | "$\eta^2$" | dummy |

**Figure 14.4:** Topological models for ($\eta^3$-allyl) metal systems.

New MM3+-based force fields for ($\mu^3$-allyl)palladium and palladium olefin complexes with various co-ligands, which are based on the points-on-a-sphere approach and do not require dummy atoms to define the connectivity, have been developed and validated with experimentally observed and quantum-mechanically computed data[454].

## 14.3   Transition Metal Phosphine Compounds

Metal-phosphine bonds can generally be modeled in much the same way as any other metal-heteroatom bond. The fact that phosphines participate in $\pi$-backbonding (filled $d_\pi$ (metal) $\rightarrow$ empty $d_\pi$ or $\sigma^*$ (phosphorous) interaction) is only of importance for generic force field parameterization schemes, and half-integer bond orders have been used to describe the effect of $\pi$-backdonation[127]. In the usually adopted empirical force field formalism, $\pi$-bonding effects, like most of the other structural/electronic effects, are accommodated by the general parameter-fitting procedure. A possible complication due to the $\pi$-bonding between the metal center and the ligands is the hindered rotation around these bonds because this is often neglected in molecular mechanics studies of metal complexes.

More important is the question of how the *trans* influence can be included in a general parameterization scheme. This is not a simple problem to solve, and at present, in the few examples reported, the *trans* influence has not been parameterized in a general way. That is, different parameter sets have been used for ligands *cis* and *trans* to the group causing the *trans* influence (see Section 12.2)[127,198]. Recent developments with electronically doped force fields (see also Section 12.2) suggest that a computation with a single set of parameters can lead to very accurate results[77].

On this restricted basis, transition metal phosphines have been modeled with accuracies similar to those of other metal-heteroatom systems[127]. A number of phosphine complexes are of interest in the area of asymmetric synthesis. Due to the requirement that the geometry and the force field of the relevant intermediates have to be known for a thorough study (see Section 8.4) most of the reports in this area are only of qualitative nature. Some of the investigations are based on molecular graphics analyses, and the evaluation of the predicted selectivities have been based on van der Waals terms alone [241,455,456].

There are detailed experimental studies on the conformational properties and dynamics of [($\eta^5$-(C$_5$H$_4$R)Fe(CO)(PPh$_3$)(L)] (R = H, Bu$^t$; L = COMe, I) and these have been compared with molecular mechanics calculations (Fig. 14.5) [443,445,446]. The energy profile associated with rotation along the substituted cyclopendadie-nyl-Fe axis of [($\eta^5$-C$_5$H$_4$Bu$^t$)Fe(CO)(P(OMe)$_3$)(I)] has been calculated with a sim-plified parameterization scheme involving a rigid cyclopentadienyl ligand. It leads to the minimum energy conformer with an angle $\alpha$(Bu$^t$-Fe-CO, projection in the cyclopentadienyl plane; see rotation (a) in Fig. 14.5) of ca. 25°, which is in good agreement with the experimental ($^1$H- and $^{13}$C-NMR) data [443]. The rotation around the COMe-Fe axis was investigated using infrared spectroscopy and, here too, the molecular mechanics calculations are in good agreement with the experi-mental data. That is, the two possible rotamers have an energy difference of 3.8 (calculated) or 5.5 kJ mol$^{-1}$ (experimental) with a calculated barrier of 31.8 kJ mol$^{-1}$ [445]. Rotations around the phosphine-Fe and the phosphorous-phenyl axes of three derivatives were studied with variable temperature $^{13}$C{$^1$H}-NMR spec-troscopy, and the line shape analysis led to activation energies of the rotations (ca. 50 and 20 kJ mol$^{-1}$ for the two rotations), in good agreement with the predic-tions based on molecular mechanics calculations [446].

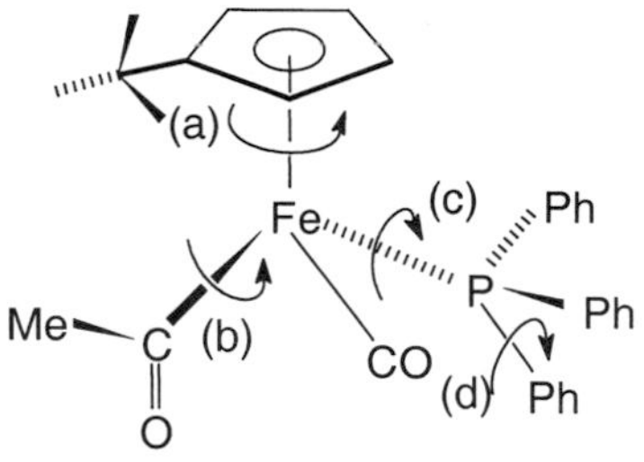

**Figure 14.5:** Dynamics of [$\eta^5$-(C$_5$H$_4$R)Fe(CO)(PPh$_3$)(L)] (R = H, Bu$^t$; L = OMe, I).

The conformational space of tripodal phosphine ligands coordinated to transi-tion metal ions has been analyzed extensively, and fascinating new techniques, in-volving neural networks and genetic algorithms have been used to optimize the force fields and analyze the data [203,457−461].

Phosphine complexes have been central to the development of many aspects of organometallic chemistry. The steric requirements of the phosphine ligands have often been varied to influence the chemistry that takes place at the metal center. In order to quantify the steric demands of different phosphine ligands, the concept of "cone angle" has been developed, and for such a simple concept has proven to be remarkably successful [462−464]. This concept has been expanded by a thorough mathematical treatment, resulting in the seat-ligand-fitting model [465−467]. The concept of the cone angle measure does have limitations and these become parti-cularly significant when the three substituents on the phosphorus atom are not all the same. A molecular mechanics method has been developed for deriving a more quantitative estimate of steric size[142,468−470]. This method involves the calculation of a term referred to as the "ligand repulsion energy", which is the energy arising when the phosphine under study binds to a "standard" [Cr(CO)$_5$] core.

## 14.4   Metal–Metal Bonding

Molecular mechanics calculations have been used to describe low-[471−473] and high-order metal-metal bonds[473,474]. These will be discussed here although many relevant examples are classical rather than organometallic coordination compounds. The force constant for Rh-Rh single bonds has been determined from the vibrational frequency of the metal-metal bond and the strain-free value was fitted using a conventional force field for the ligand systems and a series of experimentally determined structures (Fig. 14.6). The Rh-Rh bond distances cover a range of ca. 0.25 Å, and the experimentally observed trends are reasonably well reproduced by the calculations (Table 14.2).

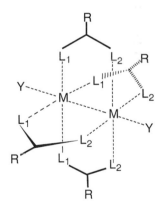

**Figure 14.6:** Metal–metal bonded dinuclear complex.

The experimental data and calculations indicate that the metal-metal distances are the result of the steric demands of the bridging ligands rather than the electronic demands of the metal centers. This is supported by the fact that the analogous Cu-Cu, Pd-Pd and Cd-Cd systems, where no metal-metal bonding exists, lead to a similar agreement between calculated and experimental data, consistent with the geometry being the result of the ligand bridges rather than the Rh-Rh bond. Interestingly, a reassignment of the Rh-Rh vibration (289 cm$^{-1}$ [475,476] vs. 170 cm$^{-1}$ [477]), with a revised set of force field parameters for the Rh-Rh bond ($k$, $r_0$; 2.73 mdyn Å$^{-1}$, 2.43 Å vs. 0.88 mdyn Å$^{-1}$, 2.53 Å) and accompanying changes of

**Table 14.2:** Experimentally determined and calculated metal–metal distances for dinuclear metal complexes (see Figure 14.6) [471,472].

| Metal | Bond order | $L_{1,2}$ | Y | R | M-M$^{obs.}$ [Å] | M-M$^{calc.}$ [Å] [471] | M-M$^{calc.}$ [Å] [472] |
|---|---|---|---|---|---|---|---|
| Rh | 1 | O,O | O | $CH_3$ | 2.386 | 2.39 | 2.39 |
| | | O,O | O | $C(CH_3)_3$ | 2.371 | | 2.37 |
| | | O,O | S | $CH_3$ | 2.406 | 2.38 | 2.40 |
| | | S,O | O | $CH_3$ | | 2.54 | |
| | | S,O | S | $CH_3$ | 2.550 | 2.53 | |
| | | S,S | O | $CH_3$ | | 2.65 | |
| | | S,S | S | $CH_3$ | | 2.64 | |
| Pd | 0 | O,O | | $CH_3$ | 2.546 | 2.50 | |
| | | S,S | | $CH_3$ | 2.754 | 2.85 | |
| Cu | 0 | O,O | O | $CH_3$ | 2.614 | 2.60 | |
| Cd | 0 | O,O | $PPh_3$ | $CH_3$ | 3.452 | 3.39 | |

the parameters for the bridging ligands [472] led to no dramatic changes in the quality of the predictions. This is another example showing that the quality of force field parameters depends on good fitting of carefully chosen experimental data, and that they are not necessarily related to physically meaningful parameters (see Sections 3.3 and 3.5, and Chapter 16). In this case the influence of the ligand geometry is probably more important than the metal preference. This complicates the determination of a unique set of force field parameters for metal-metal bonded molecules.

Unique sets of force field parameters for M-M multiple bonds (homonuclear pairs of Cr, Mo, W and Re; quadruple and triple bonds) were derived from the intersection of curves of $r_0$ as a function of $k$ for bridged and unbridged dimetallic compounds [474]. This technique for obtaining a set of force field parameters, when all other parameters are well defined, relies on the fact that all discrete sets of $k$ and $r_0$ describing a given type of compound lie on a smooth line. Lines defined by similar compounds with different ligand systems (variation of the ligand strain) cross at the point defining the unique set of $k$, $r_0$ for the metal-metal bond in question (Fig. 14.7). The parameters deduced for the metal-metal multiple bonds are given in Table 14.3. The parameters derived for Cr, Mo, W and Re triple and quadruple bonds are consistent with observed M-M stretching frequencies. The force constants were used to determine the contribution of δ-bonds and to determine the barrier to rotation about the quadruple bond [474]. This is an interesting concept with possible applications in other areas of force field calculations.

If the type of bonding and the electronic structure of metal-metal-bonded systems rather than exclusively structural aspects are the aim of a study, then molecular mechanics alone is not appropriate. Recent studies with broken-symmetry DFT approaches have yielded structural results that are in excellent agreement

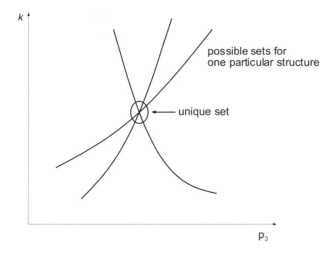

**Figure 14.7:** Fitting of unique sets of force field parameters $k, p_0$ ($p$ = distance, angle etc.).

**Table 14.3:** Force constants for metal–metal multiple bonds [472].

| Metal–metal bond | Bond order | $k$ [mdyn Å$^{-1}$] | $r_0$ [Å] |
|---|---|---|---|
| Cr–Cr | 4 | 1.87 | 1.79 |
| Mo–Mo | 4 | 2.75 | 2.04 |
| W–W | 4 | 3.65 | 2.20 |
| W–W | 3 | 1.00 | 2.23 |
| Re–Re | 4 | 4.15 | 2.11 |
| Re–Re | 3 | 3.50 | 2.15 |

with experimental data, where these have been available, and provided an insight into the bonding and electronic structure of metal-metal-bonded dinuclear systems [478–482].

## 14.5   Carbonyl Cluster Compounds

A number of examples, where two or more isomers of metal carbonyl clusters with identical metal core geometries and different ligand dispositions have been structurally characterized, reveal that ligand arrangements in carbonyl clusters can be variable. A smooth continuum encompassing terminal, bent semibridging, and symmetrically bridging carbonyl groups has been proposed [483], and a reaction trajectory for terminal – $\mu_2$ – terminal exchange has been described [484,485]. A molecular mechanics model was developed where CO ligands are allowed to float on a equipotential surface, maintaining the C-O vector perpendicular to the surface [67,68]. In this approach, the CO ligands are assumed to be bonded to the metal cluster as a whole and therefore are smoothly transformed from terminal to $\mu_2$ and $\mu_3$ bridging positions. The dominant, and therefore structure determining, contribution to the strain energy described in this model arises from non-bonded interactions. Specific electronic effects were not included. Therefore, the model is only appropriate for small distortions from a given geometry or in the absence of excessively crowded geometries [68]. Not least, the model has revealed that steric interactions are only rarely structure-determining in carbonyl cluster compounds.

# 15 Compounds with s-, p- and f-Block Elements

## 15.1 Alkali and Alkaline Earth Metals

The s- and f-block elements present an unusual challenge in the molecular mechanics field because the metal-ligand interactions in both cases are principally electrostatic. Thus, the most appropriate way to model the M-L bonds is with a combination of electrostatic and van der Waals nonbonded interactions. Indeed, most reported studies of modeling of alkali metal, alkaline earth metal and lanthanoid complexes have used such an approach.

The types of alkali and alkaline earth metal complexes subjected to molecular mechanics modeling fall into four categories; crown ethers [119,486–491], cryptands [492,493], spherands [494,495], and other biologically important ligands, such as ionophores and cyclic antibiotics [496–499].

The aims of studies on the first three groups have been broadly similar – principally elucidation of the origins of metal ion selectivity and investigation of the conformational variability. All of the studies were reasonably successful in reproducing observed structures.

### 15.1.1 Crown Ethers

The first molecular mechanics studies of alkali metals were of the $Na^+$, $K^+$, $Rb^+$ and $Cs^+$ complexes of 18-crown-6 (Fig. 15.1) [486]. The AMBER software system was used and the $M^+$...O interactions were modeled using a combination of van der Waals and electrostatic terms. The solvation energies were found to make a larger contribution than did the intrinsic, molecular-mechanics-calculated, com-

**Figure 15.1:** Metal complex of 18–crown–6.

plexation energies. Consequently, in a later study, molecular mechanics and dynamics studies were carried out with the complexes and cations surrounded by clusters of solvent molecules [487].

Ionophores such as crown ethers are of interest as membrane transport agents for use in electrochemical sensors such as ion-selective electrodes. In order to be useful in this role the ionophore must transport one cation and not others and, thus, the issue is one of metal ion selectivity (see also Chapter 9). Molecular mechanics has been used to investigate the selectivity of 18-crown-6 [488–490] and its potential as a membrane transport agent. Again, a combination of van der Waals and electrostatic terms were used to model the interaction between the metal ion and the ligand. Ab-initio methods were used to establish the energetics of the metal-ligand interactions and to parameterize the pairwise molecular mechanics terms [489].

There has also been a study of alkaline earth metal complexes of 18-crown-6 [491]. In agreement with experiment, it was found that the selectivity in the presence of phosphate and water followed the order $Ba^{2+} > Sr^{2+} > Ca^{2+} > Ra^{2+} \gg Mg^{2+}$ [119].

Ligand reorganization necessary for coordination represents an energy cost. This cost depends on the degree and type of ligand substitution, and it was shown that the cost can be predicted by molecular mechanics calculations [286]. The reorganization energy also depends on the size of the metal ion bound and, thus, variation of ligand substitution on crown ethers can be used to tune the selectivity [500].

### 15.1.2   Cryptands

Cryptands such as [222] (Fig. 15.2) have been subjected to molecular mechanics analysis using two different approaches. In the first of these the conformational preferences of [222] alone and in the presence of the alkali metal cations were investigated using the AMBER force field and van der Waals and electrostatic terms to model metal-ligand interactions [492]. In the second study, on the $K^+$ complexes of a lariat ether (an open-chain analogue of the cryptands), metal-ligand interactions were modeled using a more classical Hookes law type function [493]. This necessitates the derivation of ideal M-O and M-N bond lengths and associated force constants. It was found that this method more reliably reproduced the M-O and M-N bond lengths [493] but this conclusion was based on only two structures, and only one was studied by the two methods. The use of a Hookes law

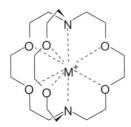

**Figure 15.2:** Metal complex of a cryptand.

function has the disadvantage of being less general; but in order to establish which of the two approaches is more reliable, a wide range of complexes would need to be studied using each of them.

### 15.1.3 Spherands

The AMBER-based approach used to model cyclic polyethers and cryptands has also been applied to the study of the $Li^+$, $Na^+$ and $K^+$ complexes of three spherands (Fig. 15.3)[494]. Experimentally determined metal ion selectivities were successfully reproduced. A similar AMBER-based model, used for molecular mechanics and dynamics of a cyclic urea-based spherand was also successful in reproducing its metal ion selectivity[495]. A number of new conformations of the spherand, including the global energy minimum, were located using molecular dynamics[495].

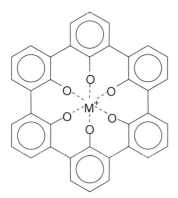

**Figure 15.3:** Metal complex of a spherand.

### 15.1.4 Biologically Relevant Ligands

Polylactones (for an example, see Fig. 15.4) are synthetic analogues of naturally occurring ionophores such as enniatin (species that transport ions across biological membranes). Molecular mechanics calculations have been used to predict the stability and selectivity with respect to $Li^+$, $Na^+$ and $K^+$, of a series of new polylactones[496]. Metal-ligand interactions were again modeled using a combination of van der Waals and electrostatic terms.

Similar studies were carried out on other enniatin analogues[497], enniatin itself[498] and valinomycin[499]. The $Ca^{2+}$ complexes of a series of $C_3$ symmetric trispeptides were also modeled, and it was shown that some of these have the ability to form chiral $Ca^{2+}$ complexes[501].

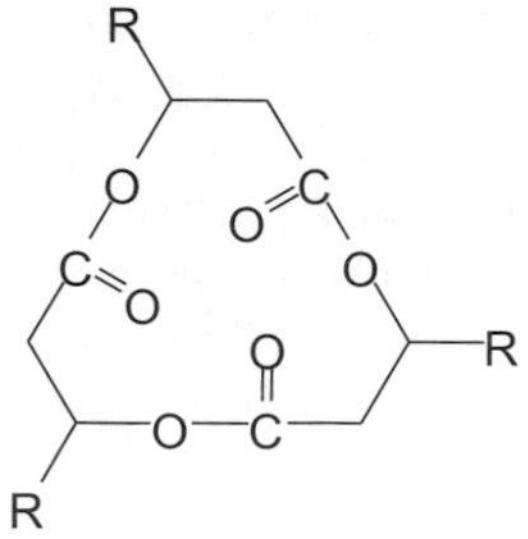

**Figure 15.4:** A polylactone ligand.

## 15.2   Main Group Elements

In comparison with organic chemistry, main group and transition metal compounds have similarities because of the large variety of structures, arising from the comparably large number of different elements and oxidation states. This variability leads to difficulties in establishing a general force field and is probably a contributing reason for the relatively small number of published molecular mechanics studies in the area of main group chemistry. Although the principal problems arising in the area of transition metal compounds – stereoelectronic effects due to partly filled d-orbitals – are absent, sterically active lone pairs often cause complications that are not easily accounted for in a molecular mechanics model. The VALBOND approach, based on valence bond theory has proven particularly successful at tackling both normal valent and hypervalent p-block compounds[78,81]. The published results indicate that the accuracy in terms of structural data is generally equivalent or superior to that of MM2 or MM3 studies, and the VALBOND approach, in contrast to conventional molecular mechanics, does not require extensive parameterization for each new class of compounds.

The classical molecular mechanics studies involving main group elements have so far concentrated mainly on establishing force fields and reproducing structures and, to a lesser extent, conformational equilibria, infrared frequencies, and heats of formation. In this respect the results obtained thus far have (not unexpectedly) been satisfactory. The outcome is that, with specialized force fields, small main group molecules can be modeled with a similarly high accuracy to that obtainable for small organic compounds. The systems treated include organo-silicon[502,503,504] and organo-phosphorous compounds[505] (see also Section 14.3), as well as organo-selenium[506], -tellurium[506], -germanium[507], -tin[507] and -lead[507] compounds. The force fields have usually been derivatives of MM2 or MM3[49,50,446], and the additional parameters have been fitted to structural, spectroscopic, and thermodynamic data. The reported structural parameters[49,50,502,503,505–507], heats of formation[502,503], moments of inertia and infrared frequencies[506], and dipole moments[507] are in good agreement with experimental data (average deviation in bond distance of ca. 0.01 Å, in valence angle of ca. 2° and in infrared frequency of ca. 10 cm$^{-1}$).

A combination of molecular mechanics calculations and electric birefringence (electrooptical Kerr effect) measurements of group IVB aryl compounds has been used to study conformational effects in these molecules [504].

Structure optimization of main group molecules with generic force fields has the same advantages and problems as similar calculations of organic and organometallic compounds: on one hand there is no need to fit a specialist force field, on the other is the expectation of lower accuracy [89,126]. The structural results are especially poor for molecules where electronic effects are important, e.g., those with hypervalent or dative bonds. An exception to this is the VALBOND force field which generally leads to excellent results (see above) [78,81].

An area of increasing interest is the selective complexation of Sn$^{2+}$ and more particularly Pb$^{2+}$ for the treatment of heavy metal poisoning. Molecular mechanics has been extensively applied to the problem of metal ion selectivity (see Chapter 9) but there have been few studies of lead or tin complexes. The fit of Sn$^{2+}$ to 18-crown-6 has been considered [508] as has the size selectivity of tetraazamacrocycles with respect to Pb$^{2+}$ binding [120]. The binding of Pb$^{2+}$ to porphyrin-1 has been modeled, though in this case the point of interest was the deformations caused by the metal cation [398].

There have been a number of molecular mechanics and dynamics studies of silicates and zeolites [509–511]. These present a number of unusual problems arising from the polymeric nature of the materials and the variability of the bond lengths and angles about the oxygen atoms. The early studies avoided these problems by keeping the zeolite lattice invariant and mapping the energy as included molecules were moved to various locations in the cavities [509,510]. An interesting approach to dealing with the variability of the oxygen geometry, based on a shell model, has been described [511]. In this study the lattice was allowed to vary and the fit between calculated and experimental cell constants was used as a measure of the correctness of the model [511].

Gallium(III) and indium(III) radionuclides are used widely for radiodiagnosis. As with all complexes used for such purposes, rapid and stable binding of the toxic radionuclide is essential. Molecular mechanics modeling has been used to predict the stability of gallium(III) and indium(III) complexes of bis(aminoethanethiol) ligands and to attempted a prediction of the coordination numbers of the complexes [512].

## 15.3  Lanthanoids and Actinoids

There are a number of significant motivations driving the development of molecular mechanics models of lanthanoid and actinoid complexes which, being relatively unstable, are notoriously difficult to crystallize and structurally characterize. At the same time there is strong interest in these structures. For example, lanthanoid shift reagents have been used for many years as aids to interpreting NMR spectra. In order to predict reliably the effects of shift reagents it is necessary to be able to predict the distance between the lanthanoid ion and the spectroscopically probed atom. More recently, lanthanoid complexes have been used as contrast agents, aiding in the medical applications of magnetic resonance imaging (MRI). The stabilities of such complexes is critical because, if the lanthanoid is lost from the carrier ligand, it can lead to serious toxic side-effects. Related to this application is the use of ligands to deliver radioactive metal ions to tumors to act as imaging agents or radiotherapeutic agents. Finally, the separation of the actinoids from the lanthanoids is a difficult task but one that is essential for the treatment of nuclear waste. Thus, both the structures and stabilities of lanthanoid and actinoid complexes are of interest and molecular mechanics has the potential to make a substantial contribution in both these areas.

Complexes of the lanthanoid elements are different in a number of ways to those of the transition metals. They have incomplete f subshells but the f orbitals are not significantly involved in the metal-ligand bonding. Instead the metal-ligand bonding is principally electrostatic and, because there is little interaction with the orbitals of the metal, there is little metal-imposed directionality. That is, the geometries of complexes formed by the lanthanoid elements are almost totally determined by the ligand requirements and interactions between the ligands. Thus, the use of L-M-L' angle bending terms is inappropriate in the case of the lanthanoid complexes. Instead, 1,3-nonbonded interactions are the dominant terms and need to be included. Two methods for doing this and modeling lanthanoid complexes have been described in recent years.

1,3-nonbonded interactions[513−515] were included in the same way as has been used firstly for the modeling of cobalt(III) complexes[59] and more recently for a wide range of transition metal compounds[57,58,65]. That is, standard models with M-L harmonic bond stretching terms were used, but the L-M-L' harmonic terms were deleted and L…L' nonbonded interactions were used in their place. In the first of these studies, on lanthanoid shift reagents, two seven-coordinate europium(III) and ytterbium(III) complexes, [Eu(dipivaloylmethanato)₃(quinuclidine)] and [Yb(acetylacetonato)₃(H₂O)] (Fig. 15.5), were modeled[513]. The structures of these complexes were well reproduced. In a subsequent study these complexes

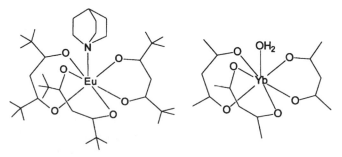

**Figure 15.5:** [Eu(dipivaloylmethanato)₃(quinuclidine)] and [Yb(acetylacetonato)₃(H₂O)].

and [Eu(dipivaloylmethanato)₃(DMSO)] were studied with the same force field, and the random incremental pulse search method was used to locate all potential coordination geometries[514].

Recently, a similar approach has been used to study eight- to twelve-coordinate aqua and nitrato complexes of the complete series of the lanthanoids[515]. Again, good agreement between observed and calculated geometries was obtained. In all three of these studies, modified versions of MM2 were used and electrostatic interactions were not explicitly included.

An alternative approach, based on the assumption that metal-ligand bonds in lanthanoid complexes are purely electrostatic in nature, has been developed and applied to MRI contrast agents[516,517]. The harmonic metal-ligand interaction is replaced by a combination of the electrostatic interaction between the positively charged metal center and the negatively charged donor atom and the van der Waals interaction between these atoms[516]. While this approach might be more "realistic" it relies on accurate values being available for the charges and the van der Waals parameters of the metal. However, the electrostatically based method has been used successfully to study the stability of a series of polyaminepolycarboxylato complexes of gadolinium(III)[516]. Molecular dynamics calculations were also used to study the mechanisms of complexation[516].

The same method was used to study a similar series of polyaminepolycarboxylato complexes of europium(III)[517]. This study included a series of dimeric complexes and the Eu...Eu distances were compared with those determined from luminescence spectra. A number of different conformations and the energy differences between them were also investigated.

Of particular importance in the field of lanthanoid chemistry is the separation of the different lanthanoid(III) cations. Molecular mechanics calculations have been used to rationalize the selectivity observed for the aqua ligand and for a range of organophosphate ligands[124]. The successful rationalization of the experimental observations suggests that it should be possible to design ligands to further improve the separation of the lanthanoid ions.

# 15.4 Conclusions

Exploration of the potential applications of molecular mechanics to compounds of s-, p- and f-block elements is starting to develop rapidly. The difficulties arising from the electrostatic bonding in the s- and f-block elements have been tackled in a number of different ways, in most cases with reasonable success, as has the problem of stereochemically active lone pairs in p-block elements.

# Part III:    Practice of Molecular Mechanics

In this part of the book we give practical advice on how to apply molecular mechanics to problems involving metal complexes. Chapter 16 is a short introduction
that gives advice on which model and force field should be chosen, how to interpret, use and describe the results and how to avoid possible pitfalls. Chapter 17 is
a tutorial. Using the software included in this book and a computer with Microsoft
Windows® the reader can learn most of the methods and techniques presented in
this book.

# 16 The Model, the Rules and the Pitfalls

## 16.1 Introduction

The basis of molecular mechanics is a classical parameterization of non-classical effects that allows the computation of molecular structures. In its original form molecular mechanics is based on harmonic potentials for bond stretching, angle bending and torsional barriers. Secondary effects, such as electrostatic, hydrogen bonding and van der Waals interactions are modeled with the usual potentials, and additional terms are sometimes added for specific electronic effects. The optimum arrangement of two atoms (bond distance), three atoms (valence angle) etc., is a function of the electron density distribution between the atoms. This is quantified in terms of the atom types which are assigned on the basis of the connectivity in a molecule and define the bond order, the hybridization and, to some extent, the charge distribution. That is, a different atom type is assigned to a given element in different environments (e.g., carbon in an alkane, in benzene, in the cyclopenta-dienyl anion coordinated to iron, in a carboxylate etc.), and this defines the electronic properties of each atom in a molecule. The assumption is that a molecular structure is a compromise between all preferences in a molecular assembly (all individual distances, angles, torsions and nonbonded contacts), i.e., the sum of all potential energy terms is at a minimum. It emerges from this, that molecular mechanics is equally applicable to simple organic compounds and metal complexes, to classical coordination compounds and organometallics, to small molecules, biomolecules, polymers and crystal lattices. The only requirement for a structure optimization by molecular mechanics is an accurate force field, and that includes an appropriate set of potential energy functions and their parameterization.

## 16.2 The Starting Model

The starting model is the set of coordinates that is input to the minimization program and defines the approximate geometry of the conformation and configuration of the molecule of interest. Most energy minimization routines require a starting model that is moderately close to an energy minimum. If the starting model is a poor approximation then the minimization procedure may diverge or oscillate wildly and perhaps refine to a minimum other than that desired. For example, if one is trying to determine the strain energies of a number of conformational isomers then the starting model will generally have to be close to the final result in order to avoid conversion of the geometry to that of one of the other conformers.

In the past, starting models were mostly derived from crystal structures of related molecules. Alternatively, a few programs were available that could be used to construct molecules. There are now many programs that enable molecules to be constructed graphically (see Appendix 3). We are unaware of programs that are specifically designed for the construction of metal complexes but most of the listed programs can, with a little effort, be used to produce reasonable starting models. Some care does need to be taken, however, because often, the bond lengths and angles imposed by such programs will be far from the ideal values for metal complexes.

When producing starting models for metal complex/bio-macromolecule adducts it is generally good practice to start with an energy-minimized macromolecule so that the reorganizations taking place on introduction of the metal complex are not too dramatic for the program to handle.

## 16.3   The Force Field

A force field is the collection of numbers that parameterize the functions used to calculate the strain energy, e.g., the force constant ($k_b$) and ideal bond length ($r_0$) in the function generally used to calculate the deformation energy ($E_b$) of a bond (see Chapter 3). There is no one set of "correct" parameters. For example, if a bond is subject to a force that lengthens it, then the energy minimized bond length can be decreased by either increasing the force constant ($k_b$) or reducing the ideal bond length ($r_0$). Because of the correlation between the parameters it is important that they are firmly based on as widely varied experimental data as possible, i.e., on spectroscopic, thermodynamic and/or structural data of an as large as possible set of compounds with the corresponding functional group. First estimates of many of the parameters can be derived directly from experiment, but some others must be developed empirically. However, the force constants derived from spectroscopy need to be fitted to experimental structures, spectra or thermodynamic data because of the fundamental difference between spectroscopic and molecular mechanics force fields (see Sections 3.3, 3.4 and 16.6).

There is a strong correlation between the parameters of different potential energy functions so that they should not be developed or refined in isolation. For example, the barrier to rotation about a bond can be modified by changing the explicit torsion angle term or by changing the nonbonded interactions. Thus, the effect of any change on a force field parameter needs to be tested extensively, i.e., some other parameters might need to be reajusted. For this reason, developing a new force field is an onerous task and one that needs to be undertaken with great care.

There are a number of molecular mechanics force fields available for organic compounds and there are now few classes of organic compounds that cannot be modeled at least reasonably well [49,106,125,135]. The same is not true for inorganic systems; while there are many force fields, they cover a limited range of metal centers and ligands. The situation is greatly complicated in the case of transition metal ions by the large number of possible oxidation states, spin states and coordination numbers that the metal center can adopt and by the way the ligand can be altered by its interaction with the metal ion. For these reasons, and because of the work involved in developing or extending a force field, it is possible that no force field will exist for the particular class of compound that you are interested in. If this is the case then in the following paragraph and in Sections 17.10–14 you will find advice on how to go about developing a new force field or extending an existing one.

Some of the commercially available molecular modeling packages include parameters for modeling inorganic systems. However, these are generally restricted to only a small number of systems. Calculations carried out using force field parameters to model molecules that differ from the type used in the development of the parameters cannot be relied upon to be accurate. Therefore, limited parameterizations are of little general use. Unfortunately, given the potential for variation in metal complex systems, a collection of parameters able to deal with a significant proportion of the complexes likely to be of interest is difficult to develop. The alternative approach of using a model that is able to deal with all complexes by predicting the force constants, as in the UFF, ESFF and DREIDING models [83,125], while useful for visualization, is unlikely to yield results accurate enough to compete with a customized force field for the particular class of complexes under study. The VALBOND model [78,79,81] (see Section 3.2.2 and Chapter 12) is an exception and leads to very accurate structural predictions in certain areas of inorganic chemistry. Also, with the CLF/MM method [77,357] a reduced parameter set can be used, since the same parameters apply for different spin states and coordination geometries. Force field parameters are generally still obtained by manual fitting to observed experimental data (often structures but also thermodynamic or vibrational spectroscopic data). However, parameters have also been fitted with optimization routines [128,518,519] and with genetic algorithms [459].

## 16.4   The Energy Minimization Procedure

Given a force field and a set of starting coordinates, the primary role of a molecular mechanics program is to calculate the geometry with the lowest strain energy. This can be achieved by a variety of mathematical methods, and these are outlined in detail in Section 4.2. The most common classes are modified first-derivative methods such as conjugate-gradient techniques, full-matrix second-derivative methods such as the Newton-Raphson procedure and block-diagonal variations of the Newton-Raphson method. Discussion hereafter will be limited to the first two of these classes. The behavior and limitations of block-diagonal Newton-Raphson refinement closely resemble those of the first-derivative methods.

Second-derivative methods are inherently better in that a mathematically verifiable minimum is usually achieved in a small number of cycles. Setting up and inverting the $3N \times 3N$ square matrix can require a large amount of computer time and memory but for small molecules (up to a few hundred atoms) this is now possible, even on a modestly configured personal computer. Given the extra reliability and certainty that second-derivative methods offer they are the method of choice whenever computational resources allow. Indeed, it has been suggested that the results obtained using other minimization procedures cannot be relied on [520]. An additional advantage of second-derivative methods is that they alone allow the imposition of mathematically strict constraints (see Section 4.3).

The disadvantage of second-derivative methods is the large amount of computer time and memory they require when dealing with large molecules. The square matrix uses more than $4.5 \times N^2$ words of storage for $N$ atoms so that modeling even a small protein or oligonucleotide uses more than a million words of storage. Also, the time required for inversion of the matrix increases in proportion to the third power of the number of atoms. For these reasons in nearly all molecular mechanics studies of macromolecules modified first-derivative methods have been used. As the availability of computer time and memory increases it will become more feasible to use second-derivative methods for energy minimization of macromolecules. For example, full-matrix Newton-Raphson methods have been used to calculate the minimum-energy geometries of oligonucleotides up to nine base-pairs long [141]. One approach to extend Newton-Raphson refinement to very large molecules is to use rigid groups, i.e., to refine parts of the molecule while other parts are kept constant (see Section 17.20).

Second-derivative methods generally require a better starting model than do first-derivative methods. If the starting model is poor then the refinement can become unstable and rapid divergence can occur. First-derivative methods are generally far more tolerant of a poor starting model, having a much wider circle of convergence. Therefore, it is often advisable to subject a starting model to a number of cycles of energy minimization using a first-derivative method and then to follow this up with a second-derivative refinement. This is also the most computationally efficient approach when dealing with moderately large molecules, and an ideal molecular modeling program might begin with first-derivative methods and then, on the basis of the status of the convergence, switch automatically to second-derivative methods to complete the refinement process.

An additional advantage of second-derivative methods is that frequencies of infrared vibrations can be calculated from the final Hessian matrix. This is only likely to be of relevance to small molecule systems where good quality spectra can be obtained. However, in such cases there is the potential to predict spectra and so characterize an unknown compound (see Section 10.1). The ability to reproduce infrared frequencies should also provide a good test of the force field parameters but little use has been made so far of this approach[13,74,108].

Unstable energy minimization processes can be controlled to some extent by "damping" the refinement. When a minimization becomes unstable the atomic shifts to be applied are generally over-estimated and the geometry calculated at the next iteration is even further from the minimum than the starting model. Application of a fraction of the calculated shift can be used to damp the process, and in some cases this will allow convergence to occur. The fraction of the shift to be applied is generally calculated using a function which is inversely related to the shift. That is, for a large calculated shift, only a small fraction of the shift is applied. Damping can increase the number of cycles needed to achieve convergence, so it should not be applied unless necessary. If the energy minimization is unstable, even with damping, then it is necessary to go back to the starting model and to check whether there are any errors or whether some aspect of the model can be improved. Alternatively, constraints, restraints or temporary modification of the force field can be used to stabilize a previously unstable refinement (see Section 4.3). For instance, if a cobalt(III) complex was being used as a starting model for the analogous cobalt(II) complex then the large change in bond lengths might make the minimization process unstable. This could be overcome by constraining the Co-L bond lengths to values intermediate between those of cobalt (III) and cobalt(II) and "stepping" from one to the other. Alternatively, the ideal Co-L bond length could be temporarily modified in the force field to a value intermediate between those of cobalt(III) and cobalt(II).

Convergence of the energy minimization process can be defined in two ways: the change in energy between two iterations (energy gradient) being less than a specified value or the largest shift in a coordinate being less than a specified value. The first is the criterion usually applied in first-derivative methods and the second is that usually used for second-derivative methods. The problem with defining convergence as being when the energy change is less than a certain amount is that if the potential energy "surface" is flat then convergence can occur far from the actual minimum. It can be argued that if the potential energy surface is flat then there are many geometries with similar energies and it is unimportant which is arrived at. However, it is possible to have convergence at some distance from a deep potential energy minimum if there is a relatively flat surface adjacent to the minimum. Thus, it is generally better to use a second-derivative method when it is important to have the true value and geometry of minimum energy. This tends to be less important for macromolecules because they certainly can adopt many closely related conformations. That is, they tend to have flat potential-energy surfaces. However, it is important to remember that minima produced by first-derivative methods are not mathematically verifiable and may not correspond to true minima at all.

## 16.5    Local and Global Energy Minima

Usually, the optimized structure is a local minimum close to the starting structure. In many molecular modeling studies it is important that the global energy minimum be located and in others that all local minima be located. To do this unequivocally, all internal coordinates (bond angles, torsion angles, etc.) would need to be systematically varied. This is a problem of $3N$-6 dimensionality for an $N$ atom molecule and is impractical for any but very small molecules. However, there are a number of methods available for scanning the potential-energy surface. Which is the most appropriate depends on the size of the molecule under consideration and on its flexibility. For relatively small molecules, systematic variation of torsion angles, the primary determinants of conformation, is feasible and is generally successful. For proteins, most of the degrees of freedom are not relevant because the secondary structure is usually not expected to alter. In such cases molecular dynamics can be used to produce ensembles of structures, and samples can be fully energy minimized. This is a non-systematic method of scanning the potential energy surface, but, given the number of degrees of freedom in a protein, only such methods are feasible. If larger structural variations are of interest then Monte Carlo methods are appropriate. For molecules of intermediate size a combination of systematic scanning and random variation is most likely to give a thorough scan of the potential-energy surface (see also Chapter 5). Various methods for conformational searching have been tested for organic molecules[187] and for transition metal coordination compounds[37,521].

There are also several reasons for mapping the potential energy surface defined by the molecular mechanics model. The most common reason is to determine the energy barrier associated with moving from one potential-energy minimum (conformation) to another. For example, the potential energy surface associated with changing from one conformation of a five-membered chelate ring to another in simple $[Co(bidentate)(monodentate)_4]^{3+}$ complexes has been calculated[180]. Alternatively, mapping can be used to locate potential energy minima on the surface.

In order to carry out a mapping study it is first necessary to decide what the mapping coordinates are to be. Generally, these will be internal coordinates, such as torsion angles. For example, in the case of a five-membered ring, two torsion angles are sufficient to define the conformation. It is a relatively simple exercise to calculate the potential-energy surface for such a ring by constraining the two torsion angles at values in the ranges that encompass all reasonable conformations. Energy is plotted as the third dimension to give a plot that clearly shows the energy minima and energy variation as a function of the torsion angles.

In order to map against an internal coordinate it is necessary to be able to set that coordinate to a particular value. This can be done by constraints or restraints (see Section 4.3). Constraints are only available when using second-derivative methods and are imposed by adding extra columns and rows to the matrices. Consequently, they are mathematically precise and at convergence the internal coordinate will be exactly equal to the value selected. Restraints are artificially large force constants that are used to drive the coordinate close to the selected value.

Given that the internal coordinate is also subject to other forces, a restraint will never force it to exactly the selected value. However, whether that is an important consideration depends on the level of accuracy required.

# 16.6 Pitfalls, Interpretation and Communication

There are two outcomes of a single molecular mechanics calculation: the structural geometry of a local or the global energy minimum and the energy value of that minimum.

The use of the energy minimized structures falls into two classes; those where the structure has been determined by crystallographic or spectroscopic means and those where the structure is unknown. Where the structure has been determined crystallographically, comparison between the energy minimized and crystal structures is generally used solely as a test of the molecular mechanics force field. However, as molecular mechanics models become more reliable other comparisons will become more valid. For example, conformational disorder is a common occurrence in crystal structures of chelate complexes. In such cases the crystal structure represents an average of two or more conformational isomers and the energy minimized structures may represent the only way of accessing the structures of the individual isomers [182,207,219]. When more than one chelate ring is disordered it is often impossible to establish what combinations of ring conformations are stable. The application of a combination of molecular mechanics with spectroscopy to such problems has become an important application of molecular modeling (see Chapters 7 and 10).

The limitations of molecular mechanics, when electronic effects are important, have been outlined in detail elsewhere in this book (Section 3.6 and Chapter 12). However, this problem can be turned around and molecular mechanics can be used to help in the unravelling of steric and electronic effects. This can be done by careful analysis of the differences between energy minimized and crystal structures and attributing these differences to electronic effects. For instance, two ligands may be suspected of having different $\pi$-bonding abilities but may also have substantially different steric demands. Both factors will influence the metal-ligand bond length, and unravelling the relative importance of the two factors is difficult. Molecular mechanics, since it relates solely to steric effects, can be used to factor out the steric contribution and so allow the estimation of the electronic contribution. This has been done for the determination of inductive effects by alkyl substitution of amines, where the increase in bond strength (increased basicity) is opposed by a increasing repulsion because of the increasing bulk of the ligands [90,166].

Structures determined by spectroscopic means are generally underdetermined in that the number of variables is similar to or greater than the number of observations. In such a case, the energy minimized structures can be used to aid in the refinement of the structure or as a check of the structure [164]. Applications of molecular mechanics modeling to the determination of solution structures is described in detail in Chapters 7, 8 and 10. Such approaches are fundamental to the determination of macromolecular structures by NMR spectroscopy.

Where the structure has not been determined or cannot be determined experimentally, by diffraction techniques or spectroscopy, then molecular mechanics can be used as a predictive tool. This is of particular value when the structure of a compound is the main target, such as in modeling associated with rational drug design (see Chapter 13).

When the goal is visualization rather than quantitative analysis, somewhat less well refined molecular mechanics models have a valid role to play. For example, the force fields for proteins and nucleic acids, and for their metal complexes are in general less rigorous than those for small molecules. Despite this, much useful information has been obtained from structures produced using these force fields. However, it is important not to over-interpret the results in such cases. In particular, the minimized strain energies obtained from such models should be considered to be qualitative estimates at best.

There is some thermodynamic information in strain energies. However, strictly this is only true if entropic terms are included, and this is not usually the case. It is necessary to evaluate in each case whether or not the neglect of entropic terms is justifiable. The point has been made a number of times earlier in this book that values of minimized strain energies obtained from molecular mechanics models of metal complexes should be taken as having no absolute meaning. In the case of simple alkanes the force fields have been refined sufficiently for the strain energy to be an accurate indication of the potential energy in a molecule, arising from the stresses associated with bringing atoms close together[49,64,106]. However, in the case of most metal complexes, the force fields are still some way from this point. This is due to the fact that most of the transition metal complex force fields are entirely based on structural data and may therefore only be used successfully for the accurate predictions of structures. However, in force fields where parts of the parameterization schemes have been fitted to thermodynamic data, it is reasonable to expect good agreement of relative strain energies with experimentally determined stabilities (see Section 3.3). This is obvious for the much-modeled hexaaminecobalt(III) complexes where clear correlations are evident between the total strain energy and the stresses involved in the formation of the complex[65,135,151]. However, even in these examples it would be risky and unwise to assume that the strain energies did have absolute meaning.

Instead, the comparison of strain energies should be limited to sets of isomers. When the atom connectivities are the same, electronic factors and other omissions and errors in the force field can be assumed to be constant and therefore will cancel when differences between strain energies are considered. The values of the strain energies reveal which isomer is the most stable (has the lowest strain energy) and what percentage of each isomer should be observed in an equilibrated system[65]. The methods for calculating these percentages are described in Chapter 8.

Binding energies or docking energies are often calculated from minimized strain energies in order to estimate binding strength in macromolecule/substrate interactions. Where such interactions directly involve a metal ion, the approximations likely to have been made in the development of the force field mean that such estimations have questionable validity.

Molecular mechanics is a purely interpolative approach. Therefore, full details of the force field have to be presented together with the results and their interpretation, in order to allow the reader to fully reproduce them. These include the functional forms of the potentials, the parameter sets that have been used for the computations, the data for which they were derived and the data for which they were validated[522]. Guidelines for the publication of research results from force field calculations have been published[523].

# 17 Tutorial

The twenty tutorial lessons will help you to appreciate the scope and limits of molecular mechanics, the importance of the set of potential energy functions and their parameterization, and the ease with which basic principles can be visualized and accurate structural and thermodynamic data obtained. Apart from structure optimization, the computation of isomer distributions, and parameter optimization, some specific examples are used to help you to appreciate the scope of inorganic molecular mechanics and further the understanding of some of the examples discussed in Part II. These include the mapping of a potential energy surface, the computation of ligand cavity sizes and of reduction potentials, and a structural study based on spectroscopic data. Each lesson begins with a short discussion of relevant theoretical aspects, followed by a carefully guided computational tutorial and some additional exercises. The first seven lessons provide experience with the software and the basic tasks of molecular mechanics modeling, i.e., building of structural models, visualization, stucture optimization and conformational analysis. The next seven lessons concentrate on the force field, i.e., on the potential energy functions and their parameterization. Exercises involving strain energy minimization with constraints and rigid groups are the basis of the last six lessons, and these lead to applications which involve reduction potentials, the size and plasticity of ligand cavities and spectroscopic data.

The tutorial is based on the combination of HyperChem® for the building and viewing of structures and MOMEC for the computational work. MOMEC is a molecular mechanics program and force field that has been specifically designed and extensively used for problems related to transition metal coordination compounds. It has a number of functions and modules that are essential for inorganic molecular modeling and one of the most extensive force fields. It is our advice that a force field should only be used for types of compounds for which it has been carefully parameterized and validated (see Section 16.6). Therefore, MOMEC does not add guess-values for missing parameters but a message is given for each missing parameter. Also, a simple force field editor allows the parameter set to be quickly changed and optimized, and parameters for new atom types can be added easily. Publications related to the MOMEC force field are available from our homepage, where they are updated routinely.

The tutorial has been written for the full, commercially available programs MOMEC and HyperChem® but most tutorials can be completed with the MOMEClite/Hinview® set of programs provided on the CD. Readers who have licences for the two programs should use the full versions. MOMEClite has most of the functionality of MOMEC but the atom types are restricted to those which are used in the tutorial and, unlike the full version, it is not possible to add new atom types. Also, the number of atoms in a molecule is restricted but is sufficient for the tutorial. Limitations imposed by MOMEClite/Hinview and the procedure for installation of the programs are outlined in the file "Readme.doc" on the enclosed CD.

In this tutorial we use MOMEC97 and HyperChem® $6.0^{[51,52]}$, but most of the exercises can also be completed with earlier versions and the programs provided on the CD. The results shown in tables, figures and screen photographs in this tutorial have been produced using MOMEC97 and, where nothing else is specified, the original MOMEC97 force field. We note that the results of force field calculations can also depend on the starting structure and the optimization control parameters such as damping factors, cut-off limits and the termination conditions. The reason is that, for highly flexible molecules, the final structure might be one of several local minima. Hence, you might obtain results that are slightly different to those given in this tutorial. In those cases, we suggest that you change some of the parameters that control the minimization process. It is always a good idea to check whether the results of a structure optimization correspond to a stable low energy structure. For clarity, the optimization parameters that we have used have not been specified in each of the examples. In most of the cases, however, we have used the parameters shown in Section 17.2, Fig. 17.2.1, and we have checked that the results presented are valid.

## 17.1  Building a Simple Metal Complex

**Theory**

Molecular mechanics force fields usually employ modified atomic symbols to specify the chemical environment of a particular atom and these are referred to as atom types. For example, $sp^3$ hybridized nitrogen atoms might be given the atom type NT, and a nitrogen atom in an aromatic ring such as pyridine might be assigned the atom type NP. MOMEC uses an atom type nomenclature that is similar but not identical to that of the *AMBER* force field available in HyperChem®. To avoid problems in MOMEC97 calculations you should first open MOMEC for each of the tutorial sessions. Under the *Execute* menu bar you can then open *HyperChem* (i.e. do not open HyperChem without or before MOMEC). This ensures that the MOMEC atom types are active. Make sure that the *AMBER* force field has been selected under *Molecular Mechanics* in the *Setup* window of HyperChem. If, for some reason, you wish to change to HyperChem atom types at any stage of a MOMEC-HyperChem session, you can do that with the option *Switch Atom Types* under the *Tools* menu bar of MOMEC97.

Many atoms have valences that are higher than those allowed for by the standard HyperChem atom types. It therefore is necessary to enable the *Allow Ions* options under the *Build* menu.

**Practice**

In this lesson we will build $[Co(NH_3)_6]^{3+}$.

1. Turn on the draw tool ⊕ in HyperChem and set the default atom to nitrogen by either double clicking on the draw tool or by selecting *Default Element* under the *Build* menu and then choosing nitrogen from the periodic table that appears.

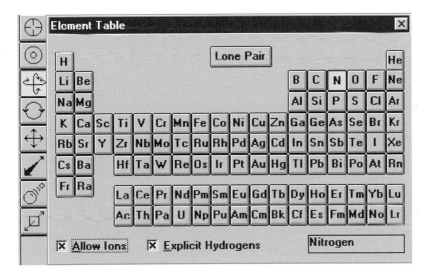

**Fig. 17.1.1**

2. Draw six bonds starting at a common point – there is no need to worry about the geometry at this point. Note: if the *Explicit Hydrogens* option in the *Build* menu is on, this creates six hydrogen atoms and it is therefore easier if this is turned off. If not, clicking on each atom in turn will convert it to a nitrogen atom.
3. Set the default atom to cobalt.
4. Change the central atom to cobalt by clicking on it.
5. Set the default atom to hydrogen.
6. Add three hydrogen atoms to each nitrogen atom.
7. Turn on the select tool ⊙.
8. Select the cobalt atom by clicking on it with the left hand mouse button, then go to the *Build* menu, choose *Constrain Geometry* and select *Octahedral*.

**Fig. 17.1.2**

9. Deselect the cobalt atom by clicking on it with the right hand mouse button.
10. In the *Build* menu check that *Explicit Hydrogens* has been selected and then select the *Model Build* option.
11. You should now have the desired molecule. However, in HyperChem version 2 an error results in one of the six $NH_3$ groups being planar rather than tetrahedral. This can be repaired as follows:
   - Select the offending nitrogen atom with the select tool ⊙.
   - In the *Build* menu choose *Constrain Geometry* and select *Tetrahedral*.
   - Select the cobalt atom and the three hydrogen atoms attached to the planar nitrogen atom.
   - In the *Build* menu choose the *Model Build* option.
12. For the refinement of the structure (Section 17.2) it is necessary that the appropriate atom types be assigned. Therefore, in the HyperChem menu *Display* choose the *Labels ...* option and check the *Atoms/Type* button. If the MOMEC types have been chosen as described above then the atom types will have been set correctly (CO3, NT, H). If not, then select *MOMEC types* in the *Tools* box of MOMEC and the *AMBER* force field method in the *Molecular Mechanics* window under the *Setup* menu of HyperChem. Then choose *Calculate Types* in the *Build* menu (you have to recompile the atom types if you have previously used another force field in HyperChem (*Compile Type Rules* in the *Build* menu)). Depending on the metal atom used it might also be necessary to assign a particular type (e. g. CO3 instead of CO2). This can be done in *HyperChem* by selecting the atom and then going to the *Set Atom Type* option under the *Build* menu. If you can't find the desired atom type, see the HyperChem manual for details on how to modify the CHEM.RUL file. Alternatively the *atom type* can be modified in MOMEC by selecting the *Set Metal Type* options in the *Tools* menu. However, if your programs are installed correctly, it will not be necessary to worry about these things for the present exercise, and we will cover these manipulations in detail in later lessons.

Save the structure by selecting the *Save* option under the *File* menu of HyperChem. When saving the file choose the path that you want to use in your MOMEC calculations.

## 17.2   Optimizing the Structure

### Theory

What will MOMEC do when you tell it to optimize your $[Co(NH_3)_6]^{3+}$ structure? It will first calculate the total strain energy $U_{total}$ from the starting structure that you have created in Section 17.1, and this is based on the equation below.

$$E_{total} = \sum_{molecule} (E_b + E_\theta + E_\phi + E_{nb} + F_c + E_{hb} + E_\delta + \ldots) \qquad (17.2.1)$$

*Bond stretch deformation:*

$$E_b = 1/2\, k_b\, (r_{ij} - r_0)^2 \qquad (17.2.2)$$

$r_{ij}$ = bond distance, $k_b$ = force constant, $r_0$ = ideal bond distance.

*Valence angle deformation:*

$$E_\theta = 1/2\, k_\theta\, (\theta_{ijk} - \theta_0)^2 \qquad (17.2.3)$$

$\theta_{ijk}$ = valence angle, $k_\theta$ = force constant, $\theta_0$ = undeformed valence angle.

Alternative for coordination angles (option: multiple harmonics):

$$E_\theta = 1/2\, k_\theta\, (\theta_{ijk} - 90°)^2 \quad \text{if} \quad 0° \le \theta_{ijk} \le 135°$$
$$E_\theta = 1/2\, k_\theta\, (\theta_{ijk} - 180°)^2 \quad \text{if} \quad 135° \le \theta_{ijk} \le 180° \qquad (17.2.4)$$

*Twist angle deformation:*

$$E_\gamma = 1/2\, k_\gamma\, (\gamma_{ij} - \gamma_0)^2 \qquad (17.2.5)$$

$\gamma_{ij}$ = angle between two planes, $k_\gamma$ = force constant, $\gamma_0$ = ideal twist angle.

*Torsion angle deformation:*

$$E_\phi = 1/2\, k_\phi\, (1 + \cos(m\,(\phi_{ijkl} + \phi_{offset}))) \qquad (17.2.6)$$

$\phi_{ijkl}$ = torsion angle, $k_\phi$ = force constant, $m$ = muliplicity of the torsion, $\phi_{offset}$ = offset of optimum torsional angle.

*Out-of-plane deformation:*

$$E_\delta = 1/2\,k_\delta\,\delta_{ijkl}^2 \tag{17.2.7}$$

$k_\delta$ = force constant, $\delta_{ijkl}$ = the angle between vector $jk$ and the plane through $ij$ and $l$.

*Nonbonded interaction (also used for 1,3-interaction):*

$$E_{nb} = ae_{ij}^{-bd} - c \cdot d_{ij}^{-6} \tag{17.2.8}$$

$d_{ij}$ = distance between two interacting atoms,
$a$ = $2014\,(\varepsilon_i \cdot \varepsilon_j)^{1/2}$, $b$ = $12.50/(vndr_i + vndr_j)$,
$c$ = $(2.25\,(\varepsilon_i \cdot \varepsilon_j)^{1/2}\,(vndr_i + vndr_j)^6)/144$

*H-bond interaction:*

$$E_{hb} = f \cdot d_{ij}^{-12} - g \cdot d_{ij}^{-10} \tag{17.2.9}$$

$d_{ij}$ = distance between the hydrogen donor and the hydrogen acceptor,
$f, g$ = atom based parameters.

*Electrostatic interaction:*

$$E_\varepsilon = q_i q_j / \varepsilon\, d_{ij} \tag{17.2.10}$$

$q_i$ and $q_j$ = partial atomic charges in electrons, $\varepsilon$ = dielectric constant (default: $\varepsilon = 4\,d_{ij}$).

   MOMEC also allows single point calculations, i.e., there will be no structural optimization after the calculation of the strain energy. The result of this calculation depends (i) on the structure that you have generated, (ii) on the functional forms of the terms in the above equations and (iii) on the parameters used in these functions. The latter two together define the force field, and this is easily modified in MOMEC. For the moment, we do not address this aspect and carry out our calculations with the default set of functions and parameters provided.

   Following the initial calculation of strain energy, MOMEC will adjust the coordinates of your structure and recalculate the total strain energy. The idea is to repeat this procedure, a cycle, until a minimum in the strain energy is reached. The results are a minimized total strain energy and the corresponding optimal structure. MOMEC offers three minimization procedures, the Fletcher-Reeves algorithm, which is a conjugate-gradient method that is particularly useful for the optimization of starting structures that are far from the energy minimum, the full-matrix Newton-Raphson procedure which reaches the optimum structure in few cycles if the starting structure is reasonably close to the energy minimum, and a combination of both methods, starting with the conjugate-gradient method (Fig. 17.2.1). Other parameters that can be varied and are important for the optimization are the termination criteria and the damping factor. You can force MO-MEC to stop after a certain number of cycles or when the mean rms difference in cartesian coordinates is smaller than a given value. The damping factors reduce the computed coordinate shifts for each cycle, and this may help to prevent a rather unstable geometry from changing to a different conformation or a poor starting structure from "exploding".

Other parameters included in the *Optimization Controls* menu of the *Setup* bar are, the *Interactions*, i.e., the type of potentials that are activated for the computation, the *Result refresh period* which determines when intermediate results are displayed (*Summary file* with strain-energy terms and the structure in the HyperChem window) and parameters that are related to the cutoff for the van der Waals interactions. For your first minimization use the default parameters displayed in Fig. 17.2.1. Note that the optimization parameters might influence your results. Also, the number of cycles for the result refresh period might influence the final result due to small rounding errors. Note also that in the *Interactions* box of the window shown below, all relevant interactions need to be activated.

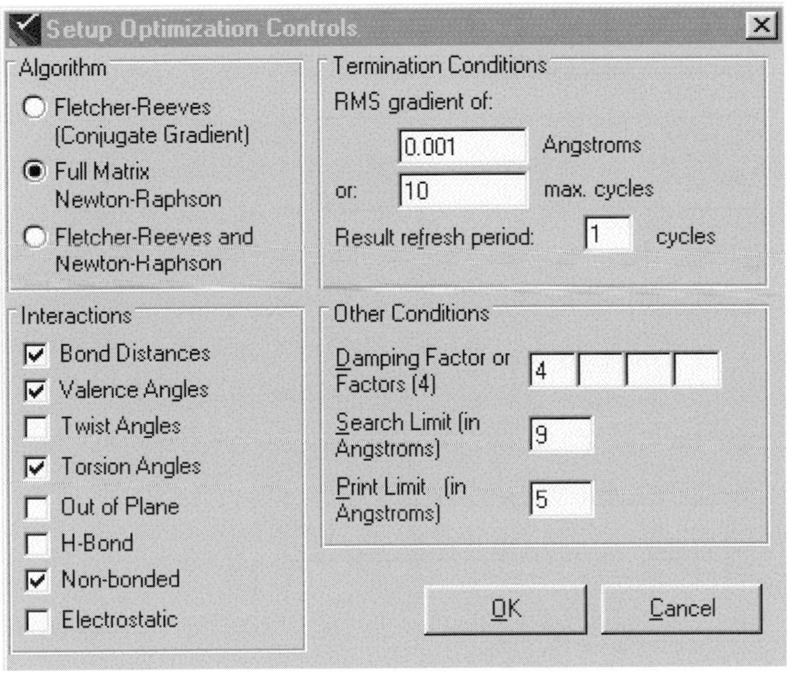

**Fig. 17.2.1**

Activating interaction types that are not relevant (e.g. *out-of-plane* in the present case) will only marginally increase the time taken for minimization but it will lead to *Force Field Messages* (see Fig. 17.2.5; in that example, we have activated the *out-of-plane* interactions).

In conclusion, the results that emerge from a force field calculation depend on a number of factors, (i) on the input structure (molecular mechanics usually does not switch between different conformations), (ii) on the force field, i.e., the type, specific form and parameterization of the potential energy functions, and (iii) on the energy minimization procedure employed.

### Practice

We will now refine the structure that you built and saved in Section 17.1. Before starting refinement we have to setup all the files.

(i) The *Files* window in the *Setup* menu (Fig. 17.2.2) allows you to specify the coordinates that are used as structural input and where and how the results of the calculation will be saved.

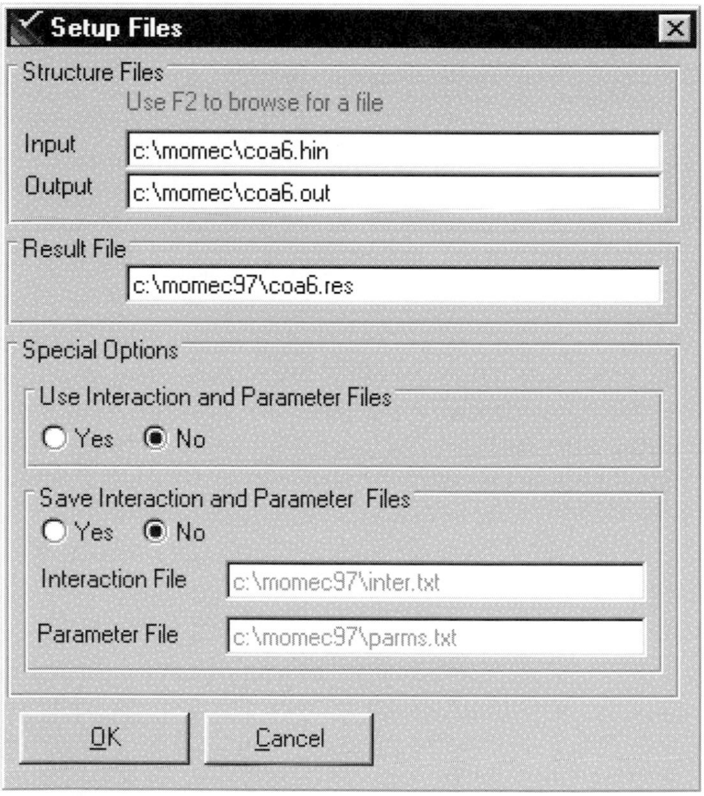

**Fig. 17.2.2**

(ii) In the *Force field* menu under *Setup* (Fig. 17.2.3) you can choose the parameter set that you want to use for the strain energy minimization.

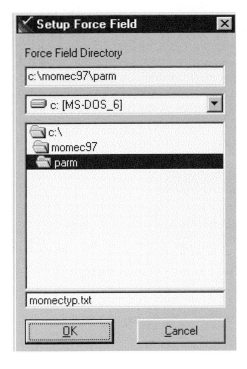

**Fig. 17.2.3**

(iii) In the menu item _Listing_ under _Setup_ (Fig. 17.2.4) you can choose the type of data tables that you want to save as result files as shown below.

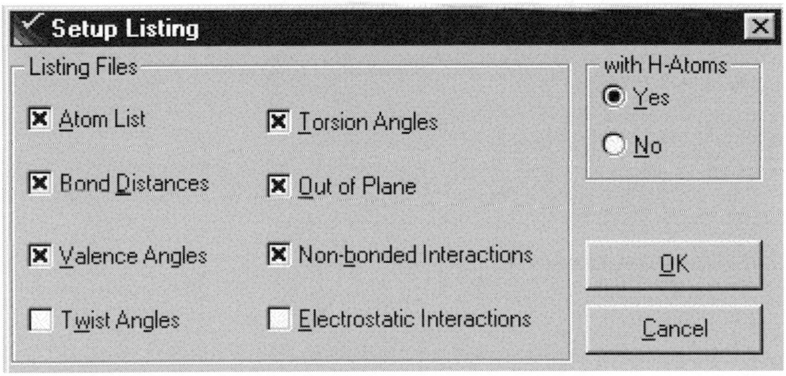

**Fig. 17.2.4**

When all parameters are set you can start the refinement by choosing the *Geometry Optimization* option in the *Execute* bar. At the end of the refinement the *Summary File* will be as shown in Fig 17.2.5. Check that convergence has been reached; if it has not, copy the output file to the input file and repeat the refinement procedure. If necessary, increase the number of cycles.

The total strain energy computed with the force field provided is 32.62 kJ mol$^{-1}$, and the average Co-N distance is 1.955 Å. You can measure the computed Co-N distance in HyperChem, or you can have a look in the data tables (*Edit/View/Listing Files/Bond Distances*). Note that, due to the free rotation around the six Co-N bonds there are many possible conformers; you might obtain one that is slightly different from the one described here. Various conformers and corresponding minimized strain energies and optimized structural parameters can be obtained by modifying the structures in HyperChem (*Edit/Set Bond Torsion*) after selecting the corresponding torsion angle with the select tool ⊙.

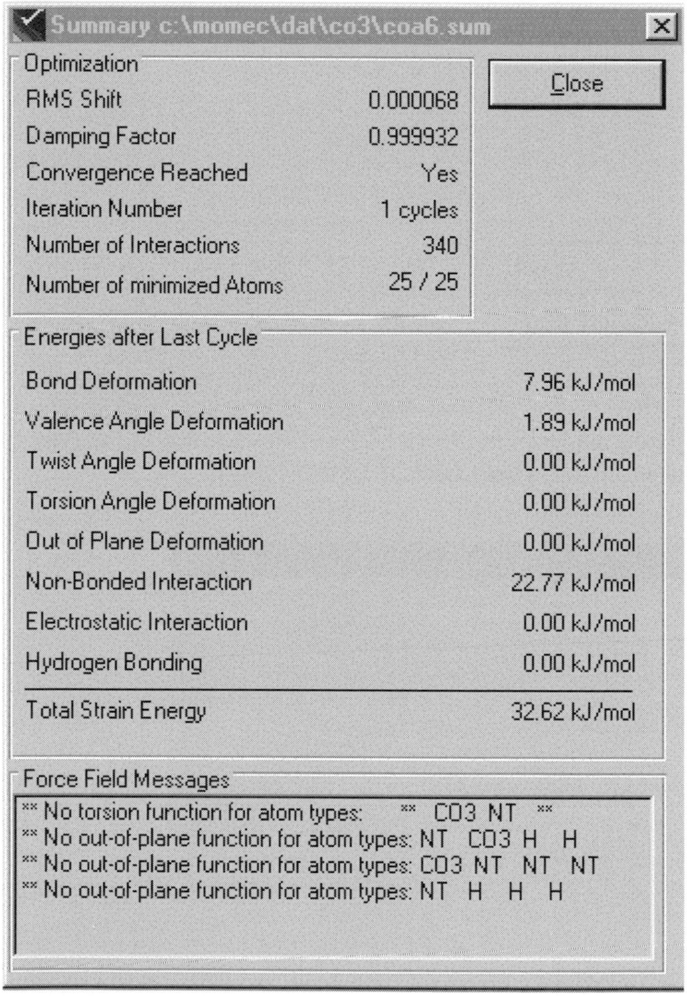

**Summary** c:\momec\dat\co3\coa6.sum

Optimization

| | | |
|---|---|---|
| RMS Shift | 0.000068 | |
| Damping Factor | 0.999932 | |
| Convergence Reached | Yes | |
| Iteration Number | 1 cycles | |
| Number of Interactions | 340 | |
| Number of minimized Atoms | 25 / 25 | |

Energies after Last Cycle

| | |
|---|---|
| Bond Deformation | 7.96 kJ/mol |
| Valence Angle Deformation | 1.89 kJ/mol |
| Twist Angle Deformation | 0.00 kJ/mol |
| Torsion Angle Deformation | 0.00 kJ/mol |
| Out of Plane Deformation | 0.00 kJ/mol |
| Non-Bonded Interaction | 22.77 kJ/mol |
| Electrostatic Interaction | 0.00 kJ/mol |
| Hydrogen Bonding | 0.00 kJ/mol |
| Total Strain Energy | 32.62 kJ/mol |

Force Field Messages

```
** No torsion function for atom types:    ** CO3  NT  **
** No out-of-plane function for atom types: NT   CO3  H    H
** No out-of-plane function for atom types: CO3  NT   NT   NT
** No out-of-plane function for atom types: NT   H   H   H
```

**Fig. 17.2.5**

# 17.3   **Building a Set of Conformers**

**Theory**

One of the strengths of molecular mechanics is the ability to readily and reliably calculate the strain energies of a set of isomers. Since the atom types and bond types are the same for each isomer many of the factors that make the comparison of strain energies of different compounds invalid do not apply, and reasonably accurate and meaningful energy differences can be obtained. These strain energy differences can be used to calculate theoretical isomer distributions that can then be compared with experimentally determined data. Studies of this kind formed the basis of much of the early application and development of molecular mechanics as applied to metal complexes. More recently, there has been a focus on enantioselective interactions between chiral molecules and chiral complexes, and these too represent systems of isomers. Thus, molecular mechanics can be used to predict enantioselectivities and to aid in the design of enantioselective complexes or targets (see Chapter 8).

The example presented here, $[Co(en)_3]^{3+}$ was one of the seminal compounds in the development of molecular mechanics for metal complexes. Each of the three chelate rings can adopt two enantiomerically related conformations that are designated $\lambda$ and $\delta$. Also, the complex itself is chiral with the chirality defined as $\Lambda$ or $\Delta$ by the direction of the screw looking down the three fold axis of the complex. This gives rise to eight isomers; $\Lambda(\delta\delta\delta)$, $\Lambda(\delta\delta\lambda)$, $\Lambda(\delta\lambda\lambda)$, $\Lambda(\lambda\lambda\lambda)$, $\Delta(\delta\delta\delta)$, $\Delta(\delta\delta\lambda)$, $\Delta(\delta\lambda\lambda)$, $\Delta(\lambda\lambda\lambda)$. Since these form two enantiomerically related and energetically indistinguishable sets we will only consider the $\Delta$ isomers hereafter. These can also be designated as $lel_3$, $lel_2ob$, $ob_2lel$ and $ob_3$ by virtue of the orientation of the C-C vector of the chelate ring parallel to (lel) or at an angle to (ob) the three-fold rotation axis of the complex.

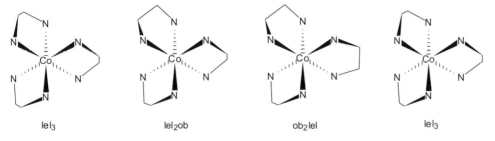

|  |  |  |  |
|---|---|---|---|
| $lel_3$ | $lel_2ob$ | $ob_2lel$ | $lel_3$ |

**Fig. 17.3.1**

**Practice**

*Building [Co(en)₃]³⁺*

1. Turn on the draw tool ⊕ and set the default atom to nitrogen by either double clicking on the draw tool or by selecting *Default Atom* under the *Build* menu and then choosing nitrogen from the periodic table that appears.
2. Draw six bonds starting at a common point – there is no need to worry about the geometry at this point.
3. Set the default atom to cobalt.
4. Change the central atom to cobalt by clicking on it.
5. Turn on the select tool ⊙.
6. Select the cobalt atom by clicking on it with the left-hand mouse button, then go to the *Build* menu and choose *Constrain Geometry* and select *Octahedral*.
7. Deselect the cobalt atom by clicking on it with the right-hand mouse button.
8. In the *Build* menu check that *Explicit Hydrogens* has been selected and then select the *Model Build* option.
9. It is good practice to save your file at each stage by selecting the *Save* option under the *File* menu. Note that, if you need to go back to a previous version, then you will need to reimpose any constraints you have been using as these are not saved. Thus, in this case you would need to select the cobalt atom and *Constrain Geometry* to *Octahedral* under the *Build* menu.
10. Set the default atom to carbon.
11. Add two carbon atoms between each of three *cis* pairs of nitrogen atoms completing the chelate rings. Be careful not to link a pair of *trans* related nitrogen atoms.
12. In the *Build* menu turn off *Explicit Hydrogens* and then select the *Add H & Model Build* option. This should generate sensible chelate rings but each nitrogen atom will only have one hydrogen atom attached.
13. Set the default atom to hydrogen.
14. Add one more hydrogen atom to each nitrogen atom.
15. In the *Build* menu turn on *Explicit Hydrogens* and then choose the *Model Build* option.
16. You should now have the target molecule. HyperChem will occasionally fail to produce a sensible geometry. If this happens try again, following carefully the sequence of steps outlined, as we have found this is the most reliable procedure.
17. Save the structure by selecting the *Save* option under the *File* menu. Before deciding on a file name you should check which conformer has been generated by orienting the molecule so you are looking down the three fold axis and counting how many C-C bonds are parallel to this axis and how many are at an angle (see diagrams above). HyperChem generally produces the lel₂ob conformation but this can vary. When saving the file choose the path that you want to use in your MOMEC calculations. Note, it is also possible to generate an "envelope" conformation in which the N-C-C-N torsion angle is approx. 0°. If this occurs, repeat the procedure.

*Changing Conformations*

Changing the conformation of a ring involves a chiral inversion and this can be achieved by reflecting all atoms of the ring through the plane defined by the cobalt atom and the two nitrogen atoms of the ring.

1. Select the cobalt atom and the two nitrogen atoms of the ring to be inverted, then under the *Select* menu choose the *Name Selection* option. This will give you a number of choices for a name or the option of inserting your own choice of name. Naming the selection as *Plane* is adequate unless you wish to have more than one plane defined at any one time.

Fig. 17.3.2

2. Deselect these atoms by clicking on the right mouse key with the select tool ⊙ highlighted and the cursor positioned away from the molecule.
3. Select all carbon and hydrogen atoms of the ring to be inverted. You can select the nitrogen atoms as well but this has no effect.
4. To invert the ring choose the *Reflect* option from the *Edit* menu. This should invert the ring producing the alternative chirality. All bond lengths and angles should be maintained – if they are not then you have made a mistake and should repeat the process. It is informative to orient the molecule so you are looking down the three-fold axis before you do the inversion. Then, as the inversion takes place you will see the change in conformation from lel to ob or vice versa. It may also help to use a 3D *Rendering* style such as *Balls and Cylinders*.
5. Other rings can now be inverted. When defining a new plane you will be asked whether you wish to overwrite the existing plane. It is best to do so as potential confusion is then avoided.
6. Before saving the new structure any named selections should be removed or they may interfere with the MOMEC refinement process. To do this go to the *Select* option under the *Select* menu, choose any named selections by clicking on them and then clicking on the *Remove* button.
7. Save the modified structure with a new file name by choosing the *Save As* option in the *File* menu. Use the file names colllen.hin, colloen.hin, colooen.hin and cooooen.hin for the $lel_3$, $lel_2ob$, $ob_2lel$ and $ob_3$ conformers, respectively.

**Exercise**

Use the methods described in Sections 17.1 and 17.3 to build the two conformational isomers of $[Co(NH_3)_4(1,2\text{-propanediamine})]^{3+}$.

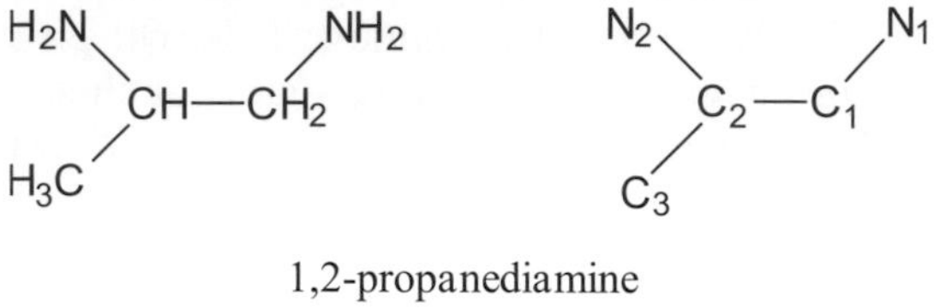

1,2-propanediamine                                    **Fig. 17.3.3**

The exocyclic methyl group should lie equatorially in one conformer and axially in the other as shown below. Note, just inverting the ring will not work in this case. Instead the positions of the methyl group and the hydrogen atom attached to $C_2$ must also be swapped. This can be achieved by defining a plane through $N_2$, $C_1$ and $C_2$ and reflecting the methyl group and hydrogen atom through this plane.

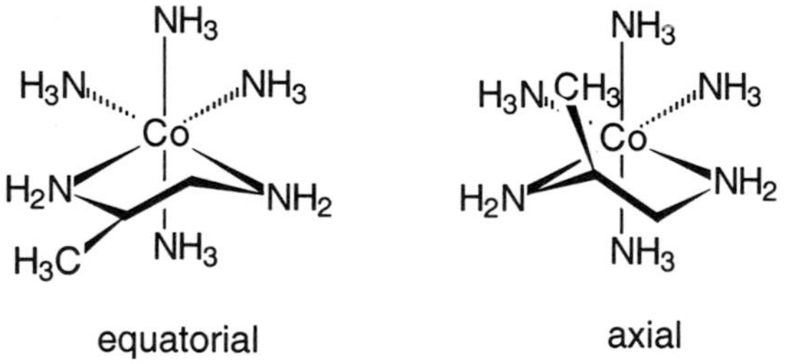

equatorial                        axial                **Fig. 17.3.4**

## 17.4 Calculating the Strain Energies and Isomer Distribution of a Set of Conformers

**Theory**

In the previous lesson we have shown how to generate the structures of the four conformational isomers of $[Co(en)_3]^{3+}$. We will now subject each of these structures to energy minimization using MOMEC and use the strain energies we obtain to calculate the isomer distribution.

The principle behind calculating the isomer distribution from the strain energies is the classical thermodynamic relationship between the difference in free energy ($G$) and equilibrium constant $K$,

$$K = e^{-\Delta G/RT} \tag{17.4.1}$$

where $R$ is the gas constant and $T$ is the temperature. The assumptions are made that the difference between strain energies ($E_{strain}$) is solely responsible for the difference in enthalpy ($H$) and that the only difference in entropy is the statistical contribution (see Chapter 8).

**Practice**

Using MOMEC you can minimize the energy of each conformer separately as described in Section 17.2. However, you could also use the batch facility to initiate a process to energy minimize all four conformers, and this procedure is now described.

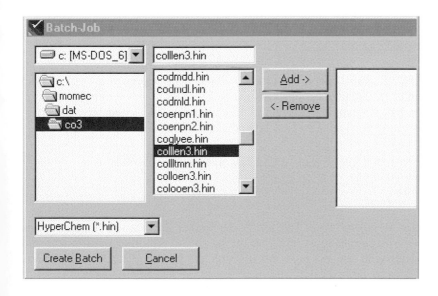

**Fig. 17.4.1**

1. Start MOMEC and under the *Edit/View* menu select *Batch-Job*. Select the four file names corresponding to the four conformers of $[Co(en)_3]^{3+}$ and add them to the batch file.
2. To start the batch job, select *Batch-Job* under the *Execute* menu. Once this is completed, check that convergence was reached for each conformer. You can do this with the menu item *Batch-Job Results* under *Edit/View*.
3. Collect the total strain energy for each conformer by looking at the *Summary* under the *Edit/View* menu for each file.

Once you have the total strain energies of the four conformers, energy differences can be calculated. These differences can be calculated by subtracting the strain energy of the lowest energy conformer from each of the others. This results in the strain energy of the lowest energy conformer being defined as zero. Before these differences can be used to calculate conformer proportions it is necessary to correct for statistical factors. The $lel_2ob$ conformer can be formed in three ways because any one of the three chelate rings could adopt the ob orientation. The same is true of the $ob_2lel$ conformer but the $lel_3$ and $ob_3$ conformers can only be formed in one way. Thus, even if the four conformers had identical strain energies, they would be observed in a $1:3:3:1$ ratio. The easiest way of correcting for this effect is to subtract the energy contribution associated with a $3:1$ preference from the strain energies of the $lel_2ob$ and $ob_2lel$ conformers. This contribution can be calculated using the expression,

$$\Delta G = -RT \ln(K) \tag{17.4.2}$$

the inverse of Eq. 17.4.1. Here $K$ is the ratio expected on the basis of statistical effects. If the temperature is 298 K (25 °C) then the equation becomes:

$$\begin{aligned}
\Delta G &= -8.3143 \text{ J K}^{-1} \text{ mol}^{-1} \cdot 298 \text{ K} \cdot \ln(3) \\
&= -2722 \text{ J mol}^{-1} \\
&= -2.72 \text{ kJ mol}^{-1}
\end{aligned}$$

Once the corrected strain energy differences have been calculated they can be used to determine the isomer proportions using Eq. 17.4.1. For example, the difference between the statistically corrected energies of the $lel_3$ and $lel_2ob$ conformers is $-1.65$ kJ mol$^{-1}$. Thus, Eq. 17.4.1 becomes:

$$\begin{aligned}
K &= \exp((-[-1650] \text{ J mol}^{-1})/(8.314 \text{ J K}^{-1} \text{ mol}^{-1} \cdot 298 \text{ K})) \\
&= \exp(0.6660) \\
&= 1.95
\end{aligned}$$

Thus, the ratio of $lel_2ob$ to $lel_3$ is 1.95 (with the former dominating because it has a lower energy). You should now calculate the ratios for the other conformers and use these to determine the proportions of each conformer. Check your results against those given in Table 17.4.1.

**Table 17.4.1**

| Conformer | $lel_3$ | $lel_2ob$ | $ob_2lel$ | $ob_3$ |
|---|---|---|---|---|
| Total strain energy [kJ mol$^{-1}$] | 45.12 | 46.19 | 46.97 | 46.94 |
| Strain-energy difference | 0.0 | 1.07 | 1.85 | 1.82 |
| Statistical correction | 0.0 | −2.72 | −2.72 | 0.0 |
| Corrected energy difference | 0.0 | −1.65 | −0.87 | 1.82 |
| Conformer ratio | 1 | 1.95 | 1.42 | 0.48 |
| Conformer percentage | 21% | 40% | 29% | 10% |

**Exercise**

Use the methods described in Sections 17.2 and 17.4 to energy minimize the two conformational isomers of $[Co(NH_3)_4(1,2\text{-propanediamine})]^{3+}$ built in Section 17.3 and shown in Fig. 17.3.4.

Compare the strain energies of the two conformational isomers and calculate the proportion of each expected in solution at room temperature. Using HyperChem measure the contacts between the methyl group and the remainder of the complex. Consider how this contributes to the difference in strain energies.

## 17.5   Constructing and Optimizing a Set of Isomers Automatically

**Theory**

Conformational isomers represent minima on an energy surface, and all structures and the corresponding strain energies can be obtained by a careful analysis. This can be performed manually (such as in Sections 17.3 and 17.4) or automatically. An automatic procedure may involve a systematic search (grid search methods), a stochastic search (e. g., torsional Monte Carlo or cartesian stochastic, i. e., the random kick method) or molecular dynamics (see Chapter 5 and Section 16.5). Implemented in MOMEC is a random kick stochastic search module, and this has been shown to lead to excellent results, not only for conformational equilibria, but also for distributions of configurational isomers[37].

**Practice**

In this section we will use the random kick module to search for conformers of the complex $[Co(en)_3]^{3+}$.

1. Open the $lel_3$ conformer of $[Co(en)_3]^{3+}$ in HyperChem (colllen.hin) and, using the select tool ⊙, select both carbon atoms and all eight hydrogen atoms of one of the three ligands. Save the file.

2. Open the same file in MOMEC97 and check that it is fully energy minimized by executing geometry optimization.
3. Under *Execute* select *Random Kick*. Select *Kick atoms selected in HyperChem* and set the other parameters to the values shown in Fig. 17.5.1.

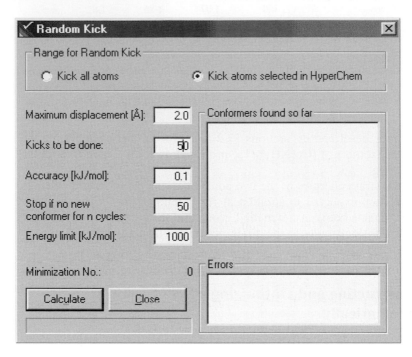

Fig. 17.5.1

4. Select *Calculate*. A series of new geometries will be generated, some of which will not be realistic. The conformers will be listed in order of increasing energy, and the energy contributions for each geometry will be listed as shown in Fig. 17.5.2. Differences in energy and/or in the relative contributions are indicative of different conformations.

| kicks | Total | Cycles | Bond | Angle | Torsion |
|---|---|---|---|---|---|
| 1 | 204.73 | 14 | 9.21 | 157.11 | 7.41 |
| 2 | 273.49 | 16 | 16.62 | 177.16 | 18.87 |
| 3 | 45.12 | 20 | 8.36 | 4.59 | 8.17 |
| 4 | 615.12 | 10 | 37.72 | 460.9 | 17.17 |
| 5 | 46.94 | 14 | 8.68 | 4.15 | 8.95 |
| 6 | 207.52 | 24 | 9.63 | 157.28 | 8.39 |
| 7 | 207.04 | 13 | 8.6 | 154.52 | 9.06 |
| 8 | 262.75 | 16 | 17.04 | 170.53 | 14.57 |
| 9 | 367.73 | 16 | 9.16 | 305.49 | 9.24 |
| 10 | 209.66 | 20 | 8.68 | 153.4 | 10.5 |
| 11 | 46.19 | 26 | 8.54 | 4.32 | 8.66 |

Result of Random Kick

Fig. 17.5.2

5. Each geometry can be viewed in HyperChem by opening the file colllen.n where n is the geometry number. One of these should correspond to the $lel_2ob$ conformer. In the example shown, conformer 3 is the original $lel_3$, conformer 11 is the $lel_2ob$ and conformer 5 is the $ob_3$.

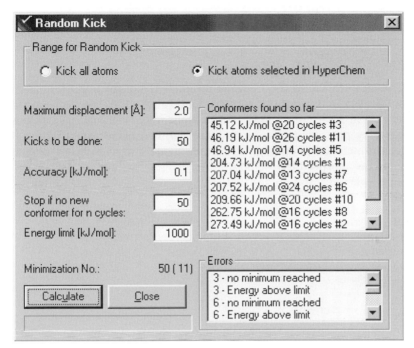

Fig. 17.5.3

**Exercise**

Repeat the procedure selecting the carbon and hydrogen atoms of two or three of the rings and search for the $lelob_2$ and $ob_3$ conformers.

# 17.6 Building More Difficult Metal Complexes

**Theory**

When building simple chelate rings coordinated to metal ions problems can arise and these become more severe with tridentate or macrocyclic ligands. These problems are not unique to HyperChem and arise because of the mathematical complexity associated with building complicated ligands. Therefore, it is necessary to adopt a variety of strategies in order to produce starting models of some complexes. These include using crystallographically determined structures as starting points, building parts of molecules separately and docking ligands to a metal framework.

## Practice

### Importing Structures from other Sources

MOMEC97 has an interface to the freeware program Babel for importing and exporting structural data in a variety of formats. Once a structure has been imported it can be modified in a variety of ways. For instance, other conformers or isomers can be generated by procedures similar to those outlined in Section 17.3 for generating the conformers of $[Co(en)_3]^{3+}$. Starting models derived from crystallographic data can be improved by the addition of hydrogen atoms or by correcting unreasonable bond lengths and angles that might be present due to disorder in the crystal structure.

### Importing Data

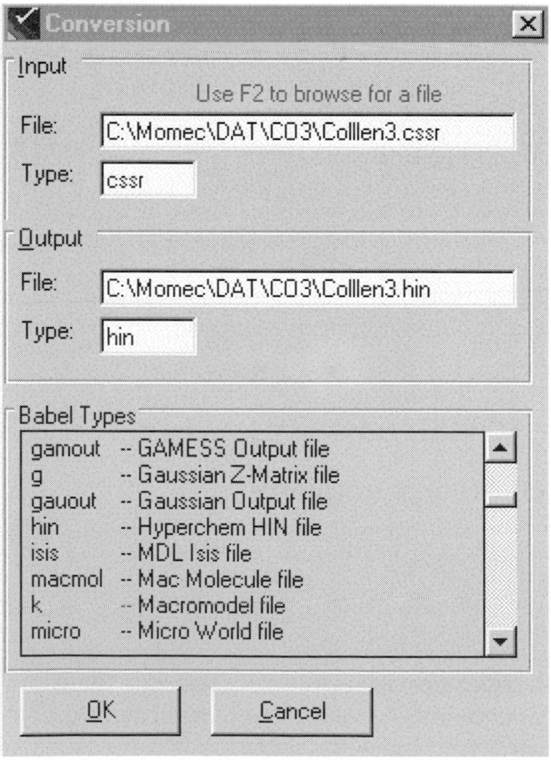

Fig. 17.6.1

1. Select the <u>C</u>onversion option from the <u>E</u>xecute menu of MOMEC.
2. Insert the file name of the foreign file in the input section. Select the file type from the *Babel input* list in *Babel Types*. For example the SHELX or CSD type would commonly be used for importing crystallographic data.
3. A default output filename will be inserted in the output section. This should be altered to have a *.hin* extension and the output file type should be set as *hin*.
4. Select <u>O</u>K and the file will be converted.

### Building Fragments

The example we will consider in this section is the construction of the enantiomers of the anti-cancer active complex [PtCl$_2$(*trans*-chxn)] and its geometric isomer [PtCl$_2$(*cis*-chxn)] (chxn = cyclohexane-1,2-diamine).

*trans*                    *cis*                    **Fig. 17.6.2**

Starting by simply drawing a six-membered ring with nitrogen atoms on adjacent carbon atoms, followed by *Model Build* in HyperChem, may or may not produce the desired isomer, and if the Pt atom is added then boat rather than chair conformations are often produced. Thus, it is best to proceed in a stepwise fashion by building the cyclohexane ring first, adding the amine groups in the desired orientations and then docking this with a square planar platinum framework.

1. Set the default atom to carbon and draw a six-membered ring.

2. Under the <u>B</u>uild menu select the *Add H & <u>M</u>odel Build* option (note: turn off the <u>E</u>xplicit Hydrogens option first). This will generate a cyclohexane ring with a chair conformation. This can be saved as say chxn.hin and used in the future for building similar molecules. Over time a library of commonly used ligand fragments can be generated and this greatly facilitates the building of new molecules.

3. Set the default atom to nitrogen and turn the *Explicit Hydrogens* option on. Locate two equatorial hydrogen atoms on adjacent carbon atoms and transform them into nitrogen atoms by clicking on them with the *draw* tool ⊕. Select *Model Build* under the <u>B</u>uild menu. This will generate *trans*-1,2-cyclohexanediamine with no hydrogen atoms on the amines.

4. Set the default atom to platinum and add a platinum atom between the two nitrogen atoms and connected to both. Select *Model Build* under the <u>B</u>uild menu.

5. Set the default atom to hydrogen and add two hydrogen atoms to each nitrogen atom. Select the hydrogen atoms and choose *Model Build* under the *Build* menu. This demonstrates a useful technique for building difficult systems – that of building only part of the molecule. Selecting the hydrogen atoms and choosing *Model Build* generates the correct geometry for the amine groups. This procedure should generate a 1,2-cyclohexanediamine ligand bound at the appropriate distances to a platinum atom. If the procedure fails at this step with a message "Unable to build only part of a ring or high-coordination system" then set the default atom to carbon and change the nitrogen atoms to carbon atoms and repeat the *Model Build* command with the amine hydrogen atoms still selected. Then set the default atom to nitrogen and change the amine donor groups back to nitrogen atoms.

6. Under the *Build* menu choose the *Calculate Types* option.

7. The file should be saved as, say, pttrchxn.hin. *Cis*-1,2-cyclohexanediamine can be generated similarly by transforming one axial and one equatorial hydrogen atom on adjacent carbon atoms and repeating steps 4 to 6.

8. Having generated the ligands we can now generate the metal center. Open a new workspace. Set the default atom to chlorine and draw four bonds from a common point. Change the default atom to platinum and transform the central atom.

9. Turn on the select tool ⊙ and select the platinum. Under the *Build* menu choose *Constrain Geometry* and select *Square Planar*.

10. Deselect the platinum atom by clicking on it with the right-hand mouse button.

11. In the *Build* menu check that *Explicit Hydrogens* has been selected and then select the *Model Build* option. This will generate a square planar $[PtCl_4]^{2-}$ complex ion.

12. Set the default atom to nitrogen and transform two *cis* related chlorine atoms to nitrogen atoms. Repeating the *Model Build* procedure will generate $[PtCl_2N_2]$ with a *cis* geometry. Starting with two nitrogen and two chlorine ligands at step 8 will sometimes generate the desired *cis* geometry but will equally often produce the *trans* geometry. With more complex systems, the stepwise procedure is essential.

13. Save this file as ptcl2n2.hin.

*Docking Molecules*

There are two ways to dock molecules; manually using translation and rotation functions that act on one and then both molecules or automatically using the *Overlay* option under the *Display* menu. Here we will use the automatic and less cumbersome method.

1. First, the two files must be merged. This is achieved by opening one file, say ptcl2n2.hin and saving it as say pttrcxcl.hin so that the starting file is not acciden-

tally overwritten. Next choose the _Merge_ option under the _File_ menu. This will give you the chance to select a second file for merging. Selecting ptcschxn.hin will result in both the metal center and the ligand appearing on the screen.

2. Turn on the select tool ◎ and on each molecule in turn select the platinum atom and the two nitrogen atoms in that order. This defines two planar groups that can be matched up. Under the _Display_ menu choose _Overlay_. If _Overlay_ is not highlighted then you have not selected the groups correctly – repeat the procedure, remembering to do one molecule then the other and following the same sequence of atoms in each case.

3. Step 2 will have generated a complex with the ligand in the correct position but with duplicate platinum and nitrogen atoms. These can be deleted by selecting them with the select tool ◎ on, then hitting the _Del_ key on the keyboard. Since the excess atoms might be very close together zooming in on them or rotating the molecule may help. The alternative of selecting the draw tool ⊕ and clicking on the excess atoms with the right-hand button of the mouse may result in other atoms being deleted and, therefore, is not recommended in this case. Deleting the excess atoms will result in the disappearance of some bonds. These can be reinserted by selecting the draw tool ⊕ and drawing new bonds between the appropriate atoms with the left-hand mouse button depressed.

4. The target molecule has now been constructed and can be saved as say pttrcxcl.hin.

5. The other enantiomer of this complex can be generated using the methods described in Section 17.3. First, the Pt atom and the two N atoms are selected and used to define a plane using _Name Selection_. Next, this plane is deselected, all ligand atoms are selected and the ligand is inverted using the _Reflect_ option in the _Edit_ window. The configuration of chiral centers can be displayed in HyperChem by selecting _Labels_ under the _Display_ menu and selecting _Chirality_.

6. Repeat steps 1 to 4, using the _cis_ isomer of the 1,2-cyclohexanediamine ligand and save the file as say ptcscxcl.hin.

**Exercise**

Use the methods described in this lesson to generate a model of [Pt(1,3-cyclohexanediamine)$Cl_2$].

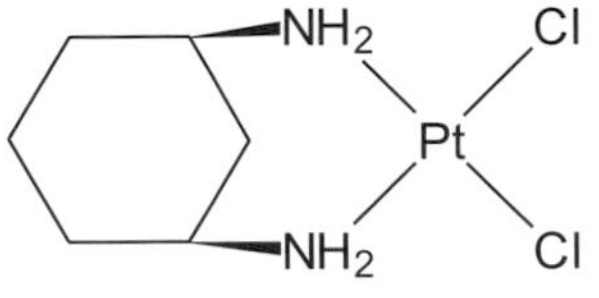

Fig. 17.6.3

## 17.7   Analyzing Structures

**Theory**

The *Structure* module of MOMEC enables you to analyze structures that have been saved as *.hin files. These can be structural data files from experimental work, from a data base (e.g. the CSD) or computed structures such as those optimized with MOMEC. The geometric parameters accessible include the calculation of a least-squares plane (defined by three or more points), the distance of atoms from this plane, the angle between a vector such as a metal-ligand bond and a plane, that between two planes, e.g., for the measure of a trigonal twist angle or a tetrahedral twist angle. In this lesson, we will analyze the structures of the four conformers of $[Co(en)_3]^{3+}$ considered in Sections 17.3, 17.4 and 17.5.

**Practice**

To prepare for this lesson you should have a MOMEC session started with lel$_3$-$[Co(en)_3]^{3+}$ as the input file in *Setup/Files* (note that the relevant file for *Structure* is the *Input* file; thus, you should choose the *.out file as *Input* for the following manipulations). Double click on the *Structure* option in the *Execute* window of MOMEC to open the *Structure* module. The window displayed in Fig. 17.7.1 will appear. Click on the *Define* button. The HyperChem window with the selected input file will open. *Structure* will ask you to select at least 3 atoms to define a plane. Choose the *Select* tool ⊚ of HyperChem, and select the Co and 4 N atoms that lie in one of the three perpendicular coordination planes. Confirm your choice by clicking the *OK* button that appears on the *Structure* window. Your manipulations and results are stored in the *log* screen of the *Structure* window, and the plane function, rms shift and distances of the atoms of the plane are displayed.

There are a number of ways to check the distortion of a hexacoordinate complex from octahedral geometry. You may, for example, measure the angle between the reference plane that you have chosen and another of the three tetragonal planes: click on the *Angle between two planes* button and *Select* Co and 4 N's of a second plane when the program asks you to define the second plane. After confirming your choice, the equation of the new best plane with the corresponding rms shift and the angle between the two planes (89.8°) are displayed.

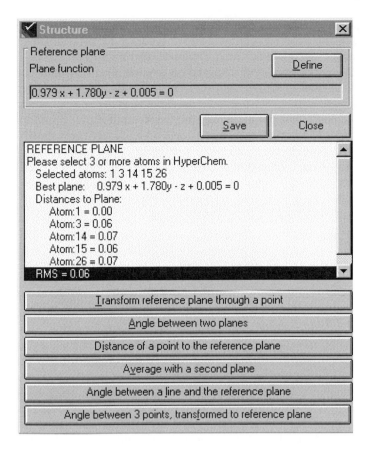

**Fig. 17.7.1**

Another measure of the distortion of $lel_3$-$[Co(en)_3]^{3+}$ from octahedral geometry is the deviation of the angle between a Co-N bonding vector and the plane that you have defined above: Click on the *Angle between a line and the reference plane* button and follow the instruction to *Select* the line, i.e., a Co-N bond, on the HyperChem window. After confirmation of your choice, the value of 87.5° will be displayed.

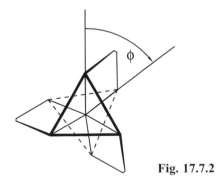

**Fig. 17.7.2**

Possibly the most important geometric parameter for tris-chelate complexes is the trigonal twist angle $\phi$ (Fig. 17.7.2). You can measure $\phi$ by defining the average of two opposite trigonal planes of your octahedral complex: *Define* a new reference plane (one N of each of the three ligands, all disposed *cis* to each other), then choose the *Average with a second plane* option, followed by selecting the other three N atoms. This new plane, appearing in the *log screen* is defined as the new reference plane. Obviously, in this $D_3$ symmetrical molecule, the Co lies in this plane, and all N atoms are at the same distance from the plane, i.e., 1.14 Å. You can measure this value with the *Distance of a point to the reference plane* item. To measure the angle $\phi$, you click on the *Angle between three points, transformed to the reference plane* button, you then select Co and two N's within one chelate ligand (located on opposite sides of the reference plane), and you get the trigonal twist angle $\phi = 53.8°$.

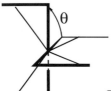

**Fig. 17.7.3**

An important geometric parameter for four-coordinate systems is the tetrahedral twist angle $\theta$ (Fig. 17.7.3). We can use this parameter to inspect the geometry around one of the coordinated amine donors. *Define* a new reference plane with one of the N donors and two of the atoms bound to it (e.g., the Co center and one of the H's). Then, click on the *Angle between two planes* button and follow the instruction to define the other plane (the same N donor again and the other two substituents, i.e., the second H and the C atom bound to the N). The value of the tetrahedral distortion ($\theta = 88.5°$) will then be displayed (the value of $\theta$ obviously depends on the choice of planes).

An important feature of five-membered chelate rings is the amount of puckering. You can check this by computing the angle between the plane defined by the Co center and two N donors of one of the chelate rings, and the line defined by the two C atoms of this ring. For $lel_3$-$[Co(en)_3]^{3+}$ this angle is 27.3°.

**Exercise**

In Sections 17.3 and 17.4 you have optimized the structures of all four conformers of $lel_3$-$[Co(en)_3]^{3+}$. Using the three other conformers, measure the modes of distortion described in this lesson.

## 17.8   Potential Energy Functions I: Bond Length, Valence Angle, Torsion Angle, Twist Angle and Out-of-Plane Deformation Functions

**Theory**

Distorting a bond length or valence angle from its "ideal" value costs energy and the function that gives the energy cost associated with a given degree of distortion is called a potential energy function. The true relationships are complex and are unsymmetrical in that contraction by a given amount will cost a different amount of energy to expansion by the same amount. The basis of molecular mechanics is that mathematically simple functions are used to mimic the true relationship between distortion and energy. In the case of bond length, valence angle, twist angle and out-of-plane deformations, quadratic potential energy functions are typically used, and these have the following form:

$$E_p = \frac{1}{2} k (p - p_o)^2 \qquad (17.8.1)$$

Here $k$ is the force constant that relates the energy that arises from a given degree of distortion, $p_o$ is the ideal value for the parameter and $p$ is the actual value of the parameter. Eq. 17.8.1 is a Hookes Law function and can also be used to describe the energy required to deform a spring. In that case $k$ corresponds to the strength of the spring and $p_o$ corresponds to the ideal length of the spring.

A quadratic function defines a symmetric parabola and therefore cannot exactly reproduce the true relationship between the distortion of a bond length or valence angle and the energy needed to effect that distortion. However, a central assumption in the application of simple molecular mechanics models is that distortions from ideal values are small and in such cases it is only necessary that the potential energy function be realistic in the region of the ideal value. This is shown in Fig. 17.8.1, where a quadratic curve is compared to a Morse potential that is believed to more accurately reflect the relationship between bond length distortion and energy cost.

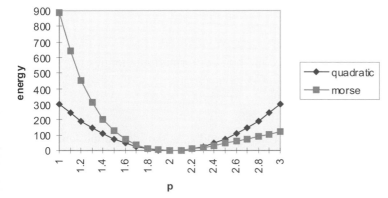

Fig. 17.8.1

## Practice

We will now examine the effect of varying the parameters in Eq. 17.8.1*. In the comparison shown in Fig. 17.8.2, $k = 1.0$ and $p_o = 2.0$. If $k$ is decreased to 0.5 the curve becomes flatter and if $k$ is increased to 2.0 the curve becomes steeper. You can see from these curves that with a larger value of $k$, a given degree of distortion requires a greater amount of energy.

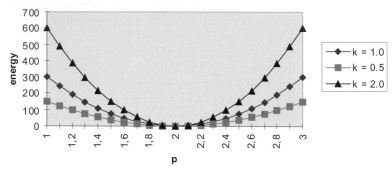

Fig. 17.8.2

Changing the value of $p_o$ to say 1.5 or 2.5 does not change the shape but shifts the minimum down or up, respectively, as shown in Fig. 17.8.3.

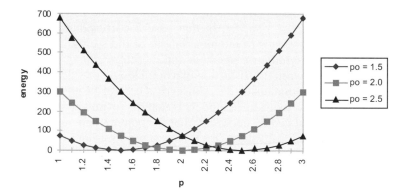

Fig. 17.8.3

*Bond Length Deformation*

The form of the equation used to calculate bond deformation energies is,

$$E_b = \frac{1}{2} k_b \left( r_{ij} - r_0 \right)^2 \tag{17.8.2}$$

where $k_b$ is the force constant or bond strength and $r_0$ is the ideal bond length.

---

\* Note that plotting of potential energy functions can be done with a conventional mathematics or table calculation program such as Mathematica®, Maple®, Excel®, etc.

To show how we use Eq. 17.8.2 to calculate strain energies we will insert some typical values. For instance, the MOMEC97 force constant for a Pt(II)-N(amine) bond is 2.54 mdyn $\text{Å}^{-1}$ and the ideal bond length is 2.03 Å. The units for the energy term (mdyn $\text{Å}^{-1}$) are not SI units but they can be converted to kJ $\text{mol}^{-1}$ by multiplying by 602.5 (see Appendix 2). The units of mdyn $\text{Å}^{-1}$ for a stretching force constant are employed as input for MOMEC and the conversion of the output energies takes place within the program. Thus, the equation used to calculate the energy required to deform a Pt-N bond of this type is:

$$E_b = \frac{1}{2} \cdot 602.5 \cdot 2.54 \cdot (r - 2.03)^2 \qquad \text{or}$$

$$E_b = 301.25 \cdot 2.54 \cdot (r - 2.03)^2 \qquad\qquad (17.8.3)$$

For example, if the Pt-N bond is stretched to 2.06 Å the energy required is:

$$
\begin{aligned}
E_b &= 301.25 \cdot 2.54 \cdot (2.06 - 2.03)^2 \text{ kJ mol}^{-1} \\
&= 301.25 \cdot 2.54 \cdot 0.0009 \text{ kJ mol}^{-1} \\
&= 0.689 \text{ kJ mol}^{-1}
\end{aligned}
$$

*Valence Angle Deformation*

The energy associated with distortion of valence angles is calculated in a similar way but it is important to note that ideal valence angles are entered into MOMEC in radians. For example, for an aliphatic C-C-C group the ideal angle is 109.4° which is inserted as 1.911 and the force constant is 0.45 mdyn Å $\text{rad}^{-2}$. Thus, the energy associated with distortion of such an angle to 115° is calculated as:

$$
\begin{aligned}
E_\theta &= 301.25 \cdot 0.45 \cdot (\theta - 1.911)^2 \text{ kJ mol}^{-1} \\
&= 301.25 \cdot 0.45 \cdot (2.007 - 1.911)^2 \text{ kJ mol}^{-1} \\
&= 301.25 \cdot 0.45 \cdot 0.00924 \text{ kJ mol}^{-1} \\
&= 2.8 \text{ kJ mol}^{-1} \qquad\qquad (17.8.4)
\end{aligned}
$$

MOMEC offers a number of possibilities for modeling valence angles, and these are of particular interest for the computation of angles around metal centers. However, all of these approaches can be used for any of the angles in a molecule. The choice includes:

- harmonic potential energy functions of the type discussed above
- multiple harmonic potentials, e.g., for octahedral and square planar compounds, where minima of 90° and 180° are found
- 1,3-nonbonded interactions, i.e., the angular geometry is established by ligand-ligand repulsion
- plane twist functions, e.g., to enforce square planar geometry for four-coordinate compounds

Note that a combination of various types of potentials can be activated. For example, the coordination geometry for octahedral transition metal compounds can be modeled by 1,3-nonbonded interactions in combination with a multiple harmonic function. This is the approach used in the MOMEC97 force field for a number of metal complexes. Also, for four-coordinate compounds one or both of these potentials can be combined with a plane twist potential that enforces square-planar geometry.

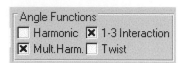

**Fig. 17.8.4**

The type of potential is chosen in the *Edit/View; Force Field; Atom Type Parameters* window of MOMEC. Note that the corresponding functions also need to be activated in the *Setup/Optimization controls* menu.

### Out-of-plane Deviations

Out-of-plane deviation functions are used when a group of four atoms is most stable when they are coplanar. This is the case for groups involving multiple bonds such as alkenes or conjugated systems such as carboxylate groups. The deviation is usually defined as the angle between the vector from the central atom to the fourth atom and the plane defined by the other three atoms.

**Fig. 17.8.5**

A quadratic term is used to calculate the energy for this deformation but the function differs from those above because the ideal value is zero (corresponding to planarity) and therefore does not appear in the equation.

$$E_\delta = \frac{1}{2} k_\delta \, (\delta)^2 \qquad (17.8.5)$$

Here too, $k_\delta$ is a force constant that determines the amount of energy that must be expended to achieve a certain distortion.

### Torsion Angle or Dihedral Angle Functions

Torsion or dihedral angles are angles involving a sequence of four atoms and can be defined as the angle between the plane defined by the first three atoms in the sequence and that defined by the last three atoms in the sequence. Alternatively, it can be viewed as the "twisting" angle about the central bond – the angle through which one bond must be rotated in order to overlay the other bond. This is shown schematically in Fig. 17.8.6 in lateral (a) and Newman (b) projections.

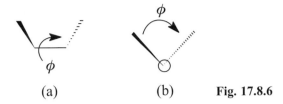

(a)                    (b)          **Fig. 17.8.6**

If the four atoms (and three bonds) are coplanar and the two terminal atoms are cis to each other then the torsion angle is defined as zero. Torsion angles have a sign that is determined from the direction of rotation needed to bring the front bond into superposition with the rear bond. If the direction of rotation is clockwise, as shown in part (b) of Fig. 17.8.6, then the torsion angle is said to be positive and if it is anti-clockwise then the angle is negative.

Rotation about the central bond costs energy and there are two components to this; one arises from the repulsion between the first and fourth atoms in the sequence and the other is intrinsic to the bond. The former component is included in the nonbonded interactions of the molecular mechanics model but the other component must be explicitly included by means of a torsion angle potential energy function.

The simplest example is ethane; the torsional strain energy of ethane is at a minimum when the torsion angles are 60°, 180° and 300° and at a maximum when they are 0°, 120° and 240°, and these are the staggered and eclipsed conformations, respectively.

staggered conformation   eclipsed conformation       **Fig. 17.8.7**

The three fold symmetry of ethane along the central bond gives rise to a periodicity in the energy as a function of the torsion angle. Thus, a sinusoidal curve describes this relationship, and the energy can be calculated using a function of the type:

$$E_\phi = \frac{1}{2} k_\phi \left(1 + \cos\left(3\,\phi_{ijkl}\right)\right) \tag{17.8.6}$$

As before, this is a harmonic function, where $k_\phi$ is the force constant that relates the energy that arises from a given degree of distortion to the torsion angle. The factor of three gives the appropriate multiplicity and the result is the curve shown in Fig. 17.8.8.

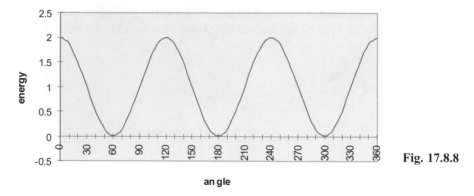

Fig. 17.8.8

an gle

As you can see in Fig. 17.8.8, this function gives energy minima at 60°, 180° and 300° and maxima at 0°, 120° and 240°. Increasing the value of $k_\phi$ increases the height of the curve. This equation is suitable when the atoms forming the central bond are both tetrahedral, giving rise to the three-fold symmetry seen in ethane but, where this is not the case, a more complex function is required as shown in Eq. 17.8.7.

$$E_\phi = \frac{1}{2} k_\phi \left(1 + \cos\left(m\left(\phi_{ijkl} + \phi_{\text{offset}}\right)\right)\right) \tag{17.8.7}$$

In this equation, $m$ is the multiplicity, the number of times the curve is repeated in one complete rotation about the central bond (3 in the example of ethane), and $\phi_{\text{offset}}$ is used to shift the energy minimum. For example, in the case of a torsion angle about a bond between an sp$^3$ hybridized atom and an sp$^2$ hybridized atom the periodicity is 6 (arising from the atoms having three-fold and two-fold symmetry, respectively), and a function of the type:

$$E_\phi = \frac{1}{2} k_\phi \left(1 + \cos\left(6\phi_{ijkl}\right)\right) \tag{17.8.8}$$

produces a curve with minima at 30°, 90°, 150° etc., as shown in Fig. 17.8.9.

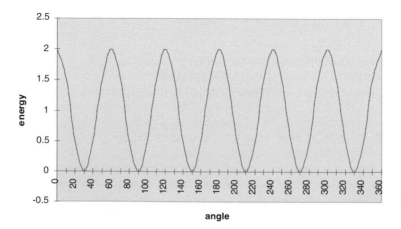

angle

Fig. 17.8.9

However, the true minima are at 60°, 120°, 180° etc. Thus, it is necessary to apply an offset of 30° or 0.524 radians, giving equation 17.8.9 and the curve shown in Fig. 17.8.10.

$$E_\phi = \frac{1}{2} k_\phi \left(1 + \cos\left(6\left(\phi_{ijkl} + 0.524\right)\right)\right) \tag{17.8.9}$$

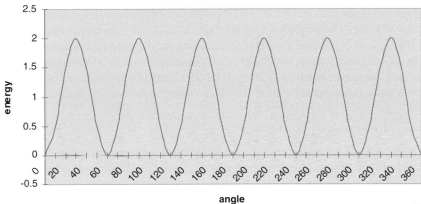

**Fig. 17.8.10**

Note that the twist angle potential will be discussed and used separately in Section 17.14.

**Exercise**

Calculate the strain energy that results from distorting one of the oxygen atoms in a carboxylate 7.5° out of the plane defined by the other three atoms, using Eq. 17.8.5, and given that $k_\delta = 0.5$ mdyn Å rad$^{-2}$.

Determine the appropriate multiplicity that would be used to calculate the torsion deformation energy arising from rotation about a Co-N bond in $[Co(NH_3)_6]^{3+}$. (Note: it is generally assumed that the energy barrier to such a rotation is so low that it need not be included in the force field).

Use Eq. 17.8.9 to calculate the strain energy arising from a torsion angle of 40°, given that $k_\phi = 2$. Compare the value that you obtain with the value estimated from the plot shown in Fig. 17.8.10.

## 17.9    Potential Energy Functions II: Nonbonded Interactions

**Theory**

Pairwise interactions between atoms not connected by a formal bond can have a number of components. There is always what is usually referred to as the van der Waals interaction which has attractive and repulsive components. If both atoms are charged or carry a significant partial charge then there will be an electrostatic component. Lastly, if one atom has a lone pair of electrons capable of accepting a hydrogen atom and the other atom is a hydrogen atom which is part of an acidic group then a hydrogen bond can form. The molecular mechanics treatment of these three components is very different and thus we will consider each one separately.

*van der Waals Interactions*

The van der Waals interactions are repulsive at short and attractive at long distances. The energy minimum is at the sum of the van der Waals radii. The repulsive component arises from overlap of electron clouds and mutual repulsion of the nuclei, the attractive component arises from interactions between dipoles and multipoles. A number of functions have been used to mimic these components but the most popular fall into two groups, the Lennard-Jones potential (shown in Eq. 17.9.1 in the $6-12$ form) and the Buckingham potential (Eq. 17.9.2).

$$E_{vdw} = \frac{A}{d_{ij}^{12}} - \frac{C}{d_{ij}^{6}} \qquad (17.9.1)$$

$$E_{vdw} = Ae^{-Bd} - Cd^{-6} \qquad (17.9.2)$$

In both equations, $d$ is the separation between the atoms. The Lennard-Jones potential is simpler and computationally less demanding and is therefore favored for models of macromolecules such as proteins and DNA. The Buckingham function more closely resembles the energy relationship and is preferred when higher accuracy is required. The latter function is available in MOMEC and we will concentrate on this.

 $A$ and $C$ in both Eqs 17.9.1 and 17.9.2 define the steepness of the repulsive and attractive components, respectively. $B$ in Eq. 17.9.2 defines the minimum of the curve or the van der Waals radius and is calculated using:

$$B = 12.50/(vdWr_i + vdWr_j) \qquad (17.9.3)$$

Here, $vdWr_i$ and $vdWr_j$ are the van der Waals radii of atoms $i$ and $j$, respectively. It is generally assumed that $A$ and $C$ are related and can be calculated from a single value. This is the basis of the Hill equation and the common factor is referred to as the polarizability, $\varepsilon$. $A$ and $C$ are calculated from $\varepsilon$ using:

$$A = 2014 \, (\varepsilon_i \cdot \varepsilon_j)^{1/2} \tag{17.9.4}$$

$$C = (2.25 \, (\varepsilon_i \cdot \varepsilon_j)^{1/2} (vdWr_i + vdWr_j)^6)/144 \tag{17.9.5}$$

**Practice**

For a H···H interaction in the MOMEC force field the van der Waals radius is 1.44 Å and the $\varepsilon$ value is 0.044 for each hydrogen atom. These give rise to $A$, $B$ and $C$ values of 48.34, 4.34 and 0.214, respectively, and Eq. 17.9.2 becomes:

$$E_{vdw} = 48.34 \, e^{-4.34d} - 0.214 \, d^{-6} \tag{17.9.6}$$

If the separation is 2.4 Å the energy is calculated to be 0.197 kJ mol$^{-1}$. Shown in Fig. 17.9.1 is the energy versus separation curve for this H···H interaction.

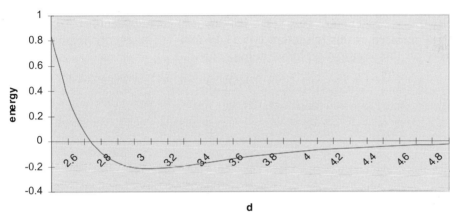

**Fig. 17.9.1**

Note, that the minimum of this curve lies at the van der Waals radius sum of 2.88 Å. Below this value the energy increases rapidly as the atoms come into close contact. Above the van der Waals radius the energy asymptotes to zero at an infinite separation. Above about 2.6 Å the energy is negative which means the attractive component dominates and the net van der Waals interaction is stabilizing, below 2.6 Å the interaction is destabilizing.

**Theory**

*Hydrogen Bonds*

Hydrogen bonds occur when an atom with a lone pair of electrons, capable of accepting a hydrogen atom, is in the vicinity of a hydrogen atom that is part of an acidic group. Modeling hydrogen bonds is difficult because there is a continuum of bond strengths ranging from interactions that are primarily electrostatic to those where the hydrogen atom is shared equally between the donor and acceptor atom. Thus, hydrogen bonds are usually modeled using an empirical function that reproduces the hydrogen bonding distance and returns the correct amount of energy at this distance. However, it cannot always be expected to reliably reproduce the relationship between energy and distance nor can it always predict the hydrogen bonding separation. The equation usually used is similar to the Lennard-Jones potential in that it has attractive and repulsive components but the exponents give a steeper curve.

$$E_{hb} = Cd^{-12} - Dd^{-10} \tag{17.9.7}$$

**Practice**

The variables $C$ and $D$ are chosen to give the desired separation and energy. For example, if $C = 234.0$ and $D = 66.8$, the minimum in energy occurs at 2.05 Å, and the energy released at this minimum is 5.11 kJ mol$^{-1}$. The energy versus separation curve for this example is shown in Fig. 17.9.2.

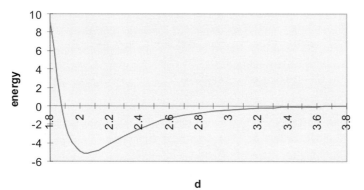

Fig. 17.9.2

**Theory**

*Electrostatic Interactions*

Nearly all atoms in complex molecules carry a residual charge; those that have a charge with the same sign will repel one another and those that have oppositely signed charges will attract one another. The equation that relates the separation of the atoms ($d$) and the charges ($q_i$, $q_j$) they carry to the energy of the attraction or repulsion is well known:

$$E_\varepsilon = (q_i \cdot q_j)/(\varepsilon \cdot d) \tag{17.9.8}$$

In this equation, $\varepsilon$ is the dielectric constant, which is 1 in a vacuum but is substantially larger in solution or the solid state, and one of the difficulties in molecular mechanics studies is deciding on an appropriate value for the dielectric constant. Values between 2 and 8 are typically used but more recently, distant dependent values, such as $4d$, have been shown to best reproduce protein crystal structures[146] and have become increasingly popular. The other difficulty in applying electrostatic interactions lies in deriving the residual charges on the atoms. For small organic molecules this is done by quantum mechanics calculations and for proteins and DNA it is assumed that calculations on the component parts can be transferred to the macromolecule. However, reliable quantum mechanics calculations on even moderately large metal complexes are still rather expensive. Thus, the inclusion of electrostatic interactions in models of metal complexes remains problematic. Additionally, in most metal complexes the electrostatic interactions are not believed to be significant because the major residual charge lies on the metal atom and this produces a symmetrical field about the metal. Consequently, the tendency is to omit electrostatic interactions, unless there is a clear need to include them. This may, for example, be necessary in oligonuclear complexes where more than one metal center is present. Electrostatic interactions are also important in macromolecules such a proteins and DNA and therefore, when modeling such molecules, whith metals present, these interactions should be included.

**Practice**

Consider the repulsion between two atoms, each with a residual charge of 0.2 electrons and separated by 3.0 Å. Eq. 17.9.8 becomes:

$$E_{\varepsilon} = (0.2\,\text{e} \cdot 0.2\,\text{e})/(\varepsilon \cdot 3.0\,\text{Å})\tag{17.9.9}$$

However, the value returned will not have units of kJ mol$^{-1}$ and a multiplier of 1390.0 must be applied (see Appendix 2). If $\varepsilon$ is taken to be 4 then we calculate an energy of 4.63 kJ mol$^{-1}$, and if $\varepsilon$ is taken to be $4d$ then this becomes 1.54 kJ mol$^{-1}$. The greater the value of $\varepsilon$ the smaller the electrostatic energy that is calculated – because the higher the dielectric constant, the more the electrostatic interaction is damped. Shown in Fig. 17.9.3 are curves for electrostatic interactions between these atoms for $\varepsilon = 4$ and for $\varepsilon = 4d$.

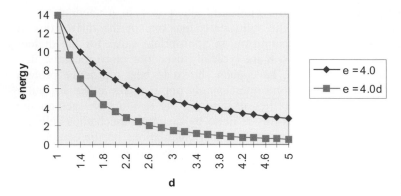

Fig. 17.9.3

Note that the energy varies much more slowly as a function of the interatomic separation than is the case for van der Waals or hydrogen bonding interactions. For example, if the interatomic separation is increased to 4.0 Å the electrostatic interaction energies referred to above become 3.48 and 0.869 kJ mol$^{-1}$, respectively. Note also, that the energy varies more rapidly if $\varepsilon = 4d$ because the relationship is inverse squared with respect to separation, rather than a simple inverse.

**Exercise**

Consider the three-atom system shown in Fig. 17.9.4 with the data listed below.

O    H⋯⋯⋯O    **Fig. 17.9.4**

The partial charge on each O atom is $-0.2$ e and the partial charge on the H atom is 0.1 e. The O⋯H separation is 1.9 Å and the O⋯O separation is 2.8 Å. The $\varepsilon_i$ values for H and O are 0.046 and 0.052, respectively and the van der Waals radii are both 1.44 Å. The $C$ and $D$ values for an O⋯H hydrogen bond are 234.0 and 66.8, respectively.

Use this information to calculate the van der Waals, hydrogen bonding and electrostatic contributions to the energies of the O⋯O and O⋯H interactions.

# 17.10   Force Field Parameters I: Developing a Force Field for Cobalt(III) Hexaamines – Normal Bond Distances

**Theory**

It is important to note that a force field is a highly correlated entity, and all parts of it, that is, the potential energy functions and the set of parameters, are interdependent: if one function or one parameter is changed, others may have to be redetermined. It is clear then that force field parameters do not necessarily have a real physical significance. The important message is: do not add force field parameters from a different force field, even if that force field is well established and produces high quality results, without carefully checking and refining them in your force field. In cases, where different types of functions are used, e.g., the Lennard-Jones instead of the Buckingham potential for van der Waals interactions, or if stretch-bend cross terms are added, it is obvious that quite different parameter sets have to be used. However, even for parameterization schemes based on identical potential energy functions, these may differ considerably.

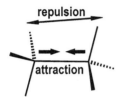

**Fig. 17.10.1**

It is relatively easy to develop a force field that models the structure of a simple molecule, such as ethane (Figure 17.10.1). The computed C-C bond distance is influenced by the van der Waals repulsion of the protons (H···H), the stretching force constant $k_b$ of the C-C bond and the corresponding strain free distance $r_0$. Thus, increasing the H···H repulsion, leading to an elongation of the C-C bond, requires that the $r_0$ value of the C-C bond be reduced, or the force constant $k_b$ be increased, or both, to reproduce the correct C-C bond length of 1.532 Å. Obviously, with a general and constant parameterization of the van der Waals term, fitting of the C-C based parameters ($k_b$ and $r_0$) to one structure, i.e., that of ethane, will not lead to a unique combination of parameters. For a general parameter set $k_b$ and $r_0$, which can be used to model a whole series of alkanes, the development of the force field has to be based on a broader body of experimental data. Since molecular mechanics modeling is an interpolative method, the choice of data for the development of the force field determines the applicability of the parameter set.

It also emerges that parameter schemes based on thermodynamic (heats of formation, isomer distributions), spectroscopic (vibrational force constants) or structural experimental data are different and primarily useful for modeling the corresponding properties. The MOMEC force field is largely based on structural data.

However, some of the parameters have also been refined against thermodynamic and/or spectroscopic data. Thus, the hexaaminecobalt(III) force field leads to accurate structural and thermodynamic predictions. Since all parameters are highly correlated, i.e., the parameterization of the ligands and that of the chromophore, the prediction of thermodynamic properties of complexes with other metal centers is also expected to be reasonably accurate. However, this may have to be tested separately for each set of compounds.

An important point, related to force fields based on X-ray structural data is that the optimized structures are related to molecular structures in a crystal lattice. Thus, the suggestion that the computations lead to "gas phase structures" is wrong, even if the environment is not explicitly considered. Obviously, the anisotropy of a "real" environment will be lost, and the modeled structures are generally more highly symmetrical than the geometries of related experimentally determined crystal structures. Thus, for judging the quality of a computed structure, you should always compare averages of symmetrically related structural parameters of the experimental structure with the corresponding computed parameters.

You might have observed another fact related to the symmetry of computed structures: $[Co(NH_3)_6]^{3+}$ that you have refined in Section 17.2 is not as highly symmetrical as you might have anticipated, in that the N-Co-N valence angles differ from the theoretically expected $90°$ and $180°$. This is not unexpected, if you think about it: the molecular cation does not have $O_h$ symmetry, this is violated by the trigonal structure of the $NH_3$ ligands, and the resulting overall symmetry is $D_{3d}$ at most, depending on the conformer considered (frozen orientation of the six ammonia ligands). Again: use average structural parameters to compare the computed geometry with experimental data.

One of the major problems facing a molecular modeler is the development of a parameter set for a new type of compound. As a simple exercise for this lesson we consider the cobalt(III)-amine bond within the framework of the MOMEC potential energy functions and the known parameterization scheme for the organic backbone of the ligands. With the approach of modeling angular geometries with 1,3-nonbonded interactions the only metal-donor dependent and therefore unknown parameters are $k_b(Co^{III}\text{-amine})$ and $r_0(Co^{III}\text{-amine})$ (the metal-donor-backbone valence angle potential, i.e., $Co^{III}$-N-C, is assumed to be metal ion independent). With the combined model, involving 1,3-nonbonded interactions and the multiple harmonic angle function for the angular geometry involving the metal center, the parameters for the latter function are generic, i.e., they must not be fitted to experimental structural data. Once we get $k_b(Co^{III}\text{-amine})$ and $r_0(Co^{III}\text{-}$ amine) we can model any hexaaminecobalt(III) compound within the interpolation range given by the range of test structures.

The first task is to get reasonable starting values for these parameters. The undeformed (ideal) bond distance $r_0$ can be estimated from the average of known experimental bond distances of relatively undistorted, compounds. Experimental bond lengths are usually elongated by $1-10\%$ due to steric strain (e.g., mutual repulsion of the substituents – remember the picture of ethane above). A first approximation to the force constant $k_b$ can be calculated from the fundamental vibration frequency $v$, taken from the vibrational spectrum of a representative compound, e.g., $[Co(NH_3)_6]^{3+}$ (Eq. 17.10.1).

$$k_b = \mu \, (2 \, v\pi c)^2 \qquad\qquad (17.10.1)$$

Here, $v$ is the frequency of the stretching mode ($v$ = 494 cm$^{-1}$ for Co$^{III}$-N), $c$ is the speed of light ($2.998 \cdot 10^8$ m s$^{-1}$), and $\mu$ is the reduced mass ($\mu = m_1 \cdot m_2 / (m_1 + m_2)$).

The starting values of the parameters $r_0$ and $k_b$ can be added to the force field parameter files, then test structures (e.g., $[Co(NH_3)_6]^{3+}$, $[Co(en)_3]^{3+}$ etc.) can be computed and the relevant structural parameters (i.e., Co$^{III}$-N distances) are compared with experimental data. The structural differences (experimental vs. calculated) are then used to optimize the force field parameters, i.e., the parameters $r_0$ and $k_b$ are varied to minimize the structural differences. It is clear that this will be more difficult with a large set of very different structural types (extremely short and long Co$^{III}$-N bonds) but the fitted parameter set will be more generally applicable, i.e., the interpolation range will be larger. Thus, it is preferable to use as large a set of experimental structures as possible to develop a new force field. The best approach is to load all relevant structures from the Cambridge Structural Data Base (CSD) and to refine the whole set as a batch job in each cycle of force field parameter adjustment. To demonstrate the dangers of using limited structural data sets for force field fitting, we restrict ourselves in this lesson to the structures of $[Co(NH_3)_6]^{3+}$ and $lel_3$-$[Co(en)_3]^{3+}$. Get these structures from the CSD if it is available to you, and convert them to *.hin files using the *Conversion* module of *MOMEC*, available in the *Execute* window (see Section 17.6).

There are various ways to compare structures and, more specifically, to list the relevant differences between computed and experimental structural data: (i) you can use the *Overlay* module in the *ChemPlus* extension of HyperChem (*Overlay* in the basic version of HyperChem does not provide rms differences and thus is not useful for parameter fitting); (ii) we suggest that you make a table of the relevant parameters; (iii) the developers kit in HyperChem 5.0 allows for coupling HyperChem with external programs, such as Excel®, and this may facilitate the production and analysis of such tables.

**Practice**

In Sections 17.10–17.13 we will manipulate the force field. Make sure that you do not lose the parameters supplied with MOMEC. Therefore, copy all force field files (momec*.txt) with the usual Windows98® tools (e.g., Explorer) from the directory that you use for MOMEC (e.g., c:\momec or c:\momec97\parm) to a subdirectory (e.g., c:\momec\user\lesson10) and choose this subdirectory as the present path for the force field: click on the MOMEC menu item *Setup* and then on *Force Field*. With the F2 key you can choose the path (c:\momec\user\lesson10). Confirm with *OK*.

Open the menu item *Edit/View*, there *Force Field* and there *Bond Stretch Parameters*. Move the cursor to the row CO3-NT, click on this line, click on the button *Delete*, confirm and leave the *Force field* editor with *OK*. Now, your MOMEC force field does not have any parameters for $Co^{III}$-amine bonds: you are ready to develop your own parameterization! Actually, see what happens if you try to refine a molecule with a missing parameter: Open the $[Co(NH_3)_6]^{3+}$ – hin file from Section 17.1 (*Setup/Files*) or from the CSD and refine it (set the maximum number of cycles (*Setup/Optimization Controls*) to 2; *Execute/Geometry Optimization*). At the bottom of the *Summary* window the Force Field Messages will tell you that there is no bond stretch function for atom types: CO3 NT.

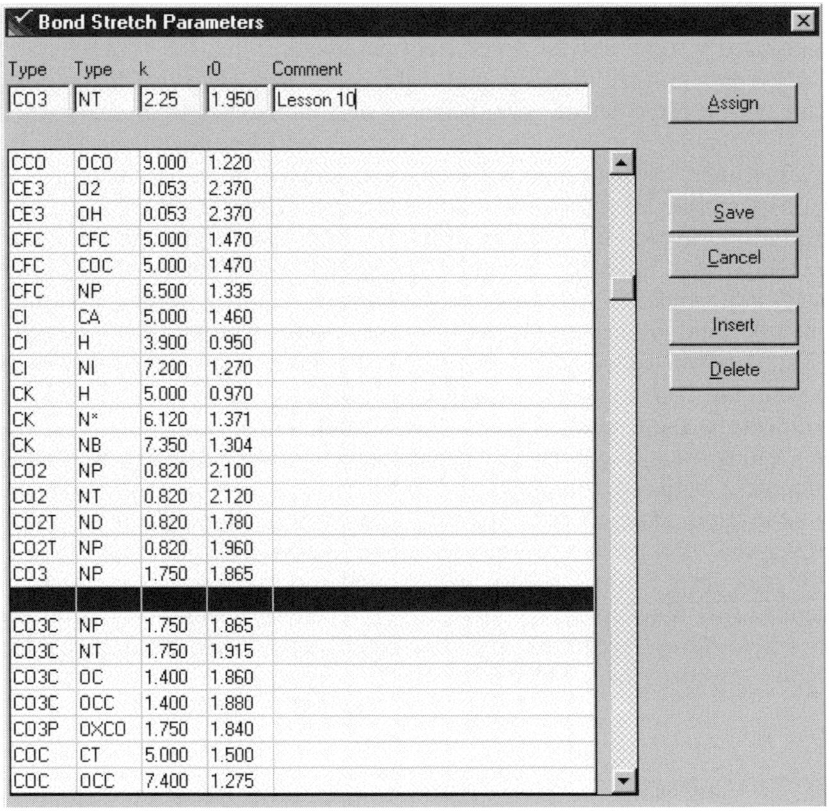

| Type | Type | k | r0 | Comment |
|------|------|------|------|---------|
| CO3 | NT | 2.25 | 1.950 | Lesson 10 |

| Type | Type | k | r0 | Comment |
|------|------|-------|-------|---------|
| CCO | OCO | 9.000 | 1.220 | |
| CE3 | O2 | 0.053 | 2.370 | |
| CE3 | OH | 0.053 | 2.370 | |
| CFC | CFC | 5.000 | 1.470 | |
| CFC | COC | 5.000 | 1.470 | |
| CFC | NP | 6.500 | 1.335 | |
| CI | CA | 5.000 | 1.460 | |
| CI | H | 3.900 | 0.950 | |
| CI | NI | 7.200 | 1.270 | |
| CK | H | 5.000 | 0.970 | |
| CK | N* | 6.120 | 1.371 | |
| CK | NB | 7.350 | 1.304 | |
| CO2 | NP | 0.820 | 2.100 | |
| CO2 | NT | 0.820 | 2.120 | |
| CO2T | ND | 0.820 | 1.780 | |
| CO2T | NP | 0.820 | 1.960 | |
| CO3 | NP | 1.750 | 1.865 | |
| | | | | |
| CO3C | NP | 1.750 | 1.865 | |
| CO3C | NT | 1.750 | 1.915 | |
| CO3C | OC | 1.400 | 1.860 | |
| CO3C | OCC | 1.400 | 1.880 | |
| CO3P | OXCO | 1.750 | 1.840 | |
| COC | CT | 5.000 | 1.500 | |
| COC | OCC | 7.400 | 1.275 | |

Buttons: Assign, Save, Cancel, Insert, Delete

**Fig. 17.10.2**

The range of force constants $k_b$ for $Co^{III}$-amine that have been used in the literature vary from 1.7 to 2.25 mdyn $Å^{-1}$ (MOMEC97 uses 1.750 mdyn $Å^{-1}$), and the reported values for $r_0$ vary from 1.905 to 1.950 Å (MOMEC97 uses 1.905 Å). In this lesson, we will fit the two missing parameters to the two structures of $[Co(NH_3)_6]^{3+}$ (Section 17.2) and lel$_3$-$[Co(en)_3]^{3+}$ (Section 17.4) with reported Co-N distances of 1.961 and 1.964 Å, respectively. Since these distances are very similar to each other, select a value close to this distance as the starting value for the ideal bond distance, i. e., 1.950 Å and a relatively stiff force constant, i. e., 2.25 mdyn Å$^{-1}$.

Return to the menu item *Bond Stretch Parameters* under *Edit/View* and *Force Field*. Go back to the row that you have deleted, click on the *Insert* button and type in the top row CO3, NT, 2.25, 1.950 and Section 17.10. Confirm these entries with *Assign* and *Save* the changes to the force field. You are now ready to refine the two molecules as you have learnt in Sections 17.2 and 17.4 (do not forget to set the number of cycles (*Setup*; *Optimization Controls*) back to 20; refine the molecules as a *Batch-Job* and see the results with *Batch-Job Results* under *Edit/View* (see Section 17.4).

**Table 17.10.1**

| Source | $[Co(NH_3)_6]^{3+}$ [Å] | $lel_3$-$[Co(en)_3]^{3+}$ [Å] |
|---|---|---|
| Experimental | 1.961 [53] | 1.964 [58] |
| *MOMEC97* | 1.955 | 1.963 |
| 2.25/1.950 | 1.980 | 1.980 |
| 2.25/1.927 | 1.964 | 1.964 |

$[Co(NH_3)_6]^{3+}$ and $lel_3$-$[Co(en)_3]^{3+}$ refine with over long bonds (1.98 Å instead of 1.96 Å each, see Table 17.10.1). The obvious thing to do now is to decrease $r_0$. You can do that as before, in the *Force Field Editor*. However, a more efficient method is to directly edit the *Parameter Array*. To do that, you have to be sure that the correct *Interaction Array* has been stored. This array can be built in the menu item *Build Interactions* under *Tools*. You can then edit the force field parameters in the *Parameter Array* field of *Edit/View*. Click the corresponding parameter (STR CO3 NT, i.e., the stretching interaction between cobalt(III) and aliphatic amine donors) and enter the new values in the *Change Values* line; *Assign* the parameter set and *Save* them.

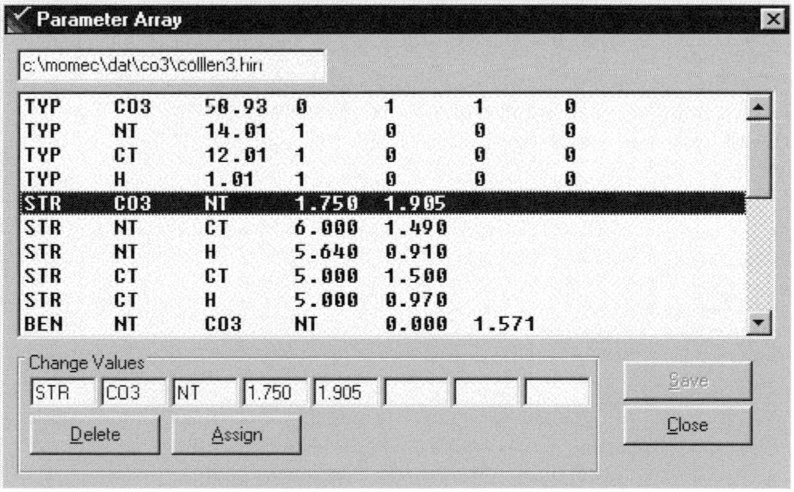

**Parameter Array**

c:\momec\dat\co3\colllen3.hin

| TYP | CO3 | 50.93 | 0 | 1 | 1 | 0 |
|---|---|---|---|---|---|---|
| TYP | NT | 14.01 | 1 | 0 | 0 | 0 |
| TYP | CT | 12.01 | 1 | 0 | 0 | 0 |
| TYP | H | 1.01 | 1 | 0 | 0 | 0 |
| STR | CO3 | NT | 1.750 | 1.905 | | |
| STR | NT | CT | 6.000 | 1.490 | | |
| STR | NT | H | 5.640 | 0.910 | | |
| STR | CT | CT | 5.000 | 1.500 | | |
| STR | CT | H | 5.000 | 0.970 | | |
| BEN | NT | CO3 | NT | 0.000 | 1.571 | |

Change Values

STR | CO3 | NT | 1.750 | 1.905

Delete    Assign    Save    Close

**Fig. 17.10.3**

In the *Setup Files* menu you must then activate the *Use Interaction Parameter Files* option.

**Fig. 17.10.4**

With $k_b$ = 2.25 mdyn $\text{Å}^{-1}$ and $r_0$ = 1.927 Å the computed bond distances are more accurate (1.964 Å each). Actually, these values are even better than those with the original MOMEC97 force field (see Table 17.10.1).

**Exercise**

You may now want to further modify your $Co^{III}$-N force field. One instructive possibility is to replace the modified "point-on-a-sphere" model (1,3-interactions around the metal center, perturbed with a multiple harmonic potential, see Section 17.8 and Section 3.2.2) by a more conventional angle function. To do this, you have to go back to the force field editor and modify the *Angle Functions* setup in the *Atom Type Parameters* menu. You can simply choose the *Mult.Harm.* option, i.e., model the angular geometry without 1,3-nonbonded interactions. A reasonable initial value for the $k_\theta$ parameter of the NT-CO3-NT valence angle (edited in the *Valence Angle Parameters* item of the *Edit/View/Force Field* menu) is 0.68 mdyn Å $\text{rad}^{-2}$ (this is a typical parameter for amine-$Co^{III}$-amine bending, used in force fields not involving the 1,3-nonbonded interactions). This parameter can obviously be refined to get better results. The resulting Co-N distances are slightly smaller since the ligand-ligand repulsion is now removed. Obviously, this effect is larger when you use the original MOMEC parameter set ($k_b$ = 1.75 mdyn $\text{Å}^{-1}$, $r_0$ = 1.905 Å). This is the reason why the ideal metal-donor distances $r_0$ are shorter in MOMEC than in force fields that do not use 1,3-interactions to model the geometry of the chromophore. Test this!

As an additional exercise you should calculate the estimated force constant from Eq. 17.10.1.

## 17.11  Force Field Parameters II: Refining the New Force Field – Very Short Bond Distances

**Theory**

The two cobalt(III) hexaamines with relatively short metal-donor distances considered here are [Co(*trans*-diammac)]$^{3+}$ (*trans*-diammac is *trans*-1,4,8,11-tetraazacyclotetradecane-6,13-diamine, see Fig. 17.11.1) and [Co(trap)$_2$]$^{3+}$ (trap is 1,2,3-triaminopropane, see Fig. 17.11.2). There are three conformers of [Co(*trans*-diammac)]$^{3+}$ ($\delta\delta$, $\delta\lambda$, $\lambda\delta$; where $\delta$ and $\lambda$ refer to the conformation of the five-membered chelate rings in the complex – see also Section 17.3). The $\lambda\delta$-conformer is the most stable form and has been characterized by an X-ray diffraction study. $\lambda\delta$-[Co(*trans*-diammac)]$^{3+}$ has very short Co-N bonds (1.937Å (four equatorial bonds), 1.946 Å (two axial bonds)), and the experimentally determined high ligand field and the strongly negative redox potential confirm that these structural features are conserved in solution[90,231,281].

There are two isomeric forms of [Co(trap)$_2$]$^{3+}$. Experimental studies and molecular mechanics calculations indicate that they have similar stabilities. The *meso* isomer has been analyzed by X-ray diffraction, and the computed structure is in good agreement with the experiment[170]. It is of interest that the structure of *meso*-[Co(trap)$_2$]$^{3+}$ is similar to that of [Co(*trans*-diammac)]$^{3+}$, and the experimentally determined Co-N distances to the amine substituent in 2-position (shorter arm, two symmetrically related *trans* bonds) are similar to the corresponding distances in [Co(*trans*-diammac)]$^{3+}$ (1.942 Å *vs* 1.946 Å), while the other four bonds are longer (1.961, 1.964 Å *vs* 1.937 Å).

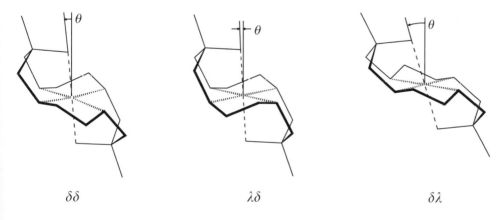

$\delta\delta$                    $\lambda\delta$                    $\delta\lambda$

**Fig. 17.11.1**

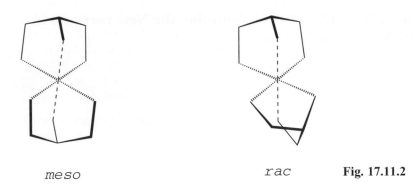

*meso*                              *rac*              **Fig. 17.11.2**

You can input the two structures from the CSD. However, it is a good exercise to build the structures using HyperChem and then refine them with MOMEC. Do not forget to set the force field back to the original MOMEC parameter set (*Setup/Force Field*/c:\momec97\parm). For [Co(*trans*-diammac)]$^{3+}$ it is sufficient to build one of the three conformers, refine it and then create the others by inversion of one or two of the five-membered chelate rings in HyperChem (see Section 17.3).

**Practice**

Once you have created the five structures and saved the *.hin files (we use the file names codmld.hin, codmdd.hin, codmdl.hin, cotrap1.hin and cotrap2.hin) you can refine them with the force field that you have developed in Section 17.10: *Setup Files/Force Field*/c:\momec\user\lesson10. Check the parameters again ($k_b$(CO3-NT) = 2.25; $r_0$(CO3-NT) = 1.927). Refine all files in the *Batch Job* mode. Relevant experimental data and those that you have calculated are assembled in Table 17.11.1.

**Table 17.11.1**

| Compound | Experimental | MOMEC97 | 2.25/1.927 |
|---|---|---|---|
| coa6 | 1.961 [Å] [53] | 1.955 [Å] | 1.964 [Å] |
| colllen | 1.964 [58] | 1.956 | 1.964 |
| codmld | 1.946$^{ax}$ [281] | 1.943$^{ax}$ 1.935$^{eq}$ | 1.954$^{ax}$ |
|  | 1.937$^{eq}$ |  | 1.946$^{eq}$ |
| codmdd | – | 1.946$^{ax}$ 1.950$^{eq}$ | 1.956$^{ax}$ |
|  |  |  | 1.959$^{eq}$ |
| codmdl | – | 1.953$^{ax}$ 1.973$^{eq}$ | 1.962$^{ax}$ |
|  |  |  | 1.978$^{eq}$ |
| cotrap1 | 1.942$^{ax}$ [170] | 1.938$^{ax}$ 1.952$^{eq}$ | 1.950$^{ax}$ |
|  | 1.961 (1.964)$^{eq}$ |  | 1.961$^{eq}$ |
| cotrap2 | – | 1.941$^{ax}$ | 1.952$^{ax}$ |
|  |  | 1.950 (1.953)$^{eq}$ | 1.959 (1.962)$^{eq}$ |

With the two new structures (codmld.hin and cotrap1.hin) it becomes obvious that our new parameter set has two deficiencies: The short bonds are generally overestimated and, therefore, the difference between long and short bonds becomes too small. Thus, the force constant of 2.25 mdyn $Å^{-1}$ is, not unexpectedly, too high and the ideal bond distance of 1.927 Å is possibly too long. Let us first consider the force constant and try the new parameter set 2.0 mdyn $Å^{-1}$/1.927 Å:

**Table 17.11.2**

| Compound | Experimental | 2.25/1.927 | 2.00/1.927 |
|---|---|---|---|
| coa6 | 1.961 [Å] [53] | 1.964 [Å] | 1.968 [Å] |
| colllen | 1.964 [58] | 1.964 | 1.968 |
| codmld | 1.946$^{ax}$ [281] | 1.954$^{ax}$ | 1.956$^{ax}$ 1.948$^{eq}$ |
|  | 1.937$^{eq}$ | 1.946$^{eq}$ |  |
| codmdd | – | 1.956$^{ax}$ | 1.959$^{ax}$ 1.961$^{eq}$ |
|  |  | 1.959$^{eq}$ |  |
| codmdl | – | 1.962$^{ax}$ | 1.966$^{ax}$ 1.983$^{eq}$ |
|  |  | 1.978$^{eq}$ |  |
| cotrap1 | 1.942$^{ax}$ [170] | 1.950$^{ax}$ | 1.952$^{ax}$ 1.964$^{eq}$ |
|  | 1.961 (1.964)$^{eq}$ | 1.961$^{eq}$ |  |
| cotrap2 | – | 1.952$^{ax}$ | 1.955$^{ax}$ |
|  |  | 1.959 (1.962)$^{eq}$ | 1.962 (1.965)$^{eq}$ |

The difference between short and longer bonds is now close to the experimental values, and it emerges that a decrease of the ideal bond distance by about 1/10 of an Angstrom should lead to satisfactory results. The parameter set 2.0 mdyn $Å^{-1}$/1.915 Å leads to the results shown in Table 17.11.3.

**Table 17.11.3**

| Compound | Experimental | *MOMEC97* | 2.00/1.915 |
|---|---|---|---|
| coa6 | 1.961 [Å] [53] | 1.955 [Å] | 1.958 [Å] |
| colllen | 1.964 [58] | 1.956 | 1.958 |
| codmld | 1.946$^{ax}$ [281] | 1.943$^{ax}$ 1.935$^{eq}$ | 1.947$^{ax}$ 1.939$^{eq}$ |
|  | 1.937$^{eq}$ |  |  |
| codmdd | – | 1.946$^{ax}$ 1.950$^{eq}$ | 1.950$^{ax}$ 1.953$^{eq}$ |
| codmdl | – | 1.953$^{ax}$ 1.973$^{eq}$ | 1.956$^{ax}$ 1.974$^{eq}$ |
| cotrap1 | 1.942$^{ax}$ [170] | 1.938$^{ax}$ 1.952$^{eq}$ | 1.943$^{ax}$ 1.956$^{eq}$ |
|  | 1.961 (1.964)$^{eq}$ |  |  |
| cotrap2 | – | 1.941$^{ax}$ | 1.946$^{ax}$ |
|  |  | 1.950 (1.953)$^{eq}$ | 1.953 (1.956)$^{eq}$ |

These results are as a whole very accurate, i. e., all the differences between experimentally determined and computed Co-N distances are well below 0.01 Å. This is especially remarkable for the very short Co-N bonds, and it is an error limit which is not generally obtained for larger data sets. Remember: our force field is still based on just four structures, and this usually is too small a basis set for anything other than a very limited set of molecules.


**Exercise**

Check the strain energies for the seven calculated structures with all the four force fields used thus far and interpret the changes. Which terms are contributing the most to the total strain energies?

Have a close look at the structures of [Co(*trans*-diammac)]$^{3+}$. The main distortion in this case is the angle $\theta$ between the best plane through the four equatorial macrocyclic N-donors and the cobalt center, and the bonds to the axial N-donors (see Fig. 17.11.1). Use the *Structure* module to measure this distortion, and compare it (i) between the three conformers and (ii) for each conformer through the four force fields.


## 17.12   Force Field Parameters III: Refining the New Force Field – Very Long Bond Distances


**Theory**

We now extend the structural basis set for cobalt(III) hexaamines with one additional structure with relatively long Co$^{III}$-N bonds. The [Co(tmen)$_3$]$^{3+}$ cation (tmen = 2,3-dimethylpropane-2,3-diamine) is a highly strained species with long Co$^{III}$-N bonds because of the four methyl substituents (see Fig. 17.12.1). The structure of the cation has been determined by an X-ray diffraction study, and the conformation in the crystal has been defined as *ob$_3$* (see Section 17.3 for the nomenclature of the conformers). Due to the elongation of the Co$^{III}$-N bonds to 1.997 Å, there is a remarkable shift in the ligand field spectra (the first d-d transition ($^1A_1 \rightarrow {}^1T_1$) is at 515 nm *vs* 470 nm for [Co(en)$_3$]$^{3+}$) and the redox potential ($-0.18$ V *vs* $+0.28$ V)[56,231].


**Practice**

Use the structure of *ob$_3$*-[Co(en)$_3$]$^{3+}$ from Section 17.4 as a starting structure to build *ob$_3$*-[Co(tmen)$_3$]$^{3+}$, and refine it with the original MOMEC force field (*Setup Files/Force Field*/c:\momec97\parm).

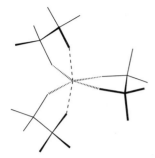

**Fig. 17.12.1**

Once you have created the structure and saved the *.hin file (we use the file name coooootmn.hin), you can refine it with the force field that you have developed in Sections 17.10 and 17.11: *Setup Files/Force Field*/c:\momec\user\lesson10. Check the parameters again ($k_b$(CO3-NT) = 2.00 mdyn Å$^{-1}$; $r_0$(CO3-NT) = 1.915 Å). Refine the structure and check the results against those in Table 17.12.1.

**Table 17.12.1**

| Compound | Experimental | *MOMEC97* | 2.00/1.915 |
|---|---|---|---|
| coa6 | 1.961 [Å] [53] | 1.955 [Å] | 1.958 [Å] |
| colllen | 1.964 [58] | 1.956 | 1.958 |
| codmld | 1.946$^{ax}$ [281] | 1.943$^{ax}$ | 1.947$^{ax}$ |
|  | 1.937$^{eq}$ | 1.935$^{eq}$ | 1.939$^{eq}$ |
| codmdd | – | 1.946$^{ax}$ | 1.950$^{ax}$ |
|  |  | 1.950$^{eq}$ | 1.953$^{eq}$ |
| codmdl | – | 1.953$^{ax}$ | 1.956$^{ax}$ |
|  |  | 1.973$^{eq}$ | 1.974$^{eq}$ |
| cotrap1 | 1.942$^{ax}$ [170] | 1.938$^{ax}$ | 1.943$^{ax}$ |
|  | 1.961 (1.964)$^{eq}$ | 1.952$^{eq}$ | 1.956$^{eq}$ |
| cotrap2 | – | 1.941$^{ax}$ | 1.946$^{ax}$ |
|  |  | 1.950 (1.953)$^{eq}$ | 1.953 (1.956)$^{eq}$ |
| coooootmn | 1.997 [56] | 1.971 | 1.973 |

It is not unexpected that the computed Co-N bond distances for [Co(tmen)$_3$]$^{3+}$ are too short. We try to correct this by reducing the steepness of the corresponding bonding potential. As a new parameter set we try $k_b$(CO3-NT) = 1.75 mdyn Å$^{-1}$; $r_0$(CO3-NT) = 1.915 Å. We now refine the whole set of structures as a *Batch Job*. The results are shown in Table 17.12.2.

**Table 17.12.2**

| Compound | Experimental | *MOMEC97* | 1.75/1.915 |
|---|---|---|---|
| coa6 | 1.961 [Å] [53] | 1.955 [Å] | 1.962 [Å] |
| colllen | 1.964 [58] | 1.956 | 1.963 |
| codmld | 1.946$^{ax}$ [281] | 1.943$^{ax}$ | 1.950$^{ax}$ |
|  | 1.937$^{eq}$ | 1.935$^{eq}$ | 1.942$^{eq}$ |
| codmdd | – | 1.946$^{ax}$ | 1.954$^{ax}$ |
|  |  | 1.950$^{eq}$ | 1.957$^{eq}$ |
| codmdl | – | 1.953$^{ax}$ | 1.960$^{ax}$ |
|  |  | 1.973$^{eq}$ | 1.978 (1.982)$^{eq}$ |
| cotrap1 | 1.942$^{ax}$ [170] | 1.938$^{ax}$ | 1.945$^{ax}$ |
|  | 1.961 (1.964)$^{eq}$ | 1.952$^{eq}$ | 1.959$^{eq}$ |
| cotrap2 | – | 1.941$^{ax}$ | 1.949$^{ax}$ |
|  |  | 1.950 (1.953)$^{eq}$ | 1.957 (1.960)$^{eq}$ |
| cooootmn | 1.997 [56] | 1.971 | 1.979 |

In terms of ob$_3$-[Co(tmen)$_3$]$^{3+}$ there is some improvement. However, the resulting Co-N distances are still too short, and the overall agreement for all five structures considered is worse. It emerges that the original MOMEC force field leads to the best overall agreement, but it has some deficiencies for compounds with very long Co-N bonds. The simple reason for that is that a harmonic potential does not reproduce well the energy arising from very distorted bonds (see Section 17.8). We have addressed this in some recent work and are currently improving MOMEC with a new stretching function [310].

**Exercise**

Construct and refine the other three conformers of [Co(tmen)$_3$]$^{3+}$ and refine them with the original MOMEC force field (c:\momec97\parm). Check the results, in particular the Co-N distances and the strain energies. Compute the conformer distribution and compare it with that of [Co(en)$_3$]$^{3+}$ (see Section 17.4).

Measure the trigonal twist angle $\phi$ of the four conformers of [Co(tmen)$_3$]$^{3+}$ (see Section 17.7 for the definition of $\phi$ and the method used to measure it with MOMEC). Check the accuracy of the MOMEC force field and those that you have developed in the last lessons by comparing the computed trigonal twist angle $\phi$ with that obtained experimentally for ob$_3$-[Co(tmen)$_3$]$^{3+}$ (43.9°) [56].

Next, examine the four structures of the corresponding cobalt(II) compounds and compare the Co-N bond distances and trigonal twist angles with those of the cobalt(III) species and with the experimentally observed data (ob$_3$-conformer: Co$^{II}$-N = 2.193 Å; $\phi = 29.5°$). The best way to quickly create the *.hin files for the cobalt(II) compounds is to use the *Set Atom Type* item in the *Tools* box of MOMEC.

# 17.13   Force Field Parameters IV: Comparison of Isomer Distributions Using Various Cobalt(III) Amine Force Fields

## Theory

The three previous lessons dealing with the development of a force field have all focused on reproducing structural details. One of the other important outcomes of molecular mechanics is the strain energies and, in particular, correlations between differences in strain energies and the amount of a given isomer or conformer that is observed experimentally (see Section 17.4). While the structures are defined by the positions of minima on potential energy surfaces, the isomer distributions are dependent on the steepness of the potentials (i. e., the first derivatives), and the vibrational frequencies (i. e., the spectroscopic force constances) are dependent on the curvature (i. e., the second derivatives). That is, a force field that has been fitted to structural data (as the one that we have developed in the preceding lessons) may lead to high quality predictions with respect to structural parameters but thermodynamic and spectroscopic parameters might be in poor agreement with experimental results (see Section 3.3).

Imagine two cobalt(III) hexaamine force fields that lead to structural predictions of similar quality, one with a relatively strong, the other with a relatively weak $Co^{III}$-$N_{amine}$ force constant. Consider two isomeric cobalt(III) hexamines, one with long, the other with short $Co^{III}$-$N_{amine}$ bonds. It is evident that the weaker force constant will generate a smaller difference in isomer abundances than will the stronger. At this point, it helps to remember that force field parameters are interrelated, i.e., if one of the parameters is changed (e. g., the force constant for the $Co^{III}$-$N_{amine}$ bond) others need to be readjusted to obtain good quality structural data (see Section 17.10). Since some of the parameters of the MOMEC force field have been fitted to thermodynamic and to spectroscopic data (this is the case for parameters of the organic part of the molecules), there is a good chance that the isomer ratios computed with a parameter set for a new type of interaction (e. g., $Co^{III}$-$N_{amine}$) with a constant parameterization for the rest of the molecule are reasonable. In other words: if the force of the organic part of the molecule is constant and fitted to thermodynamic data, the counter-force of the metal-donor bonds should also be reliable. In any case, when developing a new force field it is highly desirable to test these points, and that is what we will do in the present lesson.

## Practice

One of the classic applications of molecular mechanics modeling to a system of isomers is the application to the three geometric isomers of $[Co(dien)_2]^{3+}$. In this lesson we will examine the effect of varying the force field on the calculated energy differences between these isomers and on how well they correlate with the experimentally determined differences (see Table 8.1, Section 8.1).

The three isomers of [Co(dien)$_2$]$^{3+}$ are referred to as meridional (*mer*), unsymmetrical-facial (*u-fac*) and symmetrical-facial (*s-fac*) and are shown in schematic form in Fig. 17.13.1.

| *mer* | *u-fac* | *s-fac* | **Fig. 17.13.1** |

Build each isomer using HyperChem or import them from the CSD. Each of the chelate rings can adopt two enantiomerically related conformations referred to as $\delta$ and $\lambda$, as discussed in Sections 17.3 and 17.11. When undertaking a study of a system of isomers such as this with the aim of predicting the proportions of each isomer it is important to firstly establish which conformers are the most stable. In Table 17.13.1, strain energies of the conformers are given. Some of the conformers could not be optimized. This is not an uncommon experience and is probably a consequence of the potential energy surface being too flat in the vicinity of the local minimum corresponding to those isomers. However, in order to avoid discarding conformers unnecessarily, care must be taken to attempt refinement from a starting point as close as possible to the expected geometry and heavily damped shifts should be used to avoid jumping away from the minimum before the refinement has stabilized.

**Table 17.13.1**

| Isomer | Conformer | Strain energy [kJ mol$^{-1}$] |
|---|---|---|
| *mer* | $\lambda\delta, \lambda\lambda$ | 78.20 |
| | $\lambda\lambda, \lambda\lambda$ | 86.20 |
| | $\delta\lambda, \delta\lambda$ | 64.65 |
| | $\lambda\delta, \delta\delta$ | 78.20 |
| | $\delta\delta, \delta\delta$ | 86.94 |
| | $\delta\delta, \lambda\lambda$ | 78.77 |
| *u-fac* | $\lambda\delta, \delta\delta$ | 71.76 |
| | $\delta\delta, \delta\delta$ | 71.41 |
| | $\lambda\delta, \lambda\delta$ | 67.88 |
| | $\lambda\delta, \lambda\lambda$ | 70.19 |
| | $\lambda\lambda, \lambda\lambda$ | 71.58 |
| | $\delta\delta, \lambda\lambda$ | 75.84 |
| *s-fac* | $\lambda\delta, \lambda\delta$ | 68.74 |
| | $\lambda\lambda, \lambda\lambda$ | 75.81 |
| | $\delta\lambda, \delta\lambda$ | 86.09 |
| | $\delta\lambda, \lambda\lambda$ | 74.78 |
| | $\delta\delta, \lambda\lambda$ | 83.31 |
| | $\lambda\delta, \delta\lambda$ | 96.04 |

In the case of the *mer* and *s-fac* isomers, one conformer is substantially more stable than any other but for *u-fac*, the difference is smaller. Hereafter, we will focus on only the most stable conformer but you should be aware that this is risky. Changing the force field might reverse the stability order of the conformers or at least change the relative abundances. We leave it as an exercise for you to check the effect of the changes on the most stable conformers and to see whether there is a change in the order.

In the previous three lessons, variations in the parameters for the Co-N bond length were considered. Here we will examine the effects of changing the force constant and the ideal bond-length for the Co-N potential on the isomer distribution. Shown in Table 17.13.2 are the parameters used and the corresponding strain energies calculated for the three isomers.

**Table 17.13.2**

| Ideal bond length $r_0$ | Force constant $k_b$ | Strain energies [kJ mol$^{-1}$] (*mer–s-fac, mer–u-fac, s-fac–u-fac*) | | |
|---|---|---|---|---|
| [Å] | [mdyn Å$^{-1}$] | *mer* | *s-fac* | *u-fac* |
| 1.905 | 1.75 | 64.65 | 68.74 (4.09) | 67.88 (3.23) (0.86) |
| 1.915 | 1.75 | 61.38 | 65.71 (4.33) | 64.74 (3.36) (0.97) |
| 1.915 | 2.00 | 62.44 | 66.61 (4.17) | 65.70 (3.26) (0.91) |
| 1.927 | 2.25 | 59.39 | 63.73 (4.34) | 62.72 (3.33) (1.01) |
| 1.927 | 1.75 | 57.68 | 62.28 (4.60) | 61.18 (3.50) (1.10) |

Two points emerge from these results. Firstly, despite the small changes in the force constants, there are large changes in the magnitudes of the strain energies. For instance, the strain energy of the *mer* isomer calculated using the MOMEC parameters (1.905 Å, 1.75 mdyn Å$^{-1}$) is 64.65 kJ mol$^{-1}$ and that calculated with the ideal bond length changed to 1.927 Å is 57.68 kJ mol$^{-1}$. This very clearly demonstrates that the value of the strain energy has no absolute meaning. The reason the strain energy decreases is that expansion of the Co-N bond length reduces repulsion between the ligand atoms, reducing the size of the nonbonded contribution. Also, the ligands may relax to some extent.

The other point to emerge is that energy differences also change to some extent despite the small changes made in the force field. For example the strain energy difference between *mer* and *u-fac* changes from 4.09 kJ mol$^{-1}$ when the MOMEC force field is used to 4.60 kJ mol$^{-1}$ when the ideal bond length is increased to 1.927 Å. These strain energy differences translate to differences in the proportions of isomers that are expected. In Table 17.13.3 are listed the isomer proportions calculated from these strain energy differences. Note, when calculating these expected isomer ratios it is necessary to take into account statistical factors. Both the *mer* and *u-fac* isomers are chiral and therefore can be formed in two ways. Thus, they have 2 : 1 statistical advantage over the *s-fac* isomer, and if all three had identical strain energies we would expect the isomers to be formed in a 2 : 2 : 1 ratio. The methods for calculating isomer proportions from strain energy differences have been described in Section 17.4. Also, shown in Table 17.13.3 is the experimentally observed isomer distribution (see also Table 8.1 in Section 8.1 and the corresponding references).

Note that the isomer ratio computed using the MOMEC force field is different from that reported in Section 8.1. This difference is due to the fact that here we only consider the most stable conformer of each of the three isomers. The neglected conformers also contribute to the isomer abundance and, due to some relatively low energy conformers in the case of the *u-fac* isomer, this leads to the observed differences (see discussion above). Thus, for a more accurate computation of isomer distributions, all conformations need to be considered.

**Table 17.13.3**

| Ideal bond length $r_0$ [Å] | Force constant $k_b$ [mdyn Å$^{-1}$] | Isomer proportions [%] | | |
|---|---|---|---|---|
| | | *mer* | *u-fac* | *s-fac* |
| 1.905 | 1.75 | 73.1 | 19.9 | 7.0 |
| 1.927 | 1.75 | 75.7 | 18.4 | 5.9 |
| experimental data: | | | | |
| water | | 63 | 29 | 8 |
| methanol | | 53 | 29 | 18 |
| dimethyl sulfoxide | | 80 | 14 | 6 |
| acetone | | 74 | 17 | 9 |

The variation in the strain energy differences as a function of the force field translates into modest differences in calculated percentages. Comparison with the experimental data reveals two salient points. Firstly, the experimental data is significantly dependent on the environment. The second important point to note is that variation in the experimentally determined isomer distributions is substantially greater than the variation that results from the small changes to the force field. Consequently, it is inappropriate to fine-tune a force field on the basis of achieving agreement between observed and calculated isomer distribution. However, whenever a new force field is developed the ability of that force field to reproduce observed isomer distributions should be checked and major changes of force fields have at times been justified on the basis of substantial improvement in this ability. For example, prior to the adoption of 1,3-nonbonded interactions about the metal atom (an approach that is central to the MOMEC force field), molecular mechanics models did not reproduce the observed isomer distributions of $[Co(dien)_2]^{3+}$ even moderately well (see Section 8.1).

**Exercise**

Compute the strain energies of all conformers of the $[Co(dien)_2]^{3+}$ system with the various sets of $Co^{III}\text{-}N_{amine}$ parameters and check whether the isomer distributions change with respect to the data where only the most stable conformers have been considered.

   Examine the effect of varying force field parameters on the strain energies of other systems of isomers or conformers discussed in earlier in lessons. For instance the conformers of $[Co(en)_3]^{3+}$ or the isomers of $[Co(trap)_2]^{3+}$. Try changing valence angle, torsion angle and non-bonding parameters and also examine the effect of removing 1,3-nonbonded interactions and replacing them with valence angle terms about the metal.

## 17.14   Force Field Parameters V: Parameterizing a New Potential – The Tetrahedral Twist of Four-Coordinate Compounds

**Theory**

As one of the special features, MOMEC has a plane twist function. This has been included to limit the tetrahedral twist in four-coordinate compounds, where 1,3-nonbonded interactions lead to a preference for a tetrahedral arrangement (see Section 3.6). That is, the plane twist potential can be used to induce a square-planar arrangement or, using constraints, any intermediate structure can be enforced. The same potential can in principle be used for other structural features (see Fig. 17.14.1), such as the Bailar twist of six-coordinate complexes or for computing the rotational barrier of metallocenes. However, at present it has only been implemented in MOMEC for the tetrahedral twist and no parameters have been included as yet.

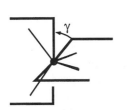 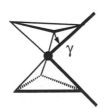 

Fig. 17.14.1

There are a number of possible models and parameterization schemes for enforcing planarity in four-coordinate compounds:

1. Using a large force constant for the multiple harmonic potential with minima at 90° and 180°. However, this may distort the chromophore in cases where the observed minimum structure is planar but with bite angles involving chelate rings that are smaller (or larger) than 90°.
2. Using dummy axial ligands to force the equatorial ligands into a planar geometry. This technique has been used widely. However, it is somewhat clumsy and might distort the backbone of the equatorial ligands.
3. Using out-of-plane potentials. This is probably the most reasonable technique from those available in older programs but it is not very intuitive and it does not allow for the constraint of specific tetrahedral twist angles.
4. The plane twist potential that we will use in this lesson.

There are many four-coordinate transition metal compounds. Apart from preferences due to the ligand geometry, ligand-ligand repulsion will tend to twist them towards tetrahedral. Depending on the electronic structure (ground state of the metal center and strength of the donors) this preference will be modulated. Thus, four-coordinate zinc(II) compounds (electronically innocent $d^{10}$ center) will generally be tetrahedral while four-coordinate palladium(II) compounds (strong field $d^8$ center) will usually be square planar. Nickel(II) has a lower ligand-field-stabilization energy and therefore less ability to enforce square planarity but usually four-coordinate nickel(II) tetraaminc compounds are also planar. Thus, the force constant for the plane twist function is dependent on the metal center and on the type of donor atoms.

In a tetrahedral coordination polyhedron there are three possible tetrahedral twist angles $\gamma$. MOMEC automatically chooses one of them (you may want to have two twist angles included; this can be done by changing the default value for twist in the momec.ini file from 1 to 2).

```
Shell=0
Print= 0 0 0 0 0 0
DefaultTwist=1

[FILES]
```
                                            **Fig. 17.14.2**

From the first donor in the atom list, MOMEC calculates the three valence angles involving the metal center and discards the largest one, i.e., the trans angle in a square planar geometry. For the other two sets of three atoms (metal center and two donors) it determines the two planes defining the tetrahedral twist and selects the combination that has the smaller twist angle. The refinement proceeds in the direction with the smallest movement, i.e., from large to small twist angles $\gamma$. Both, the selection of planes and the direction of the refinement can be changed in *Edit/View/Interaction Array* as shown in Fig. 17.14.3.

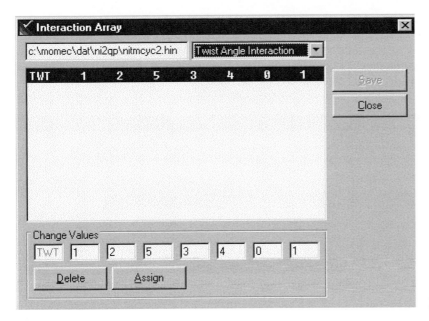

**Fig. 17.14.3**

The first digit in the boxes in Fig. 17.14.3 is the center of rotation (the metal center); the next four numbers define the two planes together with the metal center; the last number defines the direction of rotation. All these values can be changed.

## Practice

As an example we compute the structure of the platinum(II) complex of tetra-methylcyclam, $[Pt(tmtactd)]^{2+}$, see structures below. There are various isomers (configurations of the chiral amines) but the geometry with all four methyl substituents oriented on the same side of the macrocyclic plane is known to be the most stable form ($R,S,R,S$; also called *trans* I). This isomer has the two six-membered chelate rings in chair conformations, and these two chelate rings are oriented on the same side of the macrocyclic ligand plane. We will concentrate on this isomer. There are two conformers with respect to the geometry of the five-membered rings, see structures below. Generate both of them with HyperChem and save them as pttmcyc1.hin and pttmcyc2.hin.

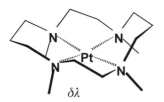

**Fig. 17.14.4**

You will need to add a value for the twist angle function. As usual, you do that with *Edit/View/Force Field/Twist Angle Parameter*. This parameter will be modified during the lesson, and this is done as in earlier lessons, by changing the parameter in the parameter array file. The first minimization will be with a parameter value of 0.0.

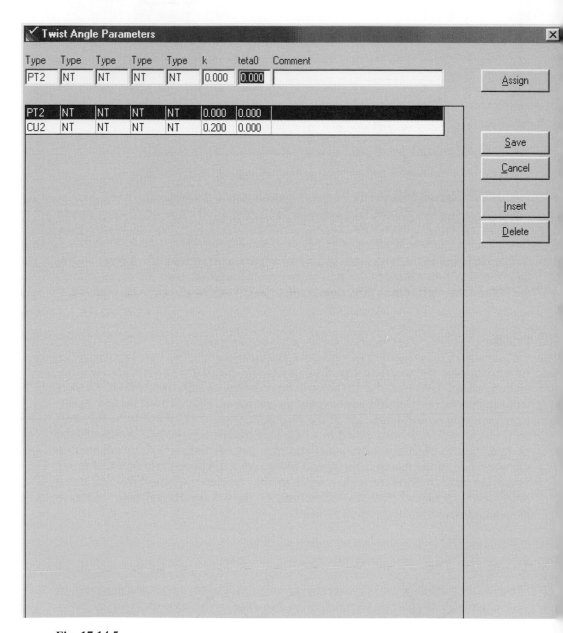

**Fig. 17.14.5**

Remember to activate the twist angle potential and the twist angle listing files in the Setup windows of MOMEC. Refinement of the two structures with these parameters leads to the results given in Table 17.4.1 (these are available in the listing files or by measuring in HyperChem and/or the *Structure* module of MOMEC):

**Table 17.14.1**

| Twist angle force constant $k_\gamma = 0.0$ |            |            |
| Parameter        | $\lambda\lambda$      | $\delta\lambda$      |
| --- | --- | --- |
| Pt-N1            | 2.053 [Å]             | 2.058 [Å]            |
| Pt-N2            | 2.053                 | 2.058                |
| Pt-N3            | 2.045                 | 2.049                |
| Pt-N4            | 2.045                 | 2.049                |
| $\gamma$         | 20.5°                 | 5.78°                |
| $E_{strain}$     | 55.6 kJ mol$^{-1}$    | 67.9 kJ mol$^{-1}$   |

Now activate the twist angle potential with a force constant of $k_\gamma = 0.1$. This gives the results shown in Table 17.14.2.

**Table 17.14.2**

| Twist angle force constant $k_\gamma = 0.1$ |            |            |
| Parameter        | $\lambda\lambda$      | $\delta\lambda$      |
| --- | --- | --- |
| Pt-N1            | 2.053 [Å]             | 2.057 [Å]            |
| Pt-N2            | 2.053                 | 2.057                |
| Pt-N3            | 2.045                 | 2.048                |
| Pt-N4            | 2.045                 | 2.048                |
| $\gamma$         | 17.3°                 | 3.7°                 |
| $E_{strain}$     | 59.5 kJ mol$^{-1}$    | 68.1 kJ mol$^{-1}$   |

There is little change in most values, particularly for the $\delta\lambda$ isomer which has little tetrahedral twist. The increase in strain energy in the $\lambda\lambda$ isomer is mainly due to a build up of torsional strain, van der Waals repulsion and twist angle strain. Minimization does not proceed as smoothly with larger values of the force constant. Thus, it may be advisable to increase the damping and/or to decrease the termination rms shift (*Setup/Optimization Controls*).

**Exercise**

Model the other Pt complexes described in Section 17.6 using a tetrahedral twist function rather than the previously used out-of-plane functions. Establish what values of the force constant $k_\gamma$ are necessary to enforce planarity in the different complexes.

## 17.15   Using Constraints to Compute Energy Barriers

**Theory**

Dynamic processes such as conformational interconversion or bond length deformation associated with changes in electronic or oxidation states have energy barriers associated with them. It is sometimes possible to obtain measures of these barriers, either directly or indirectly, but there are no experimental methods for determining the mechanisms by which these changes occur. Also, if the barriers are low it can be almost impossible to obtain experimental measures of them. Molecular mechanics calculations can be used to obtain theoretically based estimates of the barriers, irrespective of their height, and can also give mechanistic information.

Chelate rings, such as that formed when ethane-1,2-diamine binds to a metal, commonly adopt one of two enantiomerically related conformations, designated $\delta$ and $\lambda$, as described in Section 17.3.

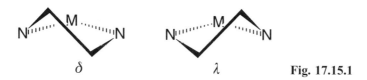

$\delta$                     $\lambda$                 **Fig. 17.15.1**

Interconversion of one conformation to the other occurs on a very rapid timescale and, therefore, it is difficult to determine the barrier using experimental methods, making it a sensible target for molecular mechanics estimation of the energy barrier. The conformations of cyclic groups such as this chelate ring can be defined by torsion angles around the ring and interconversion of the conformations involves inversion of all torsion angles; for example, the N-C-C-N torsion angle changes from about −55° to +55°. However, using constraints to "drive" this torsion angle from one extreme to the other will not produce a reliable estimate of the energy barrier. Instead, it is necessary to map the potential energy surface associated with the interconversion. In the case of a five-membered chelate ring two torsion angles are required to fully define the conformation and, therefore, assuming the interconversion does not involve substantial changes in bond lengths or bond angles, the potential energy surface is three-dimensional (see Section 7.2).

**Practice**

To map the energy profile associated with the conformational interconversion it is sensible to choose the N-C-C-N torsion, since the change in this angle is greatest and most readily visualized, and one M-N-C-C angle. These latter angles change from about 40° to −40° during the interconversion. In order to set up the calculation of the energy profile it is necessary to set up a grid of values, at least covering the ranges in these two torsion angles, and to calculate the strain energy at each point on the grid. The step size in the grid will depend on the precision required and the computational facilities available. Even a step size as coarse as 5° will require the calculation of 600 points!

So, how do we "drive" these torsion angles to the values we want and keep them there? There are two methods, constraints and restraints. Constraints are applied by additions to the matrices used in the energy minimization and are therefore mathematically precise – if a value of 35.0° is selected then on minimization the torsion angle will have exactly this value, irrespective of the strain that is induced as a result. Restraints on the other hand are just normal potential energy terms but with the minimum set at the desired value and with very large force constants to force the adoption of this desired value. Restraints are not mathematically precise; how close the minimization gets to the required value depends on the size of the artificial force constant but this is not usually a significant problem. Both constraints and restraints are available in MOMEC but it is recommended that for the purposes of mapping potential energy surfaces or computing energy barriers that constraints be used.

1. Using HyperChem build $[Co(NH_3)_4(en)]^{3+}$ by following the methods described in Sections 17.1 and 17.3.

2. Use the *Build Selections* option from the *Tools* menu to select the torsion angle. In order to do this it is necessary to open HyperChem with the *Display/Labels/Numbers* option on.

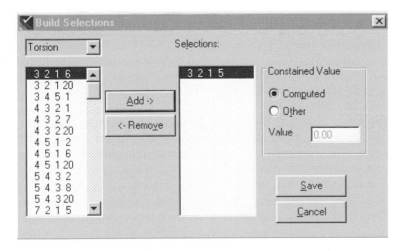

**Fig. 17.15.2**

There is an alternative method for selecting the torsion angles to be constrained and this is carried out in HyperChem (continue with step 9 if you have already added the selections):

3. Turn on the select tool ⊚ and select the N-C-C-N torsion angle of the en ring by clicking on the four atoms, being careful to follow the sequence as shown.

4. Select *Name Selection* ... under the *Select* menu. The *Other* option should be on, if not select it. In the text box name the selection as say TorNCCN and click on the *OK* button.

5. Deselect the torsion angle by clicking the right mouse button away from the molecule and then select a Co-N-C-C torsion as described above.

6. Repeat instruction 4. but this time name the selection as say TorCoNCC.

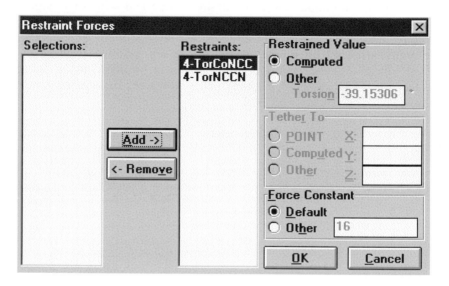

**Fig. 17.15.3**

7. Select *Restraints* ... under the *Setup* menu. The window shown in Fig. 17.15.3 will appear. The two named torsion angles should appear under the *Selections* list. Click on each of these names in turn and click on the *Add* → button, moving the two names to the *Restraints* list. Ignore the fact that this refers to restraints – that applies to minimization within HyperChem. Be warned that if other named selections appear under either of the *Selections* or the *Restraints* lists they will interfere with the refinement process and must be removed. This is done using the *Select* ... option under the *Select* menu by clicking on each extra name and the *Remove* button in turn.

8. Save the file, move to MOMEC and open the file.

9. Under the *Execute* menu select the *Energy* option. The two torsion angles selected should appear in the selection list (see Fig. 17.15.4).

10. Select the N-C-C-N torsion angle by clicking on it and then select the *Start value* by clicking on the box and typing the desired starting value of 60°. Next select the *Step* value typing a desired step size of –5° and set the *Number of Data Points* to 13. The statement "End value = 0.000" should appear. If the end value is written in red it is outside the range that is allowed in MOMEC and the parameters have to be modified. For torsion angles the range is 0° to 179.99°. Assign the chosen values to the torsion angle by clicking on the *Assign* button.

11. Repeat step 10 for the other torsion angle setting a starting value of 45° and a step size of 0°. The *Setup Energy Parameters* window should appear as shown in Fig. 17.15.4.

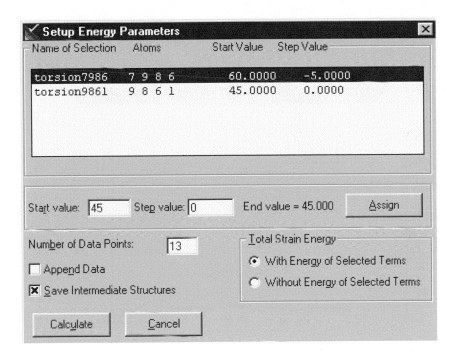

**Fig. 17.15.4**

12. Click on *Calculate*. Energy minimization with the N-C-C-N torsion constrained initially at 60°, then at 55° and so on will begin. The results will appear in tabular and graphical forms in windows as shown in Fig. 17.15.5 and 17.15.6. If there are asterisks in the *Minimum not reached* column increase the number of cycles in the optimization controls and repeat the calculations.

| Nr. | torsion7986 | torsion9861 | Energy | Minimum not reached |
|---|---|---|---|---|
| 1 | 60.000 | 45.000 | 38.93 | |
| 2 | 55.000 | 45.000 | 37.87 | |
| 3 | 50.000 | 45.000 | 38.52 | |
| 4 | 45.000 | 45.000 | 40.74 | |
| 5 | 40.000 | 45.000 | 43.14 | |
| 6 | 35.000 | 45.000 | 45.88 | |
| 7 | 30.000 | 45.000 | 48.61 | |

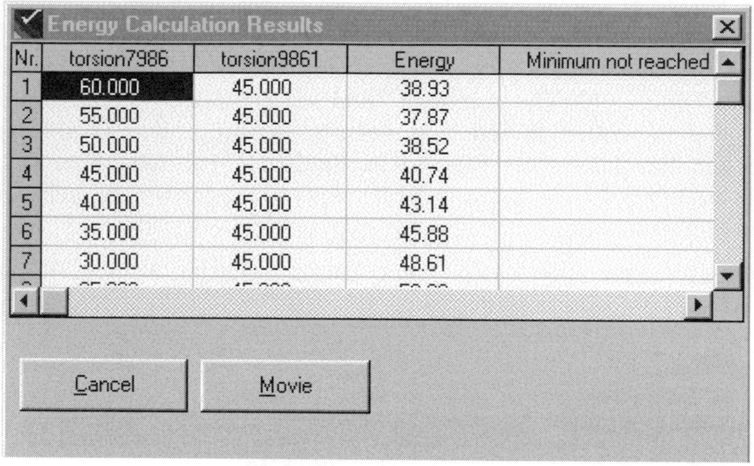

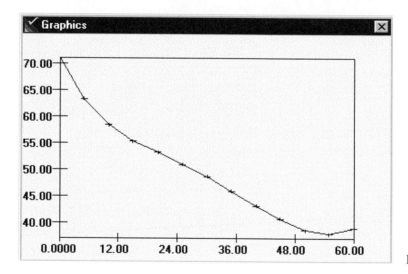

**Fig. 17.15.5**

**Fig. 17.15.6**

As mentioned above (point 10) MOMEC does not allow you to compute negative torsion angles. Since, in the present case, it is necessary to drive the N-C-C-N torsion to $-60^\circ$ we have to adopt a somewhat clumsy procedure: we have to setup the structures for both endpoints (N-C-C-N torsion of $+60^\circ$ and $-60^\circ$) by reflecting the chelate ring in HyperChem (see e.g., Section 17.3), and then drive the N-C-C-N torsion from each side to $0^\circ$, while constraining the Co-N-C-C torsion at the specified value. The results indicate that you get an identical table and plot to that above, i.e., the two traces are, as expected, symmetrically related (note that this is not true for unsymmetrical structures). A combination of the two files leads to the plot shown in Fig. 17.15.7.

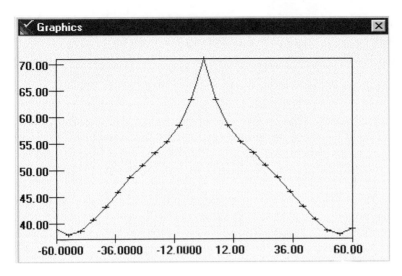

**Fig. 17.15.7**

This represents a slice through the two-dimensional potential energy surface defined by the two constrained torsion angles. The energy values and the associated torsion angles are recorded in the "*.dat" file. This file should be copied to a new file and kept.

13. Now change the *Start value* for the Co-N-C-C torsion angle to 40° and repeat the calculations. This generates the next slice of the potential energy surface and the results from the "*.dat" file should be added to the file containing the previous slice.

14. The *Start value* for the Co-N-C-C torsion angle should now be decreased a further 5° and the procedure repeated until the torsion angle has reached −45° (see note on negative torsion angles above) generating the entire potential energy surface. This can be plotted using a scientific plotting program (see Fig. 7.2 in Section 7.2).

**Exercise**

Repeat the procedure outlined above to calculate the energy profile and energy barrier associated with the conformational interconversion in $[Co(NH_3)_4(1,2\text{-pro-panediamine})]^{3+}$, built and optimized in Sections 17.3 and 17.4, respectively. Compare the results you obtain with those for $[Co(NH_3)_4(en)]^{3+}$.

## 17.16    Using Constraints to Compute Macrocyclic Ligand Hole Sizes

**Theory**

For macrocyclic and cage ligands the size of the ligand cavity (often referred to as the hole size) and its shape are important features that have been used to rationalize stabilities, redox potentials, electron transfer rates and ligand field spectra. Some of these applications will be studied in the following lessons (see Chapters 9, 10, 11).

Most applications using hole sizes are based on the idea that a metal ion that fits well into the cavity of a macrocyclic ligand will lead to a stable complex. That is, a given ligand has the potential to selectively stabilize certain metal ions in certain oxidation states. In the terminology of preorganization this means that a ligand that is highly preorganized for coordination to a given metal ion will lead to a stable product (note that cavity size is just one aspect of preorganization and that complex stability depends on many other factors in addition to preorganization (see Chapter 9)).

Let us consider specifically the metal-ligand bonding potential (Eq. 17.16.1):

$$E_{ML} = \frac{1}{2} k_{ML} \left( r_{ML} - r_0 \right)^2 \tag{17.16.1}$$

The strain induced in the metal-ligand bond will be zero, if either $k_{ML} = 0$ or $r_{ML} = r_0$. There are a number of reasons, discussed in the literature, for computing the strain induced in the ligand as a function of $r_{ML}$ with $k_{ML} = 0$, in order to accurately determine the ligand hole size. Elongation or compression of the metal-ligand distance from that corresponding to the cavity size will induce stress in the entire ligand. That is, the entire ligand and not only the M-L bonds will be deformed. Thus, the total strain energy $U_{total}$ (see Section 17.2) is plotted as a function of the metal-ligand distance $r_{ML}$ with $k_{ML} = 0$ (i.e., the stresses induced in the metal-ligand bond do not contribute to the total strain energy $U_{total}$), and this is the approach generally used in MOMEC, using the module *Energy*.

An example of such a curve is shown in Fig. 17.16.1 (the curve plotted is that of sar in the $D_3ob_3$ conformation, i.e., one of the examples that will be studied in this lesson; the structure of $D_3ob_3$-[Co(sar)]$^{3+}$ is shown in Fig. 17.16.2). Curves like this enable visualization of the size of the ligand hole and the shape of the potential energy profile.

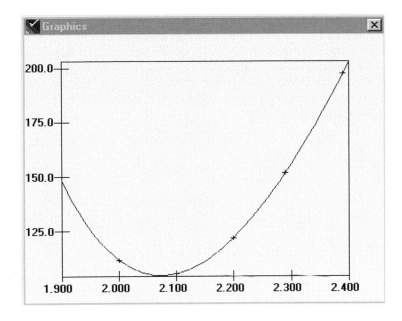

**Fig. 17.16.1**

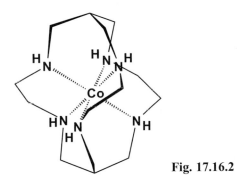

**Fig. 17.16.2**

The plot above has been produced with six identical distances from the centroid of the ligand (the metal center) to the amines. A non-spherical metal ion (e.g., a Jahn-Teller labile copper(II) ion) will induce different stresses to a symmetrical ligand. Alternatively, the ligand might be asymmetrical, i.e., the ligand itself might induce an asymmetry in the coordination sphere. Effects like these will be studied in Section 17.17.

Let us analyze the curve in Fig. 17.16.1: Spherical metal ions prefering metal-amine distances similar to those at the minimum of the curve (2.07 Å) will fit best to the ligand and consequently lead to especially stable complexes. Metal ions preferring smaller bonds (<2.07 Å, as is the case, e.g., for cobalt(III),) or larger bonds (>2.07 Å, as, e.g., for copper(I) or zinc(II)) will have elongated or compressed bonds, respectively, and relatively lower stabilities (i.e., higher strain energies).

These points are visualized in the plot shown in Fig. 17.16.3. The curve at the bottom is the same as that in Fig. 17.16.1, i.e., a metal ion independent hole size versus strain energy plot (remember that the metal-donor stretching potential is set to zero, $k_{ML} = 0$). The other curve is a strain energy versus Co-N distance plot of $D_3ob_3$-[Co(sar)]$^{3+}$. This curve is metal ion dependent and the Co-N strain contributes to the total strain energy $U_{total}$. This potential is steeper and the minimum is at a shorter M-L distance (1.96 Å vs 2.07 Å).

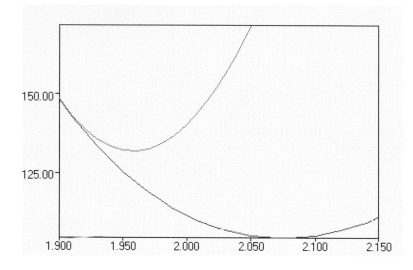

**Fig. 17.16.3**

Thus, the two curves demonstrate that the cobalt(III) ion induces some strain in the ligand and that the resulting structure of $D_3ob_3$-[Co(sar)]$^{3+}$ is a compromise between the ligand and metal ion preferences. It emerges that a metal ion with longer M-L distance preferences (a "larger" metal center) might lead to more stable complexes with the sar ligand. However, be careful with your interpretation: The stability of the complex not only depends on the strain energy that is induced in the ligand but also on the bonding energy related to the six M-L bonds. Also, different conformers of the sar ligand may have different hole sizes, and this is an aspect that will also be analyzed in this lesson.

The fact that the computed Co-N distance for $D_3ob_3$-[Co(sar)]$^{3+}$ is smaller than that preferred by the metal-free ligand indicates that the Co-N bonds are elongated in this compound (the fact that the computed and observed values for Co-N are larger than the strain-free distance $r_0$, used in our force field (1.905 Å) is irrelevant since force field parameters are not necessarily physically meaningful, see Sections 3.5, 9.2 and 10.2).

**Practice**

We will now compute the ligand hole sizes of all possible conformers of the sar-type cage ligands. We must first build the complexes with HyperChem and refine them with MOMEC. The complexes are derivatives of $[Co(en)_3]^{3+}$ (a $HC(CH_2)_3$ "cap" is added to the two trigonal faces at the bottom and the top, see the structure of $D_3ob_3$-$[Co(sar)]^{3+}$ in Fig. 17.16.2). Open the file colllen.out, refined in Section 17.4, and modify it using the methods described in Section 17.3 and Section 17.6. Refine the molecule with the MOMEC97 force field. We suggest you use a combination of Fletcher-Reeves and Newton-Raphson minimization *Setup/ Optimization Controls* since the initial structure that you have built might be rather distorted.

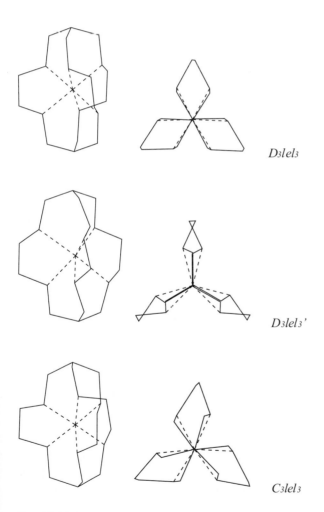

*D₃lel₃*

*D₃lel₃'*

*C₃lel₃*

**Fig. 17.16.4**

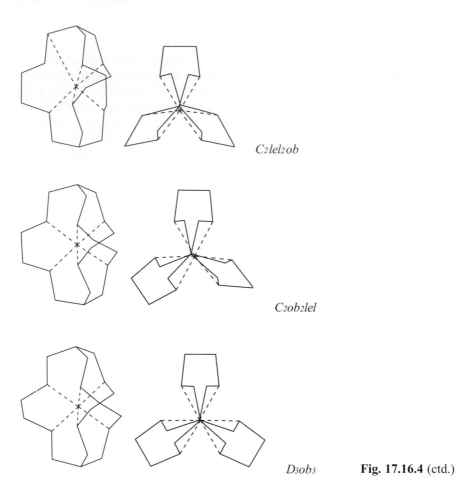

$C_2lel_2ob$

$C_2ob_2lel$

$D_3ob_3$    **Fig. 17.16.4** (ctd.)

There are six possible conformations of $[Co(sar)]^{3+}$, and these are shown in Fig. 17.16.4[89]. The plots on the left are a view perpendicular to the (pseudo) $C_3$ molecular axis through the metal center and the two carbon atoms of the "caps" on each side; the plots on the right are a view parallel to that axis. The molecule that you have drawn is either that with a $D_3lel_3$ or that with a $C_3lel_3$ conformation. To obtain the structures with *ob* conformations you can reflect the "en" type chelate rings as described in Section 17.3. Note that, due to the connectivities at the secondary amines, you should only reflect the ethylene carbon and hydrogen atoms and leave the amine protons and the $\alpha$ carbon atoms of the "caps" untouched. This will lead to some distortion. Therefore, it is again suggested that you minimize with a combination of Fletcher-Reeves and Newton-Raphson minimization. We also suggest you reflect one ring after another, and use the refined structure of the $C_2lel_2ob$ conformer as a starting point for the construction of the $C_2ob_2lel$ geometry and that for $D_3ob_3$. You can also try to create and refine all conformers automatically with the Random Kick module (see Section 17.5).

The two structures with $D_3lel_3$ and $C_3lel_3$ conformations have different orientations of the caps. The remaining conformer ($D_3lel_3'$) is only a stable energy minimum for relatively large metal ions (metal-amine distances larger than ca. 2 Å, see below). Thus, this structure can be obtained from the initial conformer ($D_3lel_3$) by constraining the six CO3-NT bonds to $\geq 2.2$ Å (see Section 17.15 for the technique used to constrain internal coordinates).

How do we create the $D_3lel_3$ conformer from $C_3lel_3$ or vice versa? There are differences in torsion angles, as shown in Table 17.16.1 and Fig. 17.16.5.

**Table 17.16.1**

| Conformer | Torsion t1 [°] | Torsion t2 [°] |
|---|---|---|
| $D_3lel_3$ | 172; 172 | 118; 118 |
| $D_3lel_3'$ | (169; 169) | (153; 153) |
| $C_3lel_3$ | 176; 162 | 107; 144 |
| $C_2lel_2ob$ | 163; 174; 88 | 141; 156; 133 |
| $C_2ob_2lel$ | 166; 95; 87 | 141; 160; 159 |
| $D_3ob_3$ | 94; 94 | 159; 159 |

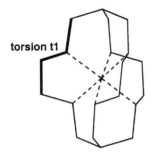
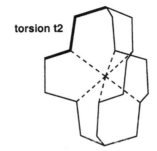

torsion t1

torsion t2

**Fig. 17.16.5**

With this information you can constrain specific torsion angles to switch between any of the six conformations. Use the *Tools/Build Selections* menu to do this, but do not forget to delete the selections with *Tools/Delete Selections* before fully refining the structure and before setting up a new conformer.

Refine the structures (except that of the $D_3lel_3'$ conformation) with the MOMEC97 force field. Check your results against those given in Tables 17.16.1 and 17.16.2 with respect to the torsion angles and the strain energies.

**Table 17.16.2**

| Conformer | $E_b$ [kJ mol$^{-1}$] | $E_\theta$ [kJ mol$^{-1}$] | $E_\phi$ [kJ mol$^{-1}$] | $E_{nb}$ [kJ mol$^{-1}$] | $E_{tot}$ [kJ mol$^{-1}$] |
|---|---|---|---|---|---|
| $D_3lel_3$ | 14.6 | 25.0 | 37.3 | 51.8 | 128.7 |
| $C_3lel_3$ | 15.4 | 22.7 | 36.2 | 53.8 | 128.0 |
| $C_2lel_2ob$ | 15.4 | 21.8 | 36.8 | 54.7 | 128.7 |
| $C_2ob_2lel$ | 14.3 | 17.6 | 39.2 | 56.7 | 127.8 |
| $D_3ob_3$ | 12.0 | 16.7 | 43.6 | 56.7 | 129.0 |

It emerges that all five structures have roughly the same total strain energies $U_{total}$ but the individual terms differ to some degree. The torsional terms can be understood from Table 17.16.1, and it might be instructive to analyze the bonding energy terms and the valence energy terms based on Table 17.16.3 (use Hyper-Chem and the *Structure* module of MOMEC to get these data; remember to use the *.out files as *Input* files in the MOMEC *Setup/Files* menu since the *Structure* module refers to the *Input* file; note also that these structural data refer to the chromophore while the strain energies refer to the entire molecule).

**Table 17.16.3**

| Conformer | Co-$N_{av}$ [Å] | N-Co-N bite$_{av}$ [°] | twist [°] |
|-----------|------------|------------------|-------|
| $D_3lel_3$ | 1.969 | 87.3 | 55.5 |
| $C_3lel_3$ | 1.970 | 85.8 | 51.6 |
| $C_2lel_2ob$ | 1.970 | 85.6 | 51.2 |
| $C_2ob_2lel$ | 1.965 | 85.8 | 54.1 |
| $D_3ob_3$ | 1.958 | 86.6 | 58.2 |

We now determine the hole sizes of the various conformers of sar. Prepare the files of the six conformers of [Co(sar)]$^{3+}$ by selecting the six Co-N bonds (*Tools/ Build Selections*) in each file to set up the constraints for the *Energy* calculations. Use the *.out files but rename them as *.hin. As outlined above, the strain energy vs. metal-donor-distance plots for the computation of the hole sizes need to be metal ion independent. Thus, you need to activate the option *Without Energy of Selected Terms* in the *Energy* setup window. Also, the donor-metal-donor valence angle term needs to be switched off, since this is also metal ion dependent. You can do that in the *Edit/View/Force Field/Atom Type Parameters* menu or in the *Edit/View/Parameter Array* window. Both options have been used before in this tutorial.

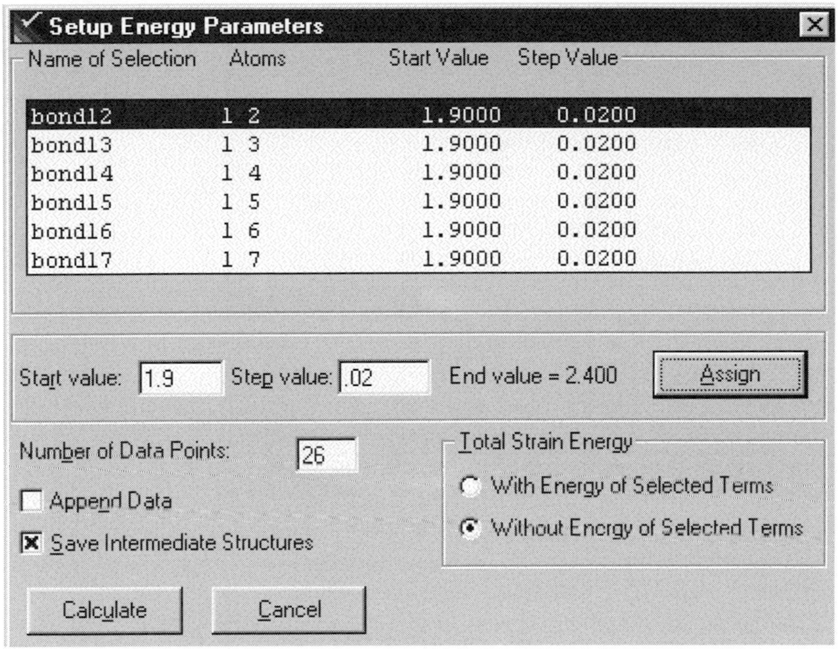

**Fig. 17.16.6**


The *Start value* and the *Step* value of each bond must be set and *Assigned* separately. We chose a starting value of 1.900 Å, a step size of 0.02 Å and 26 data points for each of the bonds. That is, we calculate the strain energy versus metal-amine distance curve in the interval from 1.900 Å to 2.400 Å. *Save Intermediate Structures* so that you can analyze them after the *Energy* calculation (these files can be deleted after the lesson since there are many of them). The option *Append Data* is used to compute additional points in the same curve, i.e., when the curve needs to be extended or when more points are needed in the region of a minimum of the curve. Start the computation with *OK*. A table with the computed data will now appear on the screen and, after the second point on the curve has been calculated, the graphic will begin to be built up. Stars in the last column of the table indicate that convergence has not been reached with the parameters chosen in the minimization options (you then need to start again with an increased maximum number of cycles). The results for the $C_3lel_3$ conformer are shown in Fig. 17.16.7.

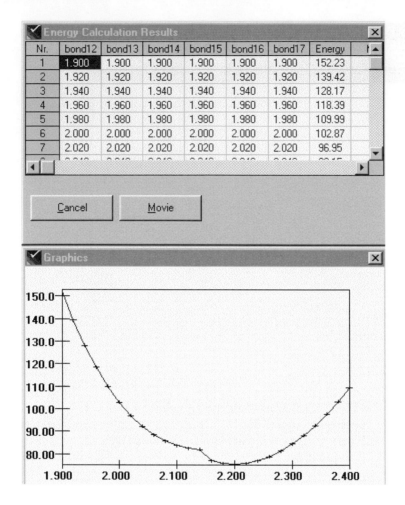

Fig. 17.16.7

Notice that at an M-N distance of 2.14/2.16 Å there is a break in the curve. Analyze the two structures by double clicking with the mouse on the corresponding lines in the *Energy* table. With the two torsion angles described above (torsion t1 and torsion t2, see Fig. 17.16.5) it emerges that the structural change is to $D_3lel_3'$ conformation. A similar conformational interconversion occurs with the $D_3lel_3$ conformer.

Compute the *Energy* plot for the $D_3lel_3'$ conformer starting at an M-N distance of 2.400 Å with a step size of −0.02 Å. When all calculations have been finished you can display all six curves in one plot with *Execute/Intersection*. Compare your results with those in Fig. 17.16.8.

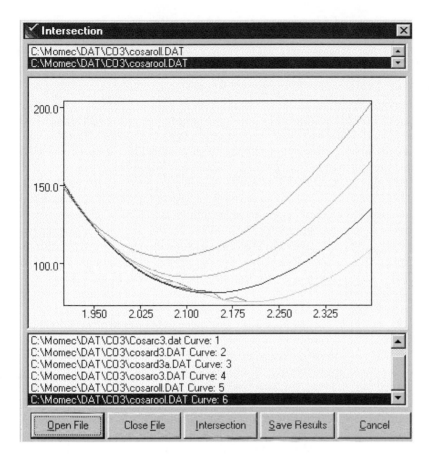

Fig. **17.16.8**

It emerges that each *ob* conformation reduces the cavity size, i.e., the $D_3ob_3$ conformation is most stable with small metal ions and the $D_3lel_3'$ conformer has the largest cavity size. This is in agreement with the computed Co-N distances of the five stable conformers (see Table 17.16.3).

**Exercise**

Determine the trigonal twist angles as a function of the M-N distance for the six conformers of sar.

Use the other two techniques for determining hole sizes for one of the structures that you have analyzed so far. These methods are:

- Using restraints (large force constants for the Co-N bonds; edit these with the *Edit/View/Parameter Array* tool) to fix the Co-N bonds.
- Changing $r_0$ in each step (using the *Edit/View/Parameter Array* tool) and record a strain energy vs $r_0$ curve.

## 17.17    Cavity Sizes of Unsymmetrical Ligands

**Theory**

You may have become aware of a fundamental problem with the computation of the ligand hole sizes of complexes in which all bond lengths are not equivalent, as is the case for the two conformers $C_2lel_2ob$-[M(sar)]$^{n+}$ and $C_2ob_2lel$-[M(sar)]$^{n+}$ (see Section 9.2). Let us have a closer look at the refined structures of the cobalt(III) complexes (open the corresponding *Listing Files/Bond Distances*): The CO3-NT distances are 1.975 Å, 1.975 Å, 1.973 Å, 1.973 Å, 1.965 Å, 1.965 Å for the former, and 1.972 Å, 1.972 Å, 1.957 Å, 1.957 Å, 1.968 Å, 1.968 Å for the latter. This bond length variation is not unexpected because the structures have only $C_2$ symmetry.

Cobalt(III)-amine bonds are relatively short. If we refine the corresponding cobalt(II) complexes we observe longer metal-amine distances and greater variation in these distances. (You can do that using the *Set Metal Type* tool in the *Tools* module of MOMEC, followed by saving the files under a new name and executing a structure optimization). The corresponding CO2-NT distances are: 2.139 Å, 2.139 Å, 2.126 Å, 2.126 Å, 2.121 Å, 2.121 Å, and 2.124 Å, 2.124 Å, 2.105 Å, 2.105 Å, 2.117 Å, 2.117 Å, for the $C_2lel_2ob$ and the $C_2ob_2lel$ conformers, respectively. The corresponding differences between the bond distances for the smaller cobalt(III) and the larger cobalt(II) complex for the $C_2lel_2ob$ conformer are 0.151 Å, 0.148 Å and 0.174 Å, those for the $C_2ob_2lel$-conformer are 0.152 Å, 0.147 Å and 0.149 Å, respectively. Thus, for asymmetrical structures it is not appropriate to vary all six bonds with the same starting values and step sizes. For highly asymmetrical ligands, neglecting this effect might lead to erroneous predictions. In this lesson we will deal with such a ligand, a disubstituted bispidine derivative. Shown in Fig. 17.17.1 is a metal complex of this ligand that enforces a distorted tetrahedral coordination geometry (see Section 9.2 [186]).

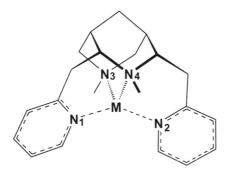
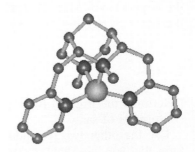

**Fig. 17.17.1**

**Practice**

Build the bispidine complex using HyperChem and refine the corresponding cobalt(III) and cobalt(II) complexes with MOMEC (note that the force fields are not optimized for tetrahedral chromophores but our aim here is just to compute relative metal-ligand distances). Note, that all these structural optimizations need to be performed with 1,3-nonbonded interactions alone for the angular geometry around the metal ions (i.e., deactivate the Mult. Harm. functions for both metal atom types, either with the force field editor, in the interaction array or in the parameter array, where you can simply set the force constant to zero). Compare your results with those in Table 17.17.1.

**Table 17.17.1**

| Metal center | M-$N_1$ [Å] | M-$N_2$ [Å] | M-$N_3$ [Å] | M-$N_4$ [Å] |
|---|---|---|---|---|
| $Co^{III}$ | 1.891 | 1.891 | 1.863 | 1.863 |
| $Co^{II}$ | 2.074 | 2.079 | 2.071 | 2.071 |

There are at least two conventional ways of determining the appropriate start and end points for the strain energy versus metal-donor distance plots:

- The variations can be determined from the slopes of plots of the M-N distances versus the ionic radii of the metal ions.
- The variations can be determined from the slopes of the plots of M-$N_{1,2}$ versus M-$N_3$ and M-$N_{1,2}$ versus M N$_4$.

The resulting values for the start and end points are used for the computation of the hole size of the bispidine derivative.

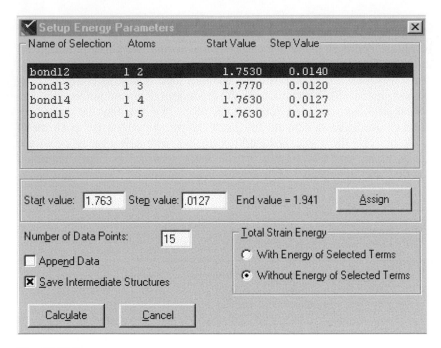

**Fig. 17.17.2**

The bispidine type ligands are very rigid. Therefore, for elongated and compressed cavity sizes achieving energy minimization is a problem and you may find that it is difficult to reach convergence. Use a large number of cycles, a high damping factor and small step values. Also, prerefine the structure at M-L values close to the starting point of the *Energy* calculation. The results are shown in Fig. 17.17.3.

More recent versions of MOMEC also offer a more elegant and accurate module for computing and analyzing cavity sizes and plasticities of asymmetrical ligands, and we suggest you use this method for all calculations of hole sizes. The *Decoupled Energy* function in the *Execute* mode is similar to that of the *Energy* module but it mathematically fixes the sum of bond distances in a molecule (e. g., the four M–N distances in metal bispidine compounds) with Lagrangian multipliers to specific values. It is possible to compute decoupled energy curves with or without the strain energies imposed by the metal-donor bonds; at this stage this has to be done by setting the metal-donor force constants to the required values or to zero, respectively.

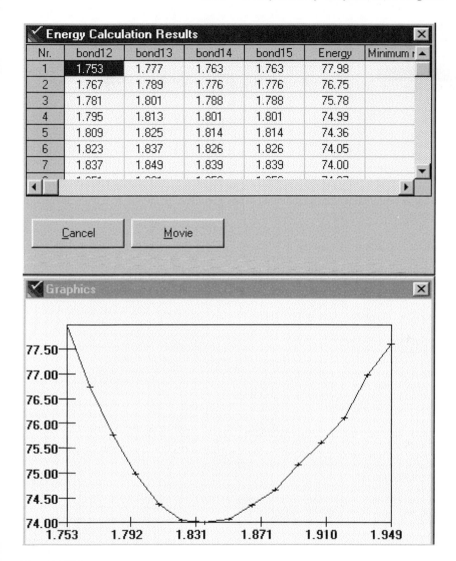

**Fig. 17.17.3**

Bond lengths to be constrained can be selected using HyperChem, as described previously, or in MOMEC using *Tools/Build Selections*. Then, go to *Execute/Decoupled Energy*, set the *Start Value*, *Step Value* and *Number* of Data Points to the desired values as shown in Fig. 17.17.4.

Select *Calculate* and the procedure will commence. The results obtained for the cobalt(III) complex are also shown in Fig. 17.17.4. Repeat the procedure for Co(II).

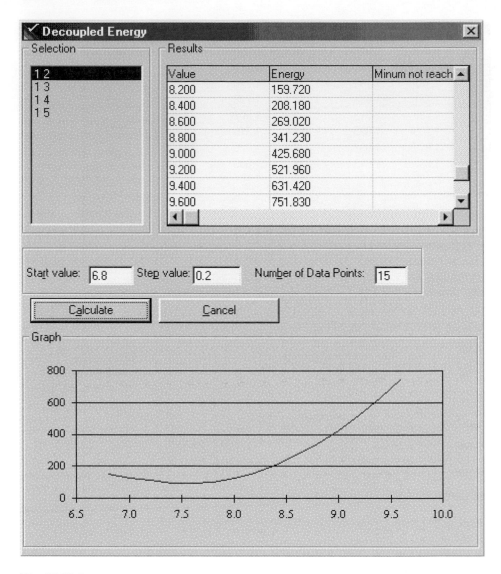

**Fig. 17.17.4**

**Exercise**

Compute the hole sizes of the two unsymmetrical conformers of the cage ligand sar ($C_2lel_2ob$ and $C_2ob_2lel$; see Section 17.16) and compare the results with those obtained in Section 17.16.

The asymmetry of the three conformers of the hexacoordinated *trans*-diammac ligand is even larger. Plot hole size curves for the three conformers of *trans*-diammac (see Section 17.11 for the corresponding structures).

# 17.18   Using Strain Energies to Compute Reduction Potentials of Coordination Compounds

## Theory

The reduction potential is related to the Gibbs energy (the free energy) of the electron transfer reaction.

$$\Delta G^{\circ} = -n \cdot F \cdot E^{\circ} \tag{17.18.1}$$

For transition metal coordination compounds $\Delta G^{\circ}$ involves terms related to the ionization potential $I$ of the free metal ion, the difference in complexation free energy $\Delta(\Delta G_{c}^{\circ})$, involving the oxidized and the reduced free metal ion, and the difference in solvation free energy $\Delta(\Delta G_{aq}^{\circ})$, involving the complexes in the two oxidation states:

$$\Delta G^{\circ} = I + \Delta(\Delta G_{c}^{\circ}) + \Delta(\Delta G_{aq}^{\circ}) \tag{17.18.2}$$

The approach of using strain energies to predict reduction potentials is based on the idea that a set of ligands that enforce large metal-donor bond distances stabilize the reduced form of the corresponding complex (more positive reduction potentials) and a ligand sphere that leads to short metal-donor distances stabilizes the smaller, oxidized form (more negative reduction potentials). However, the strain energy difference, $\Delta U_{strain}$, between the oxidized and the reduced form of the couple will only be correlated with the redox potential $E^{\circ}$ if the other terms of Eq. 17.18.2 are approximately constant or linearly dependent on $\Delta U_{strain}$. Thus, the molecular mechanics approach for the computation of reduction potentials is based on the following assumptions[231]:

- The ionization potential $I$ is only dependent on the metal center – that is, it is a constant for a series of redox couples with an identical metal center.

- The electronic contribution to $\Delta(\Delta G_{c}^{\circ})$ (metal-donor bonding) for a set of similar compounds (identical metal center, same type of donor, similar coordination polyhedra, e.g., hexaaminecobalt(III/II) couples with variable amines) is only dependent on the metal-donor distance, i.e., $\Delta(\Delta G_{c}^{\circ})$ is correlated with the strain energy difference between the oxidized and the reduced forms of the couples.

- For a series of similar redox couples (e.g., hexaaminecobalt(III/II) couples) the solvation term $\Delta(\Delta G_{aq}^{\circ})$ depends only on the relative size of the molecular cations (Born equation). That is, this term is also correlated with the strain energy difference between the oxidized and the reduced forms.

- The entropy contribution depends on the size of the molecular cations.

If these assumptions are valid, then the major contribution to $\Delta G^{\circ}$ for redox couples with identical metal centers and similar ligands is the strain energy difference between the oxidized and the reduced forms of the complex ($\Delta U_{strain}$), and the neglected terms vary roughly linearly with $\Delta U_{strain}$.

$$\Delta G^{\circ} = f \cdot (\Delta U_{strain}) + c \qquad (17.18.3)$$

where $f$ is a factor that depends on steric effects, and $c$ is a constant that includes, among others, corrections related to the ionization potential $I$ and electrode specific parameters. Therefore, if our assumptions are valid, the reduction potential is linearly dependent on the strain energy difference between the oxidized and the reduced form of the complex:

$$E^{\circ} = f' \cdot \Delta U_{strain} + c' \qquad (17.18.4)$$

where $f' = f/(n \cdot F)$. That is, for $f = 1$ (see Eq. 17.18.3) the slope $f'$ of a $\Delta U_{strain}$ vs $E^{\circ}$ correlation curve is $(n \cdot F)^{-1}$, i.e., $(96.5 \text{ kJ mol}^{-1} \text{ V}^{-1})^{-1}$, and the intercept is zero volts.

A number of studies have shown that the assumptions discussed above are valid for certain classes of redox couples, and reasonably accurate predictions for reduction potentials have been obtained for redox couples with large structural differences between the oxidized and reduced forms of the complex (see Section 11.1). This is the case for hexaaminecobalt(III/II)[231] and tetraaminecopper(II/I)[341] couples, where the metal-donor bond length differences are approximately 0.2 Å each. Note that additional assumptions are required for copper(II/I) couples, where the stoichiometries for the oxidized and reduced forms (five- or six-coordinate copper(II) versus four-coordinate copper(I)) are different. These are also discussed in Section 11.1, and it appears that they do not lead to undue inaccuracies.

Theoretically, each pair of conformers of a redox couple will lead to a specific and different reduction potential (see Fig. 17.18.1). However, if the variation in strain energy difference $\Delta U_{strain}$ is small the difference in reduction potentials will generally not be resolved experimentally. On the other hand, large strain energy differences $\Delta U_{strain}$ will usually lead to situations where the less stable conformer is not abundant enough to be observed. Thus, there has only been one report so far, where more than one reduction potential has been resolved experimentally (see Section 11.1)[310].

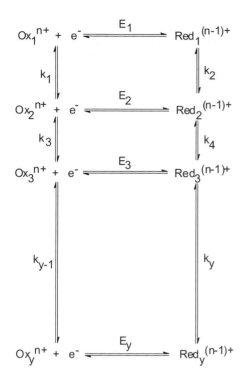

**Fig. 17.18.1**

The development of a correlation function for a class of compounds should involve as many and as variable data sets ($E°$ vs $\Delta U_{\text{strain}}$) as possible. In this lesson we will use known correlation functions to predict a limited number of reduction potentials and compare them with the known, experimentally observed values. The following correlation function has been found for hexaaminecobalt(III/II) couples:

$$E° = \frac{\Delta U_{\text{strain}}}{40.378} - 1.354 - a\left(\frac{\Delta U_{\text{strain}}}{1285.1} - 0.0525\right) \qquad (17.18.5)$$

where $a$ is the number of alkyl groups at the nitrogen donors, i.e., $a = 0$ for six ammonia ligands, $a = 1$ for five ammonia ligands and one primary amine, $a = 8$ for two secondary and four primary amines etc.[310].

**Practice**

You have built and refined a number of hexaaminecobalt(III) compounds in earlier lessons. These include $[Co(NH_3)_6]^{3+}$ (Section 17.2), the four conformers of $[Co(en)_3]^{3+}$ (Section 17.4), the three conformers of $[Co(\text{trans-diammac})]^{3+}$ (Section 17.11), the two isomers of $[Co(\text{trap})_2]^{3+}$ (Section 17.11), the four conformers of $[Co(tmen)_3]^{3+}$ (Section 17.12), and five conformers of $[Co(sar)]^{3+}$ (Section 17.16). Here, we will calculate the reduction potentials of these six compounds; we will consider all possible isomers but only the most favorable conformer of each. Make sure that all have been fully refined with the original MO-MEC97 force field. If you are not sure about that, refine the seven structures (nineteen, if you include all conformers) again as a batch-job (choose the original force field and refine the *.out files). Record all of the strain energies in a table. Check and record also the average Co-N bond distances for each compound. Try to list the compounds in the order of increasing reduction potential (remember that ligand sets that enforce short bond distances stabilize the oxidized form).

We now need to set up and refine the corresponding files for the reduced forms. Save the refined coordinates (*.out files) of each of the structures with a new name (e.g. co2a6.hin). This can be done in *Edit/View/HyperChem File* with the usual *Save as* command. Open *Tools/Set Metal Type* for each of the new files and change the *Oxidation state* from III to II. Refine the structures, calculate the strain energy differences between the oxidized and the reduced forms for the most stable conformers, calculate the reduction potentials using Eq. 17.18.5 and compare the results with those in Table 17.18.1.

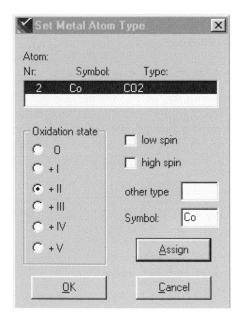

**Fig. 17.18.2**

**Table 17.18.1**

| Compound | $\Delta U_{strain}$ [kJ mol$^{-1}$] | $E^\circ_{obs}$ [V] | $E^\circ_{calc}$ [V] |
|---|---|---|---|
| [Co(*trans*-diammac)]$^{3+/2+}$ | 4.2 | $-0.63$ | $-0.75$ |
| *meso*-[Co(trap)$_2$]$^{3+/2+}$ | 29.9 | $-0.34$ | $-0.43$ |
| *rac*-[Co(trap)$_2$]$^{3+/2+}$ | 27.3 | $-$ | $-0.49$ |
| [Co(sar)]$^{3+/2+}$ | 24.3 | $-0.40$ | $-0.35$ |
| [Co(NH$_3$)$_6$]$^{3+/2+}$ | 39.8 | $-0.02$ | $-0.33$ |
| [Co(en)$_3$]$^{3+/2+}$ | 37.3 | $-0.17$ | $-0.25$ |
| [Co(tmen)$_3$]$^{3+/2+}$ | 67.8 | $+0.28$ | $+0.33$ |

A recent MM-Redox study of two isomers of a hexaamine-cobalt(III/II) cage compound has shown how powerful this technique is[310]: the structure of the two conformations has been solved with MOMEC-HyperChem in conjunction with the experimentally determined redox potentials (see Section 11.1).

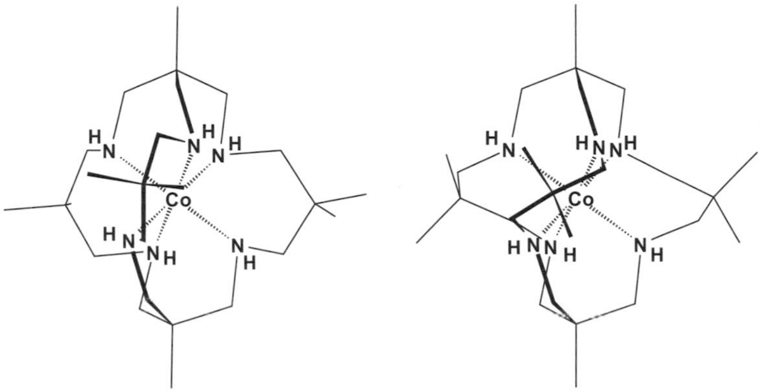

*lel*$_3$     *ob*$_3$     **Fig. 17.18.3**

This hexaamine cage ligand is derived from sar (see Section 17.16) but the three annular chelate rings are not five-membered but six-membered chelate rings and the central carbon has two methyl substituents. These chelate rings have skew-boat conformations, and their orientation is *lel*$_3$ for one and *ob*$_3$ for the other of the two conformers. Build these two structures in HyperChem and refine them as cobalt(III) and cobalt(II) compounds. Compute the reduction potentials and compare them with the data in Table 17.18.2.

**Table 17.18.2**

| Compound | $\Delta U_{strain}$ [kJ mol$^{-1}$] | $E^\circ_{obs}$ [V] | $E^\circ_{calc}$ [V] |
|---|---|---|---|
| *ob*$_3$-[Co(sixrcage)]$^{3+/2+}$ | 41 | 0.0 | $-0.1$ |
| *lel*$_3$-[Co(sixrcage)]$^{3+/2+}$ | 105 | $+0.84$ | $+0.89$ |

From the results so far it emerges that the ligand set with the largest hole size (sixrcage, *lel₃* conformation) leads to the largest (positive) reduction potential and that with the smallest hole size (*trans*-diammac) leads to the smallest (negative) potential. We will now try to design a ligand that enforces even smaller reduction potentials, and that does not yet exist. We start from *cis*-diammac and connect the two primary amine donors. This leads to a sar-type ligand with two of the six-membered chelate rings of the "caps" contracted to five-membered chelate rings. There are two isomeric forms of this cage ligand: the two contracted rings of the caps may be linked by an annular five-membered ring or not. Note also, that there are a number of conformers of each of these compounds, and we restrict ourselves to those with all annular chelate rings in lel conformation. Build the two cobalt(III) compounds in HyperChem, starting with [Co(sar)]³⁺, and refine the structures and those of the corresponding cobalt(II) complexes.

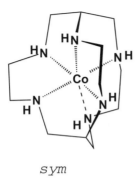 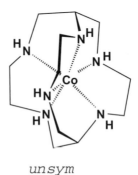

*sym*                              *unsym*                  **Fig. 17.18.4**

Compare the results with those in Table 17.18.3 which now lists all the calculations that we have done in this lesson.

**Table 17.18.3**

| Compound | $\Delta U_{strain}$ [kJ mol$^{-1}$] | $E^{\circ}_{obs}$ [V] | $E^{\circ}_{calc}$ [V] |
|---|---|---|---|
| [Co(*trans*-diammac)]³⁺/²⁺ | 4.2 | −0.63 | −0.75 |
| *sym*-[Co(diabsar)]³⁺/²⁺ | 7.7 | – | −0.61 |
| *unsym*-[Co(diabsar)]³⁺/²⁺ | 11.4 | – | −0.55 |
| *meso*-[Co(trap)₂]³⁺/²⁺ | 29.9 | −0.34 | −0.43 |
| *rac*-[Co(trap)₂]³⁺/²⁺ | 27.3 | – | −0.49 |
| [Co(sar)]³⁺/²⁺ | 24.3 | −0.40 | −0.35 |
| [Co(NH₃)₆]³⁺/²⁺ | 39.8 | −0.02 | −0.33 |
| [Co(en)₃]³⁺/²⁺ | 37.3 | −0.17 | −0.25 |
| *ob₃*-[Co(sixrcage)]³⁺/²⁺ | 41 | 0.0 | −0.1 |
| [Co(tmen)₃]³⁺/²⁺ | 67.8 | +0.28 | +0.33 |
| *lel₃*-[Co(sixrcage)]³⁺/²⁺ | 105 | +0.84 | +0.89 |

**Exercise**

In Sections 17.10–17.12 you developed and used some cobalt(III)-amine force field parameters that are different from those of MOMEC. Refine the cobalt(III) structures of Table 17.18.3 with some of these parameter sets, recompute the reduction potentials and see what the difference is. This gives you a feeling of how accurate the data need to be to make reasonable predictions in terms of the reduction potentials.

# 17.19  Using Force Field Calculations with NMR Data

**Theory**

Molecular mechanics modeling and multidimensional NMR methods are complementary and together can be a powerful tool for the study of large and flexible molecules in solution. Large molecules with many degrees of conformational freedom represent a difficult problem for molecular mechanics methods because there can be enormous numbers of local energy minima, all with similar strain energies. It is possible to calculate the strain energies of these conformations but it is virtually impossible to ensure that all conformations have been found, i.e., that the global energy minimum has been located. It is also difficult to predict which of the energy minima predominate in solution since solvation, ion pairing and electrostatic effects are often neglected or computed with simple models. Multidimensional NMR methods, on the other hand, provide information on which atoms are close to one another and this can be used to deduce which conformations predominate. However, the amount of such information available from NMR spectra is usually insufficient to fully define the geometry of the molecule. Thus, the two techniques are complementary and together can be used to generate models of molecules as they exist in solution.

The combination of molecular mechanics like methods and multidimensional NMR has formed the basis for numerous studies of peptides, proteins and DNA fragments. Paramagnetic shifts in metalloproteins have also been used to obtain structural information that is used as constraints in molecular mechanics and molecular dynamics calculations. For a number of reasons there are relatively few reported applications of combined NMR-molecular modeling studies involving metal complexes (see Section 10.4).

**Practice**

Intensities of through space correlations between atoms in multidimensional NMR are related to the separation ($d$) between those atoms by a $d^{-6}$ term. The intensities are usually standardized with respect to a correlation with a known atom-atom separation such as a geminal H···H contact and then the correlation intensities can be converted to distances. These distances can then be included as constraints in the energy minimization process, leading to a geometry with the required atom-atom separations. However, it is important to realize that factors other than atom-atom separation can influence the intensity of the correlation and, therefore, it is preferable to include the distance information based on these correlations as soft restraints rather than constraints.

At the simplest level, when there are relatively few conformations to be considered, each of them can be generated and the correlation intensities can be used as a guide to choosing between these possibilities. The relevant distances are readily available from MOMEC and/or HyperChem.

Using HyperChem build the molecule *cis*-[Pt(quinoline)(3-methylquinoline)Cl$_2$] shown here:

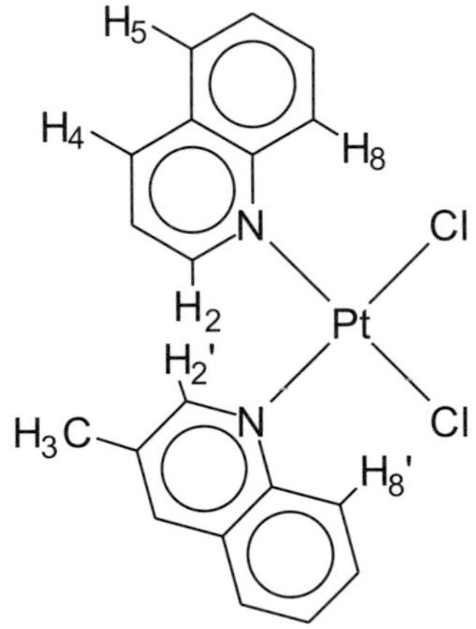

**Fig. 17.19.1**

Note that there are two conformers, one with the two H$_8$ atoms on the same side of the coordination plane (*syn* isomer) and the other with one H$_8$ atom on each side of the coordination plane (*anti* isomer). Refine and save both using the MOMEC97 force field. Section 17.14 describes how to enforce planarity of a coordination compound. Two dimensional NMR methods can be used to determine which isomer dominates – as long as interconversion of the isomers is not rapid on the NMR time scale. The data used here are hypothetical and we have assumed that one isomer dominates to the exclusion of the other and that there is no interconversion, i. e., the observed NMR spectrum is that of an isomerically pure compound.

Shown in the table below are the most intense NOESY correlations for this hypothetical molecule.

**Table 17.19.1**

| Atom pair | Relative intensity |
| --- | --- |
| $H_2 \cdots H_8'$ | 4400 |
| $H_4 \cdots H_5$ | 7400 |
| $H_2 \cdots H_2'$ | 130 |

Using the $d^{-6}$ relationship between separation $(d)$ and relative intensity $(I)$ we can deduce:

$$I_1/I_2 = (d_1/d_2)^{-6} \tag{17.19.1}$$

or

$$(I_1/I_2)^6 = d_1/d_2 \tag{17.19.2}$$

Using a model of the quinoline ligand we can measure the $H_4 \cdots H_5$ separation as being 2.49 Å. Check this in the two structures of cis-[Pt(quinoline)(3-methyl-quinoline)Cl$_2$] that you have refined. Given this distance and Eq. 17.19.2 we can calculate the $H_2 \cdots H_8'$ and $H_2 \cdots H_2'$ separations:

$H_2 \cdots H_8'$
$(7400/4400)^6 = 2.49$ Å$/d_2$
$d_2 = 2.49$ Å X$(1.68)^{1/6}$
$\quad = 2.49$ Å X $1.091$
$\quad = 2.72$ Å

and

$H_2 \cdots H_2'$
$(7400/130)^6 = 2.49$ Å$/d_2$
$d_2 = 2.49$ Å X$(56.9)^{1/6}$
$\quad = 2.49$ Å X $1.96$
$\quad = 4.88$ Å

Check the corresponding distances in the refined molecules and decide which of the two conformers is present in solution.

## Exercise

Consider the related complex with the quinoline ligand replaced by 2-methylqui-noline. Calculate the strain energies of the two isomers of the complex with this ligand and predict which isomer would be preferred. Use the energy minimized models of the two isomers and measure the closest $H \cdots H$ contacts. Calculate the intensities of the expected NOESY crosspeaks arising from these contacts.

## 17.20   Optimizing Structures with Rigid Groups

**Theory**

There are a number of reasons for carrying out refinements with parts of the molecule remaining invariant. For example, you may wish to examine the effect of increasing the rigidity of a portion of a structure or to model the interaction of a molecule with a solid surface. Unstable refinements might be stablized by minimizing part of the structure first and then refining the entire structure. For very large structures (macromolecules) it might be necessary to constrain most of the molecule because of computational limitations.

In earlier lessons we have described the use of constraints to fix an internal coordinate such as a bond length or a torsion angle to a set value or to a series of values. However, if the part of the structure to be constrained is not conveniently described by one or more internal coordinates then we need an alternative approach. Two alternatives are available in MOMEC; fixing of atomic coordinates and definition of regions of the molecule to be refined. Fixing coordinates can be useful in some cases. For instance, a bond can be forced to lie along the $x$ axis by fixing the $y$ and $z$ coordinates of the two atoms to zero. Equally, fixing all coordinates for a number of atoms could be used to generate a rigid group within the molecule. However, fixing coordinates does not lead to a decrease in the computational demands of the energy minimization. The alternative approach of defining a fixed shell does increase the efficiency and will therefore be preferred in many cases, particularly when dealing with larger molecules. In the present lesson we demonstrate how the approach of fixing coordinates can be used as a convenient technique for building a relatively large metal complex.

**Practice**

The compounds that we will study are tris-bidentate complexes with *trans*-cyclohexane-1,2-diamine (see Section 17.6). We start from $[Co(NH_3)_6]^{3+}$ (see Section 17.1). Using the usual tools in HyperChem, edit that complex so that it has six identical Co-N distances of 1.955 Å and valence angles involving Co of 90° or 180°. Select the cobalt and the six nitrogen atoms with the select tool ⊚ and save this structure; make sure that the same name also appears as the MOMEC input structure (*Setup/Files/Input*). Open the module that allows you to compute the structure with the metal center and the six donor atoms fixed: *Execute/Rigid Geometry/Fixed Coordinates*. *Read File* will mark the coordinates to be fixed, i.e., the x,y,z coordinates of the atoms that you have selected in HyperChem. The window shown in Fig. 17.20.1 will appear.

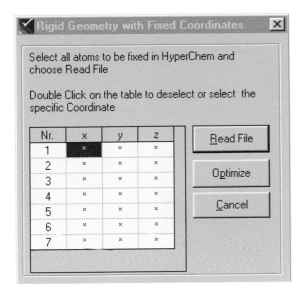

**Fig. 17.20.1**

Click on *Optimize* to start the refinement with the parameters specified in *Setup/Optimization Controls*.

Replace two H-atoms of cis-disposed amines by C-atoms, connect them and add the missing H-atoms (*Add Hydrogens* in HyperChem) to get $[Co(en)(NH_3)_4]^{3+}$. Keep all atoms except the two carbons and all hydrogens of the ethane-1,2-diamine chelate ring fixed and refine again in the rigid geometry mode.

Complete the chelate ligand to get one coordinated cyclohexane-1,2-diamine (Section 17.6) and refine the new complex, leaving only the four new carbon atoms and their hydrogen atoms not fixed.

As a last step, you can now dock the refined cyclohexane-1,2-diamine ligand with methods that we have seen in Section 17.6 to the other two chelating positions and refine the compound with the Co, the six nitrogen and the six carbon atoms of the $[Co(en)_3]^{3+}$ fragment fixed. You will get the following result:

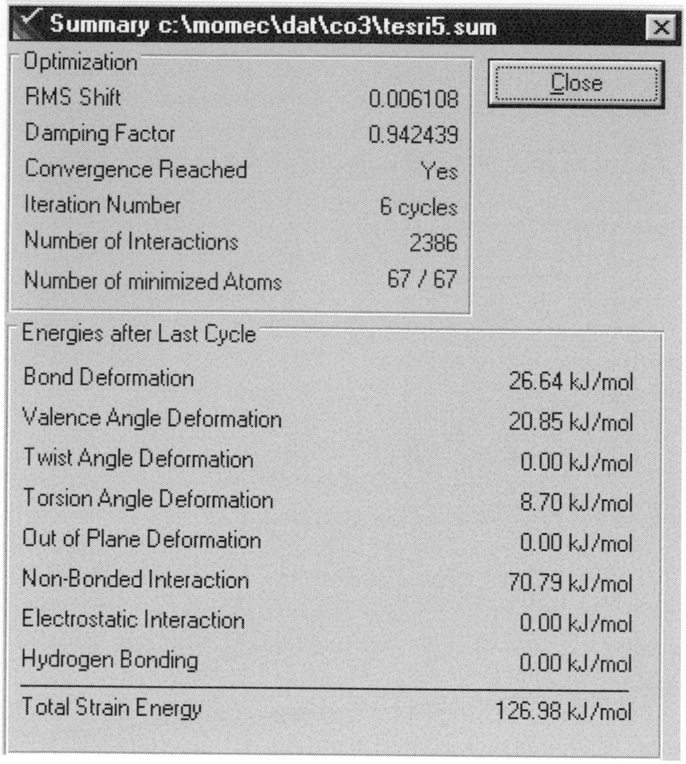

Fig. 17.20.2

Due to the restriction in terms of degrees of freedom convergence is usually reached very quickly.

Refine the same molecule without fixing any coordinates:

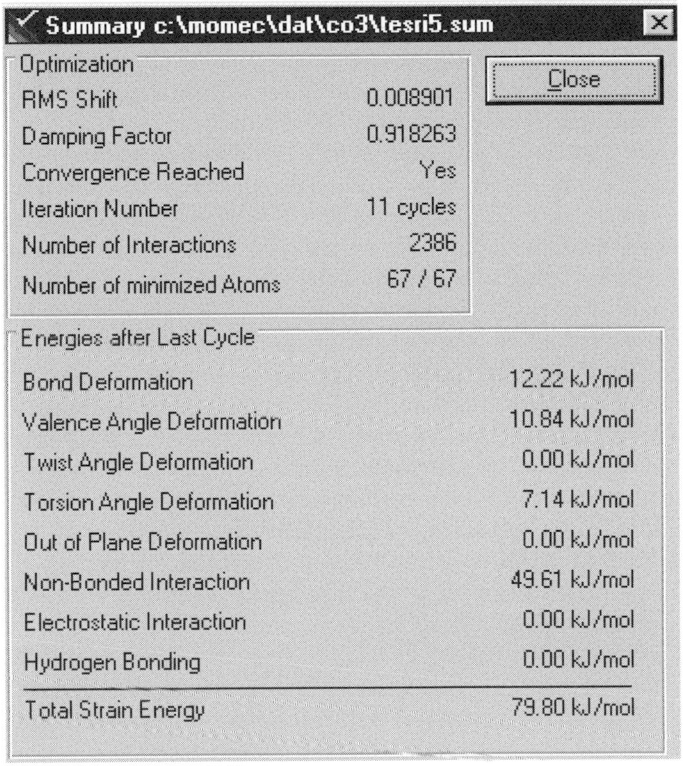

**Fig. 17.20.3**

The strain energy decreases substantially, as anticipated, but the refinement, using the same starting structure is much slower (11 instead of 6 cycles).

There are two possible applications of the techniques that we have seen here:

1. The rigid geometry module provides an elegant way for building complex structures.
2. Refinement with rigid groups allows the measurement of the stresses imposed by certain groups on the remainder of the molecule.

**Exercise**

Increase the complexity of the ligands using the same techniques. For example, you can now "cap" the chelate ligands to produce cage ligands with annular diaminocyclohexane rings (see for example Section 17.16).

# Appendices

## 1 Glossary

*ab-initio*
Based on first principles. Used for rigorous quantum chemistry, that is for MO calculations based on Slater determinants. Generally, the Schrödinger equation ($H\Psi = E\Psi$) is solved in the BO approximation (see *Born-Oppenheimer approximation*) with a large but finite basis set of atomic orbitals.

*Born-Oppenheimer approximation*
Separation of the movement of the nuclei and electrons. This is possible because the electrons move much more rapidly (smaller mass) than the nuclei. The position of the nuclei is fixed for the calculation of the electronic Schrödinger equation (in MO calculations the nuclear positions are then parameters, not quantum chemical variables). Born-Oppenheimer surfaces are energy vs. nuclear structure plots which are (n+1)-dimensional, where n is 3N-6 with N atoms (see *potential energy surface*).

*Computer Assisted Molecular Design (CAMD)*
Methods based on molecular graphics, molecular mechanics, molecular dynamics and MO calculations, used for the design of new compounds. CAMD is not equivalent but rather a subdiscipline of molecular modeling (see *molecular modeling, molecular mechanics, molecular dynamics*).

*Conformational analysis*
Experimental or theoretical delineation of the number of conformers of a system and of the population of each conformer. Also used to describe a molecular mechanics analysis of the conformers (see *molecular mechanics*).

*Conformational search*
Scanning of a potential energy surface. Only deterministic methods (that is point by point searches) are fully reliable. However, these are in practice, due to the enormous computational effort, hardly ever possible. The methods currently used include random search methods (stochastic search, for example Monte Carlo methods) and mo-

lecular dynamics (see *potential energy surface, deterministic search, Monte Carlo search, stochastic search, molecular dynamics, scanning an energy surface*).

*Constraint*
Mathematically precise fixing of internal coordinates. This is only possible with energy minimization by second derivative methods (*see restraint*).

*Deterministic search*
Systematic scanning of the potential energy surface, that is, generation of starting geometries by systematic variation of internal coordinates, followed by structure optimization (see *potential energy surface, conformational search*).

*Dummy atoms*
Non-existent atoms used for the definition of a structure if the program does not allow a specific connectivity (for example, coordination numbers greater than four) or if certain interactions and geometries (for example, Jahn-Teller distortions, planarity of a system, such as a square planar nickel(II) complex) cannot be modeled in another way.

*Electronic structure*
Arrangement and population of molecular orbitals.

*Empirical force field calculations*
Molecular mechanics or dynamics calculations (see *molecular mechanics, molecular dynamics*).

*Energy minimum*
A minimum, local or global, on a potential energy surface. In other words: any small change of the nuclear coordinates on the 3N-6 dimensional surface will lead to a loss of energy, that is, there always is a force driving the molecule back to this minimum (see *energy surface, Born-Oppenheimer approximation, saddle point*).

*Energy surface*
A multidimensional plot of energy as a function of internal or cartesian coordinates (see *potential energy surface*).

*Equilibrium parameter value*
See *ideal parameter value*

*Force constant*
Steepness of a potential energy function, for instance, spring strength in Hookes' law. Empirical force constants are different from spectroscopic force constants (see *molecular mechanics, force field*).

*Force field*
Collection of numbers that parameterize the potential energy functions. These include the force constants, the ideal distances and angles, and parameters for van der Waals, electrostatic and other terms. Since the force field parameters are dependent on the potential energy functions, the entire set of functions and parameters are sometimes referred to as "the force field".

*Free energy perturbation calculations*
Free energy perturbation calculations use Monte Carlo or molecular-dynamics approaches to calculate relative free energies (e.g., of solvation or host-guest binding) on the basis of thermodynamic cycles involving the binding processes of two related systems and the artificial mutation of one system into the other.

*Global energy minimum*
The energy minimum with the lowest energy value on an energy surface (see *energy minimum, local energy minimum*).

*Ideal parameter value*
Ideal (or equilibrium) bond lengths, valence angles etc. are the distances and angles that each atom in a molecule wants to be from each other atom (see *force field*).

*Improper torsion*
Torsional function applied to four atoms not connected by consecutive bonds. Often used to enforce planarity in conjugated systems.

*Internal coordinates*
Distances and angles. Structures can be presented in an internal coordinate system (symmetry adapted coordinates used in spectroscopy or Z-matrices, that is interatomic distances, three center angles and four center angles) instead of a global coordinate system (coordinate triples, for example cartesian, crystal, cylindrical or spherical coordinates).

*Local energy minimum*
A minimum on an energy surface that has an energy value greater than that of the global minimum (see *energy minimum, global energy minimum*).

*Mapping an energy surface*
Calculation of the shape of an energy surface. Location of minima, maxima and saddle points for the evaluation of reaction pathways (see *potential energy surface, energy minimum, saddle points, scanning of an energy surface*).

*Minimizer*
Algorithm used for the minimization of mathematical functions (steepest descent, gradient, second derivative and other methods). In molecular mechanics these are

used to minimize the strain energy and thereby optimize the molecular structure (see *molecular mechanics*).

*Molecular dynamics*
Calculation of the molecular structure at successive small time intervals using a molecular mechanics force field, with the shifts determined using Newton's laws of motion.

*Molecular mechanics*
Calculation of the molecular structure and the corresponding strain energy by minimization of a total energy, calculated using functions which relate internal co-ordinates to energy values (see *conformational analysis*).

*Molecular modeling*
Visualization and analysis of structures, molecular properties (thermodynamics, reactivity, spectroscopy) and molecular interactions, based on a theoretical means for predicting the structures and properties of molecules and complexes (see *computer assisted molecular design*).

*Molecular structure*
Three dimensional arrangement of atoms in a molecule.

*Monte Carlo search*
Random search of configurational space followed by calculation of probabilities based on a Boltzmann distribution (see *energy surface, conformational search, stochastic search*).

*PES*
see *potential energy surface*

*Potential energy function*
A function which relates the deformation of an internal coordinate to the energy cost of the deformation (see *force field*).

*Potential energy surface*
A multidimensional plot of potential energy as a function of internal or cartesian coordinates With N nuclei the surface is $F = 3N-6$ ($3N-5$ for linear molecules) dimensional, and it is represented in the $F+1$ dimensional space. Usually, one-dimensional projections are used for visualization (see *Born-Oppenheimer approximation, energy minimum, saddle point*).

*QSAR*
Quantitative-structure-activity relationship. A quantitative relationship between a specified activity and the structure of a compound. Usually used in pharmacology, where the activity refers to biological activity but also of use in coordination chemistry.

*QSPR*
Quantitative-structure-property relationship. A quantitative relationship between a specified property and the structure of a compound. Very similar to *QSAR* but of more relevance in coordination chemistry, where properties rather than activities are usually referred to.

*Restraints*
Fixation of a structural parameter by artificially large force constants to drive an internal coordinate close to a selected value (see *constraints*).

*Saddle point*
A point on an energy surface which is a local minimum with respect to some directions but a local maximum with respect to at least one other direction, that is, representing a transition state that will decay in the direction where there is no back driving force (see *energy minimum*).

*Scanning an energy surface*
Searching for all possible minima on an energy surface for the evaluation of conformers and their population (see *potential energy surface, conformational analysis, mapping an energy surface*).

*Stochastic search*
Random search of an energy surface (see *conformational search, deterministic search, Monte Carlo search, molecular dynamics*).

*Strain energy*
The energy penalty associated with deforming an internal coordinate.

*Strain*
The deformation in a molecule that results from stresses (see *stress*).

*Stress*
The interactions between atoms in a molecule that lead to distortion and strain (see *strain*).

*Unstrained parameter value*
See *ideal parameter value*

*Urey-Bradley force field*
A force field in which 1,3-nonbonded interactions are included.

## 2    Fundamental Constants, Units and Conversion Factors

*Constants*

| | |
|---|---|
| Avogadro constant | $N_A = 6.022 \ 10^{23} \ mol^{-1}$ |
| Gas constant | $R \ = 8.314 \ J \ mol^{-1} \ K^{-1}$ |
| Planck constant | $h \ = 6.626 \ 10^{-34} \ J \ s$ |
| Boltzmann constant | $k \ = 1.381 \ 10^{-23} \ J \ K^{-1} \ (= R/N)$ |
| Faraday constant | $F \ = 9.649 \ 10^4 \ C \ mol^{-1}$ |
| Charge of an electron | $e \ = 1.602 \ 10^{-19} \ C$ |
| Speed of light | $c \ = 2.997 \ 10^8 \ m \ s^{-1}$ |

*Basic SI units*

| | |
|---|---|
| length | m |
| angle | rad |
| mass | kg |
| time | s |
| temperature | K |
| energy | eV |
| amount | mol |

*Derived units and conversion factors*

| | | | |
|---|---|---|---|
| length | $1 \ Å \ = 10^{-10} \ m$ | | |
| | $1 \ pm = 10^{-12} \ m$ | | |
| | $1 \ nm = 10^{-9} \ m$ | | |
| angle | $1 \ deg = \pi/180 \ rad$ | | |
| force | $1 \ N \ = 1 \ m \ kg \ s^{-2}$ | $1 \ kp \ = 9.81 \ N$ | |
| | | $1 \ dyn = 10^{-5} \ N$ | |
| energy | $1 \ J \ = 1 \ m^2 kg \ s^{-2}$ | $1 \ erg = 10^{-7} \ J$ | |
| | | $1 \ cal \ = 4.184 \ J$ | |
| | | $1 \ eV \ = 1.6022 \ 10^{-19} \ J$ | |

*Energy units in molecular mechanics calculations*

The units for the force constants are usually given in mdyn/Å (bond length, out-of-plane), mdyn/rad (valence and torsional angle bending and rotation), and dimensions of other parameters used are as follows: van der Waals interactions: $A$ (dimensionless), $B$ ($Å^{-1}$), $C$ ($Å^6$); out-of-plane interactions: $\delta$ (Å); electrostatic interactions: $q$ (e), $\varepsilon$ (mdyn $Å^2$ $e^2$); hydrogen bonding interaction: $F$ (mdyn $Å^{13}$), $G$ (mdyn $Å^{11}$). In order to get energies in kJ/mol for each of the potential energy terms (see Eqns. 3.6–3.12) the corresponding results need to be converted. The conversion factors $f$ are as follows: $f(E_b) = 6.022 \times 10^2$; $f(E_\theta) = 6.022 \times 10^2$; $f(E_\phi) = 6.022 \times 10^2$; $f(E_{nb}) = 6.022 \times 10^2$; $f(E_\delta) = 6.022 \times 10^2$; $f(E_\varepsilon = 1.390 \times 10^3$; $f(E_{hb}) = 6.022 \times 10^2$.

# 3   Software and Force Fields

There is an increasing number of commercially available *Molecular Graphics* packages for Apple Macintosh, IBM compatible PCs and workstations (Silicon Graphics and others). Ideally, they are directly linked to *Molecular Mechanics* programs. Alternatively, the *Molecular Graphics* or the *Molecular Mechanics* programs must include file conversion tools that are able to import graphics files into the strain energy minimization programs for structure optimization, and the calculated structures into the graphics program for display and printing (if not available, these conversion tools are simple programs that can be written by the user, that is, it is relatively easy to adapt any existing force field program to an available graphics software). The *Molecular Graphics* program must be able to

-- import experimental structural data
- construct new molecules (model build option)
- freely move the molecule in space
- overlay various structures and calculate rms values
- produce three dimensional models for visualization and printing.

A compilation of available programs and force fields is given in the series *"Reviews in Computational Chemistry"*[524–527], and updated information is available on our molecular modeling homepage.

Basically, any molecular mechanics program can be used for inorganic molecules, if it allows for

- definition of coordination numbers larger than four
- definition of new atom types
- modeling of specific interactions in inorganic molecules (that is, it must include the necessary potential energy functions)
- the addition of new force field parameters and modification of existing ones.

In principle, the force field does not have to be an integral part of the program, though most of the packages used include a basic force field.

# 4   Books on Molecular Modeling and Reviews on Inorganic Molecular Modeling

**List of books on molecular modeling**

Niketic, S. R.; Rasmussen, K.
*The Consistent Force Field*; Springer: Berlin, Heidelberg, New York, ISBN 3540083448 **1977**.

Burkert, U.; Allinger, N. L.
*Molecular Mechanics*; American Chemical Society: New York, ISBN 0841205841 **1982**.

Clark, T.
*A Handbook of Computational Chemistry*; John Wiley & Sons: New York, Chichester, ISBN 0471882119 **1985**.

Warshel, A.
*Computer Modeling of Chemical Reactions in Enzymes and Solutions*; John Wiley & Sons Inc.: New York, ISBN 0471533955 **1991**.

Haile, J. M.
*Molecular Dynamics Simulation Elementary Methods*; John Wiley & Sons Inc.: New York, ISBN 0471819662 **1992**.

Cartwright, H. M.
*Applications of Articial Intelligence in Chemistry*; Oxford University Press: Oxford, ISBN 019855736 **1993**.

Zupan, J.; Gasteiger, J.
*Neural Networks for Chemists: an Introduction*; Wiley-VCH: New York, Weinheim, ISBN 3527286039 **1993**.

Grant, G. H.; Richards, W. G.
*Computational Chemistry*; Oxford University Press: Oxford, ISBN 019855740X **1995**.

Gans, W.; Amann, A.; Boeyens, J. A. C.
*Fundamental Principles of Molecular Modeling*; Plenum Press: New York, London, ISBN 0306453053 **1996**.

Leach, A. R.
*Molecular Modelling*; Longman: Edinburgh, ISBN 0582239338 **1996**.

Doucet, J.-P.; Weber, J.
*Computer-Aided Molecular Design*; Academic Press: London, ISBN 0122212851 **1996**.

Banci, L.; Comba, P.
*Molecular Modeling and Dynamics of Bioinorganic Systems*, Kluwer Academic Publishers: Dordrecht, ISBN 0792348294 **1997**.

Höltje, H.-D.; Folkers, G.
*Molecular Modeling*; VCH: Weinheim, ISBN 3527293841 **1997**.

Rappe, A. K.; Casewit, C. J.
*Molecular Mechanics across Chemistry*; University Science Books: Sausalito, ISBN 0935702776 **1997**.

Gavezzotti, A.
*Theoretical Aspects and Computer Modeling of the Molecular Solid State.* John Wiley & Sons: Chichester, ISBN 0471961876 **1997**.

Goodman, J. M.
*Chemical Applications of Molecular Modelling*; The Royal Society of Chemistry: Cambridge, ISBN 0854045791 **1998**.

Jensen, F.
*Introduction to Computational Chemistry*; Wiley & Sons Ltd.: Chichester, ISBN 0471984256 **1999**.

Lipkowitz, K.B.; Boyd, D.B.
*Reviews in Computional Chemistry*; Wiley-VCH: New York, Weinheim, Vol. 1–14 **1990–2000**.


## List of reviews in the field of inorganic molecular modeling

Brubaker, G.R.; Johnson, D.W.
*Coord. Chem. Rev.* **1984**, *53*, 1.
'Molecular mechanics calculations in coordination chemistry'.

Boeyens, J.C.A.
*Struct. Bonding* (Berlin) **1985**, *63*, 85.
'Molecular mechanics and the structure hypothesis'.

Hancock, R.D.
*Prog. Inorg. Chem.* **1989**, *37*, 187.
'Molecular mechanics calculations as a tool in coordination chemistry'.

Hambley, T.W.
*Comm. Inorg. Chem.* **1992**, *14*, 1.
'What can be learnt from computer-generated models of interactions between DNA and Pt(II) based anti-cancer drugs?'.

Comba, P.
*Coord. Chem. Rev.* **1993**, *123,* 1.
'The relation between ligand structure, coordination stereochemistry and electronic and thermodynamic properties'.

Hay, B.P.
*Coord. Chem. Rev.* **1993,** *126,* 177.
'Methods for molecular mechanics modeling of coordination compounds'.

Comba, P.
*Comm. Inorg. Chem.* **1994**, *16*, 133.
'Solution structures of coordination compounds'.

Deeth, R.J.
*Struct. Bonding* (Berlin) **1995**, *82*, 1.
'Computational modelling of transition metal centres'

Landis, C.R.; Root, D.M.; Cleveland, T., In: 'Reviews in Computational Chemistry', Lipkowitz, K.B.; Boyd, D.B., Eds.,VCH, New York, **1995**, *6,* 73.
'Molecular mechanics force fields for modeling inorganic and organometallic compounds'.

Zimmer, M.
*Chem. Rev.* **1995**, *95*, 2629.
'Bioinorganic molecular mechanics'.

Comba, P.
In: 'Fundamental Priciples of Molecular Modeling', Gans, W.; Amann, A; Boeyens, J.C.A., Eds., Plenum Press, New York, **1996**, 167.
'Modeling of structural and spectroscopic properties of transition metal compounds'.

Comba, P.; Zimmer, M.
*J. Chem. Educ.* **1996**, 73, 108.
'Inorganic molecular mechanics'.

Comba, P.
In: 'Molecular modeling and dynamics of bioinorganic systems', Banci, L.; Comba, P., Eds., Kluwer Academic Publishers, Dordrecht, **1997**, 21.
'Modeling of structures and molecular properties of transition metal compounds – toward metalloprotein modeling'.

Landis, C.R., Firman, T.K., Cleveland, T., Root, D.M.;
In: 'Molecular modeling and dynamics of bioinorganic systems', Banci, L.; Comba, P., Eds., Kluwer Academic Publishers, Dordrecht, **1997**, 49.
'Extending molecular mechanics methods to the discription of transition metal complexes and bond-making and – breaking processes'.

Deeth, R.J., Munslow, I.J.; Paget,V.J.;
In: 'Molecular modeling and dynamics of bioinorganic systems', Banci, L.; Comba, P., Eds., Kluwer Academic Publishers, Dordrecht, **1997**, 77.
'A novel molecular mechanics strategy for transition metals bound to biological molecules'.

Comba, P.
In: 'Intermolecular Interactions', Gans, W.; Boeyens, J.C.A., Eds., Plenum Press, New York, **1998**, 97.
'The importance of intra- and intermolecular weak bonds in transition metal coordination compounds'.

Cundari, T.R.
*J. Chem. Soc., Dalton Trans.* **1998**, 2771.
'Molecular Modeling of d- and f-block Metal Complexes'.

Comba, P.
*Coord. Chem. Rev.* **1999**, *182*, 343.
'Strains and stresses in coordination compounds'.

Comba, P.
*Coord. Chem. Rev.* **1999**, *185*, 81.
'Metal ion selectivity and molecular modeling'.

Comba, P.
In: 'Implications of Molecular and Materials Structure for New Technologies',
Howard, J.A.K.; Allen, F., Eds., Kluwer Academic Publishers, Dordrecht, Boston,
London **1999**, 71.
'Molecular mechanics modeling of transition metal compounds'.

Comba, P.
In: 'Implications of Molecular and Materials Structure for New Technologies',
Howard, J.A.K.; Allen, F., Eds., Kluwer Academic Publishers, Dordrecht, Boston,
London **1999**, 87.
'Modeling structural, spectroscopic and redox properties of transition metal compounds'.

# References

1 Mathieu, J.-P. *Ann. de Phys.* **(1944)** *19,* 335.
2 Corey, E. J.; Bailar Jr., J. C. *J. Am. Chem. Soc.* **(1959)** *81,* 2620.
3 Buckingham, D. A.; Marzilli, L. G.; Sargeson, A. M. *J. Am. Chem. Soc.* **(1967)** 5133.
4 Gollogly, J. R.; Hawkins, C. J. *Aust. J. Chem.* **(1967)** *20,* 2395.
5 Gollogly, J. R.; Hawkins, C. J. *Inorg. Chem.* **(1969)** *8,* 1168.
6 Gollogly, J. R.; Hawkins, C. J. *Inorg. Chem.* **(1970)** *9,* 576.
7 Gollogly, J. R.; Hawkins, C. J.; Beattie, J. K. *Inorg. Chem.* **(1971)** *10,* 317.
8 Gollogly, J. R.; Hawkins, C. J. *Inorg. Chem.* **(1972)** *11,* 156.
9 Bixon, M.; Lifson, S. *Tetrahedron* **(1967)** *23,* 769.
10 Jacob, E. J.; Thompson, H. B.; Bartell, L. S. *J. Chem. Phys.* **(1967)** *47,* 3736.
11 Bartell, L. S. *J. Chem. Educ.* **(1968)** *45,* 754.
12 Allinger, N. L.; Hirsch, J. A.; Miller, M. A.; Tyminski, I. J.; Van-Catledge, F. A. *J. Am. Chem. Soc.* **(1968)** 1199.
13 Boyd, R. H. *J. Chem. Phys.* **(1968)** *49,* 2574.
14 Snow, M. R.; Buckingham, D. A.; Marzilli, D. A.; Sargeson, A. M. *J. Chem. Soc., Chem. Commun.* **(1969)** 891.
15 Snow, M. R. *J. Am. Chem. Soc.* **(1970)** *92,* 3610.
16 Buckingham, D. A.; Maxwell, I. E.; Sargeson, A. M.; Snow, M. R. *J. Am. Chem. Soc.* **(1970)** *92,* 3617.
17 Born, M.; Oppenheimer, J. R. *Ann. Phys.* **(1927)** *84,* 457.
18 Møller, C.; Plesset, M. S. *Phys. Rev.* **(1934)** *46,* 618.
19 Pople, J. A.; Segal, G. *J. Chem. Phys.* **(1966)** *44,* 3289.
20 Pople, J. A.; Beveridge, D. L.; Dobosh, P. A. *J. Chem. Phys.* **(1967)** *47,* 2026.
21 Bingham, R. C.; Dewar, M. J. S.; Lo, D. H. *J. Am. Chem. Soc.* **(1975)** *97,* 1295.
22 Deeth, R. J. *Struct. Bond.* **(1995)** *82,* 1.
23 Anderson, W. P.; Cundari, T. R.; Drago, R. S.; Zerner, M. C. *Inorg. Chem.* **(1980)** *29,* 1.
24 Zerner, M. C. In: *Reviews in Computational Chemistry*; Lipkowitz, K. B. Boyd, D. B. Eds.; VCH: New York, Weinheim, **(1991)** *2,* 313.
25 Jensen, F. *Introduction to Computational Chemistry*; Wiley: New York, 1991.
26 Hoffman, R. *Science* **(1981)** *211,* 995.
27 Hall, M. B.; Fenske, R. F. *Inorg. Chem.* **(1972)** *11,* 768.
28 Koch, W.; Holthausen, M. C. *A Chemist's Guide to Density Functional 5 Theory*; Wiley-VCH: Weinheim, 2000.
29 Ziegler, T. *Chem. Rev.* **(1991)** *91,* 651.
30 Parr, R. G.; Yang, W. *Density Functional Theory of Atoms and Molecules*; Oxford Universtity Press: U.S.A., 1989.
31 Hohenberg, P.; Kohn, W. *Phys. Rev.* **(1964)** *136,* 3804.
32 Kohn, W.; Sham, L. *Phys. Rev.* **(1965)** *140,* A1133.
33 Johnson, K. H. *J. Chem. Phys.* **(1966)** *45,* 3085.
34 Slater, J. C. *Quantum Theory of Molecules and Solids*; McGraw Hill: New York, 1974.
35 Leach, A. R. In: *Reviews in Computational Chemistry*; Lipkowitz, K. B. Boyd, D. B. Eds.; VCH: New York, Weinheim, **(1991)** *2,* 1.

36  Saunders, M.; Houk, K. N.; Wu, Y.-D.; Still, W. C.; Lipton, M.; Chang, G.; Guida, W. C. *J. Am. Chem. Soc.* **(1990)** *112,* 1419.
37  Bartol, J.; Comba, P.; Melter, M.; Zimmer, M. *J. Comput. Chem.* **(1999)** *20,* 1549.
38  Metropolis, N.; Rosenbluth, W.; Rosenbluth, M. N.; Teller, A. H.; Teller, E. *J. Chem. Phys.* **(1953)** *21,* 1087.
39  Haile, J. M. *Molecular Dynamics Simulation: Elementary Methods*; Wiley: New York, 1992.
40  Da Cruz, M. F.; Zimmer, M. *Inorg. Chem.* **(1996)** *35,* 2872.
41  Zabrodsky, H.; Peleg, S.; Avnir, D. *J. Am. Chem. Soc.* **(1992)** *114,* 7843.
42  Pinsky, M.; Avnir, D. *Inorg. Chem.* **(1998)** *37,* 5575.
43  Lybrand, T. P. *Computer Simulation of Biomolecular Systems Using Molecular Dynamics and Free Energy Peturbation Methods*; VCH: New York, Weinheim, 1990.
44  Kollman, P. *Chem. Rev.* **(1993)** *93,* 2395.
45  Lopez, M. A.; Kollman, P. A. *J. Am. Chem. Soc.* **(1989)** *111,* 6212.
46  Hansch, C.; Fujita, T. *J. Am. Chem. Soc.* **(1964)** *86,* 1616.
47  Oprea, T. I.; Waller, C. L. In: *Reviews in Computational Chemistry*; Lipkowitz, K. B., Boyd, D. B. Eds.; VCH: New York, Weinheim, **(1997)** *11,* 127.
48  Bersuker, I. B.; Dimoglo, A. S. In: *Reviews in Computational Chemistry*; Lipkowitz, K. B. Boyd, D. B. Eds.; VCH: New York, Weinheim, **(1991)** *2,* 423.
49  Allinger, N. L. *J. Am. Chem. Soc.* **(1977)** *99,* 8127.
50  *Quantum Chemistry Program Exchange (QCPE)*; Bloomington, In., USA.
51  *HyperChem$^{TM}$*; Hypercube Inc.: 1115 NW 4$^{th}$ Street, Gainesville, Florida 32601, USA.
52  Comba, P.; Hambley, T. W.; Lauer, G.; Melter, M.; Okon, N. *MOMEC, a molecular modeling package for inorganic compounds*; CVS: Heidelberg, 1997.
53  Figgis, B. N.; Skelton, B. W.; White, A. H. *Aust. J. Chem.* **(1979)** *32,* 417.
54  Yoneda, H. *unpublished data.*
55  Whuler, P. A.; Brouty, C.; Spinat, P.; Herpin, P. *Acta Crystallogr., Section B* **(1975)** *B31,* 2069.
56  Brunner, H.; Bürgi, H.-B.; Ludi, A. *unpublished data.*
57  Bernhardt, P. V.; Comba, P. *Inorg. Chem.* **(1992)** *31,* 2638.
58  Comba, P.; Hambley, T. W.; Ströhle, M. *Helv. Chim. Acta.* **(1995)** *78,* 2042.
59  Hambley, T. W.; Hawkins, C. J.; Palmer, J. A.; Snow, M. R. *Aust. J. Chem.* **(1981)** *34,* 45.
60  Saunders, M.; Jarret, R. M. *J. Comput. Chem.* **(1986)** *7,* 578.
61  Rappé, A. K.; Casewit, C. J.; Colwell, K. S.; Goddard III, W. A.; Skiff, W. M. *J. Am. Chem. Soc.* **(1992)** *114,* 10024.
62  Charles, R.; Ganly-Cunningham, M.; Warren, R.; Zimmer, M. *J. Mol. Struct.* **(1992)** *265,* 385.
63  Tueting, J. L.; Spence, K. L.; Zimmer, M. *J. Chem. Soc., Dalton Trans* **(1994)** 551.
64  Clark, T. *A Handbook of Computational Chemistry*; Wiley: New York, Chichester, 1985.
65  Comba, P. *Coord. Chem. Rev.* **(1993)** *123,* 1.
66  Comba, P. *Coord. Chem. Rev.* **(1999)** *185,* 81.
67  Lauher, J. W. *J. Am. Chem. Soc.* **(1986)** *108,* 1521.
68  Sironi, A. *Inorg. Chem.* **(1992)** *31,* 2467.
69  Burton, V. J.; Deeth, R. J. *J. Chem. Soc., Chem. Commun.* **(1995)** 573.
70  Morse, P. M. *Phys. Rev.* **(1929)** *34,* 57.
71  Nakamoto, K. *Infrared and Raman Spectra of Inorganic and Coordination Compounds*; 4th ed.; Wiley: New York Chichester, 1986.
72  Allinger, N. L.; Yuh, Y. H.; Lii, J.-H. *J. Am. Chem. Soc.* **(1989)** *111,* 8551.
73  Lii, J.-H.; Allinger, N. L. *J. Am. Chem. Soc.* **(1989)** *111,* 8576.
74  Lii, J.-H.; Allinger, N. L. *J. Am. Chem. Soc.* **(1989)** *111,* 8566.
75  Allinger, N. L.; Geise, H. J.; Pyckhout, W.; Paquette, L. A.; Galluci, J. C. *J. Am. Chem. Soc.* **(1989)** *111,* 1106.
76  Comba, P.; Zimmer, M. *J. Chem. Educ.* **(1996)** *73,* 108.
77  Deeth, R. J.; Munslow, I. J.; Paget, V. J. In: *Molecular Modeling and Dynamics of Bioinorganic Systems*; Banci, L., Comba, P., Eds.; Kluwer: Dordrecht, 1997.
78  Root, D. M.; Landis, C. R.; Cleveland, T. *J. Am. Chem. Soc.* **(1993)** *115,* 4201.
79  Landis, C. R.; Firman, T. K.; Cleveland, T.; Root, D. M. In: *Molecular Modeling and Dynamics of Bioinorganic Systems*; Banci, L., Comba, P., Eds.; Kluwer: Dordrecht, 1997.
80  Pauling, L. *J. Am. Chem. Soc.* **(1931)** *53,* 1367.

81 Cleveland, T.; Landis, C. R. *J. Am. Chem. Soc.* **(1996)** *118*, 6020.

82 Landis, C. R.; Cleveland, T.; Firman, T. K. *J. Am. Chem. Soc.* **(1995)** *117*, 1859.

83 Mayo, S. L.; Olafson, B. D.; Goddard III, A. *J. Phys. Chem.* **(1990)** *94*, 8897.

84 Allured, V. S.; Kelly, C. M.; Landis, C. R. *J. Am. Chem. Soc.* **(1991)** *113*, 1.

85 Kepert, D. L. *Inorg. Chem.* **(1972)** *11*, 1561.

86 Kepert, D. L. *Prog. Inorg. Chem.* **(1977)** *23*, 1.

87 Kepert, D. L. *Repulsion: Inorganic Chemistry Concepts*, Vol. 6; Springer: Berlin, Heidelberg, New York, 1980.

88 Clare, B. W.; Favas, M. C.; Kepert, D. L.; May, A. S.; Taylor, N. R. *J. Organomet. Chem.* **(1994)** *478*, 111.

89 Comba, P. *Inorg. Chem.* **(1989)** *28*, 426.

90 Bernhardt, P. V.; Comba, P. *Inorg. Chem.* **(1993)** *32*, 2798.

91 Pletnev, I. V.; Melnikov, V. L. *Russ. J. Coord. Chem.* **(1997)** *23*, 188.

92 Lennard-Jones, J. E. *Proc. R. Soc. London, Ser. A.* **(1924)** *106*, 463.

93 Weiner, S. J.; Kollman, P. A.; Case, D. A.; Singh, C.; Ghio, C.; Profeta, S.; Alagona, G.; Weiner, P. *J. Am. Chem. Soc.* **(1984)** *106*, 765.

94 Cornell, W. D.; Cieplak, P.; Bayly, C. I.; Gould, I. R.; Merz, K. M.; Ferguson, D. M.; Spellmeyer, D. C.; Fox, T.; Caldwell, J. W.; Kollman, P. A. *J. Am. Chem. Soc.* **(1995)** *117*, 5179.

95 Buckingham, A. D. *Ann. Rev. Phys. Chem.* **(1970)** *21*, 287.

96 Bol, J. E.; Buning, C.; Comba, P.; Reedijk, J.; Ströhle, M. *J. Comput. Chem.* **(1998)** *19*, 512.

97 Niketic, S. R.; Rasmussen, K. *The Consistent Force Field*; Springer: Berlin, Heidelberg, New York, 1977.

98 Yao, S.; Plastaras, J. P.; Marzilli, L. G. *Inorg. Chem.* **(1994)** *33*, 6061.

99 Hambley, T. W. *Inorg. Chem.* **(1998)** *37*, 3767.

100 Cox, S. R.; Williams, D. E. *J. Comput. Chem.* **(1981)** *2*, 304.

101 Chirlian, L. E.; Francl, M. M. *J. Comput. Chem.* **(1987)** *8*, 894.

102 Abraham, R. J.; Griffiths, L.; Loftus, P. *J. Comput. Chem.* **(1982)** *3*, 407.

103 Sternberg, U.; Koch, F. T.; Möllhoff, M. *J. Comput. Chem.* **(1993)** *15*, 524.

104 Rappé, A. K.; Goddard III, W. A. *J. Phys. Chem.* **(1991)** *95*, 3358.

105 Vedani, A.; Dunitz, J. D. *J. Am. Chem. Soc.* **(1985)** *107*, 7653.

106 Lifson, S.; Warshel, A. *J. Chem. Phys.* **(1968)** *49*, 516.

107 Ermer, O. *Struct. Bond.* **(1976)** *27*, 161.

108 Doman, T. N.; Landis, C. R.; Bosnich, B. *J. Am. Chem. Soc.* **(1992)** *114*, 7264.

109 Comba, P. In: *Implications of Molecular and Materials Structure for New Technologies*; Howard, J. K. A., Allen, F. H., Eds.; Kluwer: Dordrecht, 1999, p71.

110 Pearlman, D. A.; Case, D. A.; Caldwell, J. W.; Ross, W. S.; Cheatham, T. E.; Debolt, S.; Ferguson, D.; Seibel, G.; Kollman, P. *Comput. Phys. Commun.* **(1995)** *91*, 1.

111 Cornell, W. D.; Cieplak, P.; Bayly, C. I.; Gould, I. R.; Merz, K. M.; Ferguson, D. M.; Spellmeyer, D. C.; Fox, T.; Caldwell, J. W.; Kollmann, P. A. *J. Am. Chem. Soc.* **(1995)** *117*, 5179.

112 Canales, C. R.; Zimmer, M. *J. Mol. Struc.* **(1991)** *245*, 341.

113 Bell, T. W.; Guzzo, F.; Drew, M. G. B. *J. Am. Chem. Soc.* **(1991)** *113*, 3115.

114 Endicott, J. F.; Kumar, K.; Schwarz, C. L.; Percovic, M. W.; Lin, W.-K. *J. Am. Chem. Soc.* **(1989)** *111*, 7411.

115 Yoshikawa, Y. *J. Comput. Chem.* **(1990)** *11*, 326.

116 Adam, K. R.; Antolovich, M.; Brigden, L. G.; Lindoy, L. F. *J. Am. Chem. Soc.* **(1991)** *113*, 3346.

117 Yates, P. C.; Marsden, A. K. *Comp. Chem.* **(1994)** *18*, 89.

118 Albinati, A.; Lianza, F.; Berger, H.; Pregosin, P. S.; Rüegger, H.; Kunz, R. W. *Inorg. Chem.* **(1993)** *32*, 478.

119 Hay, B. P.; Rustad, J. R. *J. Am. Chem. Soc.* **(1994)** *116*, 6316.

120 Hancock, R. D. *Prog. Inorg. Chem.* **(1989)** *37*, 187.

121 DeHayes, L. J.; Busch, D. H. *Inorg. Chem.* **(1973)** *12*, 1505.

122 Bierbach, U.; Farrell, N. P. *Inorg. Chem.* **(1997)** *36*, 3657.

123 Hay, B. P. *Coord. Chem. Rev.* **(1993)** *126*, 177.

124 Comba, P.; Gloe, K.; Inoue, K.; Krüger, T.; Stephan, H.; Yoshizuka, K. *Inorg. Chem.* **(1998)** *37*, 3310.

125 Casewit, C. J.; Colwell, K. S.; Rappé, A. K. *J. Am. Chem. Soc.* **(1992)** *114*, 10035.

126  Casewit, C. J.; Colwell, K. S.; Rappé, A. K. *J. Am. Chem. Soc.* **(1992)** *114*, 10046.

127  Rappé, A. K.; Colwell, K. S.; Casewit, C. J. *Inorg. Chem.* **(1993)** *32*, 3438.

128  Norrby, P. O.; Liliefors, T. *J. Comput. Chem.* **(1998)** *19*, 1146.

129  Norrby, P.-O.; Rasmussen, T.; Haller, J.; Strassner, T.; Houk, K. N. *J. Am. Chem. Soc.* **(1999)** *121*, 10186.

130  Schmidt, K. H.; Müller, A. *Inorg. Chem.* **(1975)** *14*, 2183.

131  Nakagawa, I.; Shimanouchi, T. *Spectrochim. Acta* **(1966)** *22*, 1707.

132  Hambley, T. W. *Inorg. Chem.* **(1988)** *27*, 2496.

133  Comba, P. *Coord. Chem. Rev.* **(1999)** *182*, 343.

134  Burkert, U.; Allinger, N. L. *Molecular Mechanics*; American Chemical Society: New York, 1982.

135  Hambley, T. W. *Inorg. Chem.* **(1988)** *27*, 1073.

136  Keck, M. V.; Lippard, S. J. *J. Am. Chem. Soc.* **(1992)** *114*, 3386.

137  Lowe, J. P. *Prog. Phys. Org. Chem.* **(1968)** *6*, 1.

138  Kim, P. H. *J. Phys. Soc. Jpn.* **(1960)** *15*, 445.

139  Hambley, T. W.; Searle, G. H. *Aust. J. Chem.* **(1984)** *37*, 249.

140  Hambley, T. W. *Inorg. Chem.* **(1991)** *30*, 937.

141  Hambley, T. W. *Comm. Inorg. Chem.* **(1992)** *14*, 1.

142  Brown, T. L. *Inorg. Chem.* **(1992)** *31*, 1286.

143  Still, W. C.; Mohamadi, F.; Richards, N. G. J.; Guida, W. C.; Liskamp, R.; Lipton, M.; Caufield, C.; Chang, G.; Hendrickson, T.; Dept. of Chemistry, Columbia University: New York.

144  Abraham, R. J.; Hudson, B. *J. Comput. Chem.* **(1984)** *5*, 562.

145  Pauling, L. *J. Am. Chem. Soc.* **(1948)** *71*, 1461.

146  Whitlow, M.; Teeter, M. M. *J. Am. Chem. Soc.* **(1986)** *108*, 7163.

147  Warshel, A. *J. Phys. Chem.* **(1979)** *83*, 1640.

148  Hall, D.; Pavitt, N. *J. Comput. Chem.* **(1984)** *5*, 441.

149  Brubaker, G. R.; Johnson, D. W. *Inorg. Chem.* **(1983)** *22*, 1422.

150  Brubaker, G. R.; Johnson, D. W. *Coord. Chem. Rev.* **(1984)** *53*, 1.

151  Bond, A. M.; Hambley, T. W.; Snow, M. R. *Inorg. Chem.* **(1985)** *24*, 1920.

152  McDougall, G. J.; Hancock, R. D.; Boeyens, J. C. A. *J. Chem. Soc., Dalton Trans.* **(1978)** 1438.

153  Hambley, T. W. *J. Chem. Soc., Dalton Trans.* **(1986)** 565.

154  Raos, N.; Niketic, S. R.; Simeon, V. *J. Inorg. Biochem.* **(1982)** *16*, 1.

155  Comba, P.; Gyr, T. *Eur. J. Inorg. Chem.* **(1999)** 1787.

156  Raymond, K. N.; Meek, D. W.; Ibers, J. A. *Inorg. Chem.* **(1968)** *7*, 1111.

157  Beattie, J. K.; Hambley, T. W. *unpublished results.*

158  Hancock, R. D. *Acc. Chem. Res.* **(1990)** *23*, 253.

159  Comba, P.; Engelhardt, L. M.; Harrowfield, J. M.; Horn, E.; Sargeson, A. M.; Snow, M. R.; White, A. H. *Inorg. Chem.* **(1985)** *24*, 2325.

160  Saigo, K. *Chem. Lett.* **(1990)** 74.

161  Jorgensen, W. L.; Tirado-Rives, J. *J. Am. Chem. Soc.* **(1988)** *110*, 1657.

162  Brubaker, G. R.; Shi, C.-X.; Johnson, D. W. *Inorg. Chim. Acta* **(1987)** *134*, 15.

163  Sabolovic, J.; Liedl, K. R. *Inorg. Chem.* **(1999)** *38*, 2764.

164  Comba, P. *Comm. Inorg. Chem.* **(1994)** *16*, 133.

165  Bernhardt, P. V.; Comba, P.; Hambley, T. W.; Massoud, S. S.; Stebler, S. *Inorg. Chem.* **(1992)** *31*, 2644.

166  Comba, P.; Hambley, T. W.; Hitchman, M. A.; Stratemeier, H. *Inorg. Chem.* **(1995)** *34*, 3903.

167  Still, W. C.; Tempczyk, A. T.; Hawley, R. C.; Hendrickson, T. *J. Am. Chem. Soc.* **(1990)** *112*, 6127.

168  Kozaki, T.; Morihashi, K.; Kikuchi, O. *J. Am. Chem. Soc.* **(1989)** *111*, 1547.

169  Hilleary, C. J.; Them, T. F.; Tapscott, R. E. *Inorg. Chem.* **(1980)** *19*, 102.

170  Comba, P.; Hambley, T. W.; Zipper, L. *Helv. Chim. Acta* **(1988)** *71*, 1875.

171  Beveridge, D. L.; Mezie, M.; Rhavishanker, G.; Jayaram, B. *J. Biosci.* **(1985)** *8*, 167.

172  Mezei, M.; Methrotra, P. K.; Beveridge, D. L. *J. Am. Chem. Soc.* **(1985)** *107*, 2239.

173  Karplus, M.; Kushick, J. *Macromol.* **(1981)** *14*, 325.

174  Levy, R.; Karplus, M.; Kushick, J.; Perahia, D. *Macromol.* **(1984)** *17*, 1370.

175  Fletcher, R. *Practical Methods of Optimization*; 2. ed.; Wiley: New York, 1980.

176  Press, W. H. *Numerical Recipes: The Art of Scientific Computing*; Cambridge University Press: Cambridge, 1986.

177 Gill, P. E.; Murray, W.; Wright, M. H. *Practical Optimization*; Academic Press: New York, 1988.

178 Carley, A. F.; Morgan, P. H. *Computational Methods in the Chemical Sciences*; Ellis Horwood: Chichester, 1989.

179 Marquardt, D. W. *J. Soc. Indust. Appl. Math.* **(1963)** *11,* 431.

180 Hambley, T. W. *J. Comput. Chem.* **(1987)** *8,* 651.

181 Bernhardt, P. V.; Comba, P.; Hambley, T. W. *Inorg. Chem.* **(1993)** *32,* 2804.

182 Bernhardt, P. V.; Comba, P. *Helv. Chim. Acta* **(1991)** *74,* 1834.

183 Wiberg, K. B.; Boyd, R. H. *J. Am. Chem. Soc.* **(1972)** 8426.

184 Thomas, M. W.; Emerson, D. *J. Mol. Struct.* **(1973)** *16,* 473.

185 van de Graaf, B.; Baas, J. M. A. *Recl. Trav. Chim. Pays.-Bas* **(1980)** *99,* 327.

186 Comba, P.; Okon, N.; Remenyi, R. *J. Comput. Chem.* **(1999)** *20,* 781.

187 Saunders, M.; Houk, K. N.; Wu, Y.-D.; Still, W. C.; Lipton, M.; Chang, G.; Guida, W. C. *J. Am. Chem. Soc.* **(1990)** *112,* 1419.

188 Kaitner, B.; Paulic, N.; Raos, N. *J. Coord. Chem.* **(1991)** *24,* 291.

189 Kurachi, S.; Ohba, S. *Bull. Chem. Soc. Jpn.* **(1992)** *65,* 3033.

190 Cundari, T. R.; Fu, W. *J. Mol. Struct. (Theochem)* **(1998)** *51,* 425.

191 Csiki, C.; Zimmer, M. *J. Biomol. Struct. Dyn.* **(1999)** *17,* 121.

192 Schweizer, W. B. In: *Structure Correlation*; Bürgi, H.-B., Dunitz, J. D., Eds.; VCH: New York, 1994; Vol. 1, p. 369.

193 Chang, G., Guida, W. C.; Still, W. C. *J. Am. Chem. Soc.* **(1989)** *111,* 4379.

194 Saunders, M. *J. Am. Chem. Soc.* **(1987)** *109,* 3150.

195 van Gunsteren, W. F. *First European Conference on Computational Chemistry*, Nancy, France, 1994.

196 Zimmer, M.; Crabtree, R. H. *J. Am. Chem. Soc.* **(1990)** *112,* 1062.

197 Comba, P. In: *Implications of Molecular and Materials Structure for New Technologies*; Howard, J. K. A., Allen, F. H., Eds.; Kluwer: Dordrecht, 1999, p87.

198 Comba, P.; Jakob, H.; Nuber, B.; Keppler, B. K. *Inorg. Chem.* **(1994)** *33,* 3396.

199 Comba, P.; Maeder, M.; Zipper, L. *Helv. Chim. Acta* **(1989)** *72,* 1029.

200 Rüegger, H.; Kunz, R. W.; Ammann, C. J.; Pregosin, P. S. *Magn. Res. Chem.* **(1991)** *29,* 197.

201 Bunel, S.; Ibarra, C.; Moraga, E.; Blasko, A.; Bunton, C. A. *Carbohydr. Res.* **(1993)** *244,* 1.

202 Won, H.; Olson, K. D.; Hare, D. R.; Wolfe, R. S.; Kratky, C.; Summers, M. F. *J. Am. Chem. Soc.* **(1992)** *114,* 6880.

203 Beyreuther, S.; Hunger, J.; Cunskis, S.; Diercks, T.; Frick, A.; Planker, E.; Huttner, G. *Eur. J. Inorg. Chem.* **(1998)** 1641.

204 Comba, P.; Hilfenhaus, P. *J. Chem. Soc., Dalton Trans.* **(1995)** 3269.

205 Comba, P.; Cusack, R.; Fairlie, D. P.; Gahan, L. R.; Hanson, G. R.; Kazmaier, U.; Ramlow, A. *Inorg. Chem.* **(1998)** *37,* 6721.

206 Comba, P.; Gavrish, S. P.; Hay, R. W.; Hilfenhaus, P.; Lampeka, Y. D.; Lightfoot, P.; Peters, A. *Inorg. Chem.* **(1999)** *38,* 1416.

207 Comba, P. *Inorg. Chem.* **(1994)** *33,* 4577.

208 Iwata, M.; Nakatzu, K.; Saito, Y. *Acta Crystallogr.* **(1969)** *B25,* 2562.

209 Rohrbaugh, W. J.; Jacobson, R. A. *Acta Crystallogr., Section B* **(1977)** *B33,* 3254.

210 Enemark, J. H.; Quinby, M. S.; Reed, L. L.; Steuck, M. J.; Walthers, K. K. *Inorg. Chem.* **(1970)** *9,* 2397.

211 Endicott, J. F.; Brubaker, G. R.; Ramasami, T.; Kumar, K.; Dwarakanath, K.; Cassel, J.; Johnson, D. *Inorg. Chem.* **(1983)** *22,* 3754.

212 Bosnich, B.; Poon, C. K.; Tobe, M. L. *Inorg. Chem.* **(1965)** *4,* 1102.

213 Moore, P.; Sachinidis, J.; Willey, G. R. *J. Chem. Soc., Chem. Commun.* **(1983)** 522.

214 Sargeson, A. M. *Pure Appl. Chem.* **(1986)** *58,* 1511.

215 Comba, P.; Sargeson, A. M. *Phosph. Sulf.* **(1986)** *28,* 137.

216 Comba, P.; Gavrish, S. P.; Lampeka, Y. D.; Lightfoot, P.; Peters, A. *J. Chem. Soc., Dalton Trans.* **(1999)** 4099.

217 Bernhardt, P. V.; Comba, P.; Hambley, T. W.; Lawrance, G. A. *Inorg. Chem.* **(1991)** *30,* 942.

218 Royer, D. J.; Schievelbein, V. H.; Kalyanaraman, A. R.; Bertrand, J. A. *Inorg. Chim. Acta* **(1972)** *6,* 307.

219 Boeyens, J. C. A.; Fox, C. C.; Hancock, R. D. *Inorg. Chim. Acta* **(1984)** *87,* 1.

220 Wüthrich, K. *Acc. Chem. Res.* **(1989)** *22,* 36.
221 Comba, P.; Hörmann, A.; Martin, L. L.; Zipper, L. *Helv. Chim. Acta* **(1990)** *73,* 874.
222 Bernhardt, P. V.; Comba, P.; Gyr, T.; Várnagy, K. *Inorg. Chem.* **(1992)** *31,* 1220.
223 Bernhardt, P. V.; Comba, P.; Hambley, T. W.; Sóvágó, I.; Várnagy, K. *J. Chem. Soc., Dalton Trans.* **(1993)** 2023.
224 Niketic, S. R.; Rasmussen, K. *Acta Chem. Scand.* **(1981)** *A35,* 213.
225 Bernhardt, P. V.; Comba, P.; Hambley, T. W.; *Inorg. Chem.* **(1993)** *32,* 2804.
226 Dwyer, F. P.; Sargeson, A. M. *Nature* **(1960)** *187,* 1022.
227 Keene, F. R.; Searle, G. H. *Inorg. Chem.* **(1974)** *13,* 2173.
228 Bond, A. M.; Keene, F. R.; Rumble, N. W.; Searle, G. H.; Snow, M. R. *Inorg, Chem.* **(1978)** *17,* 2847.
229 Dwyer, M.; Searle, G. H. *J. Chem. Soc., Chem. Commun.* **(1972)** 726.
230 Yoshikawa, Y. *Bull. Chem. Soc. Jpn.* **(1976)** *49,* 159.
231 Comba, P.; Sickmüller, A. F. *Inorg. Chem.* **(1997)** *36,* 4500.
232 Keene, F. R.; Searle, G. H. *Inorg. Chem.* **(1972)** *11,* 148.
233 Searle, G. H. *Aust. J. Chem.* **(1977)** *30,* 2525.
234 Davankov, V. A. In: *Ligand Exchange Chromatography*; Davankov, V. A., Navratil, J. D., Walton, H. F., Eds.; CRC Press: Boca Raton, 1988.
235 Bernhardt, P. V.; Comba, P.; Hambley, T. W.; Martin, L. L.; Várnagy, K.; Zipper, L. *Helv. Chim. Acta* **(1992)** *75,* 145.
236 Wilen, S. H.; Collet, A.; Jacques, J. *Tetrahedron* **(1977)** *33,* 2725.
237 Bernal, I.; Cetrullo, J. *J. Coord. Chem.* **(1989)** *20,* 247.
238 Davankov, V. A. *Chromatographia* **(1989)** *27,* 475.
239 Norrby, P.-O. In: *Transition State Modeling for Catalysis*; Truhlar, D. G., Morokuma, K., Eds.; ACS Symposium Series **(1999)**, p. 163.
240 Deeth, R. J.; Elding, L. I. *Inorg. Chem.* **(1996)** *35,* 5019.
241 Bogdan, P. L.; Irwin, J. J.; Bosnich, B. *Organometallics* **(1989)** *8,* 1450.
242 Brown, J. M.; Evans, P. L. *Tetrahedron* **(1988)** *44,* 4905.
243 Doyle, M. P.; Winchester, W. R.; Hoorn, J. A. A.; Lynch, V.; Simonsen, S. H.; Ghosh, R. *J. Am. Chem. Soc.* **(1993)** *115,* 9968.
244 Gugelchuk, M. M.; Houk, K. N. *J. Am. Chem. Soc.* **(1994)** *116,* 330.
245 Norrby, P.-O.; Akermark, B.; Haeffner, F.; Hansson, S.; Blomberg, M. *J. Am. Chem. Soc.* **(1993)** *115,* 4859.
246 Keppler, B. K.; Friesen, C.; Vongerichten, H.; Vogel, E. In: *Metal complexes in cancer chemotherapy*; Keppler, B. K., Ed.; VCH: Weinheim, 1993.
247 Bradley, D. C.; Holloway, C. E. *J. Chem. Soc (A)* **(1969)** 282.
248 Williams, D. H.; Busch, D. H. *J. Am. Chem. Soc.* **(1965)** *87,* 4644.
249 Yamaguchi, M.; Yamamatsu, S.; Furusawa, T.; Yano, S.; Saburi, M.; Yoshikawa, S. *Inorg. Chem.* **(1980)** *19,* 2010.
250 Yamaguchi, M.; Yano, S.; Saburi, M.; Yoshikawa, S. *Inorg. Chem.* **(1980)** *19,* 2016.
251 Comba, P.; Hambley, T. W.; Lawrance, G. A.; Martin, L. L.; Renold, P.; Várnagy, K. *J. Chem. Soc., Dalton Trans.* **(1991)** 277.
252 Bernhardt, P. V.; Comba, P.; Hambley, T. W.; Lawrance, G. A.; Várnagy, K. *J. Chem. Soc., Dalton Trans.* **(1992)** 355.
253 Irving, H.; Williams, R. J. P. *Nature* **(1948)** *162,* 746.
254 Irving, H.; Williams, R. J. P. *J. Chem. Soc.,***(1953)** 3192.
255 Johnson, D. A.; Nelson, P. G. *Inorg. Chem.* **(1995)** *34,* 5666.
256 Johnson, D. A.; Nelson, P. G. *J. Chem. Soc., Dalton Trans.* **(1995)** 3483.
257 Pearson, R. G. *Chemical hardness – Applications from molecules to solids*; VCH-Wiley: Weinheim, 1997.
258 Hancock, R. D. In: *Perspectives in Coordination Chemistry*; Williams, A. F., Floriani, C., Merbach, A. E., Eds.; VCH: Weinheim, 1992.
259 Kushi, Y.; Morimasa, K.; Yoshisugu, K. 35th Symposium on Japanese Coordination Chemistry, Hiroshima, 1985, p 236.
260 Hoard, J. L.; Kennard, C. H. L.; Smith, G. S. *Inorg. Chem.* **(1963)** *2,* 1316.
261 Lind, M. D.; Hamor, M. J.; Hamor, T. A.; Hoard, J. L. *Inorg. Chem.* **(1964)** *3,* 34.

262 Shimoi, M.; Saito, Y.; Ogino, H. *Bull. Chem. Soc. Jpn.* **(1991)** *64,* 2629.
263 Drummond, L. A.; Hendrick, K.; Kanagasundaram, M. J. L.; Lindoy, L. F.; MacPartlin, M.; Tasker, P. A. *Inorg. Chem.* **(1982)** *21,* 3923.
264 Lindoy, L. F. *Progr. Macrocycl. Chem.*; Izatt, R. M., Ed.; **(1986)** *33,* 53.
265 Martin, L. Y.; DeHayes, L. J.; Zompa, L. J.; Busch, D. H. *J. Am. Chem. Soc.* **(1974)** 4046.
266 Busch, D. H. *Acc. Chem. Res.* **(1978)** *11,* 392.
267 Drew, M. G. B.; Yates, P. C. J. *J. Chem. Soc., Dalton Trans.* **(1986)** 2506.
268 Hancock, R. D. *J. Chem. Soc., Dalton Trans.* **(1986)** 2505.
269 Drew, M. G. B.; Hollis, S. T. L.; Yates, P. C. *J. Chem. Soc., Dalton Trans.* **(1985)** 1829.
270 Drew, M. G. B.; Rice, D. A.; bin Silong, S.; Yates, P. C. *J. Chem. Soc., Dalton Trans.* **(1986)** 1081.
271 Haanstra, W. G.; van der Donk, W. A. J. W.; Driessen, W. L.; Reedijk, J.; Wood, J. S.; Drew, M. G. B. *J. Chem. Soc., Dalton Trans.* **(1990)** 3123.
272 Adam, K. R.; Antolovich, M.; Baldwin, D. S.; Brigden, L. G.; Duckworth, P. A.; Lindoy, L. F.; Bashall, A.; McPartlin, M.; Tasker, P. A. *J. Chem. Soc., Dalton Trans.* **(1992)** 1869.
273 Adam, K. R.; Antolovich, M.; Baldwin, D. S.; Duckworth, P. A.; Leong, A. J.; Lindoy, L. F.; McPartlin, M.; Tasker, P. A. *J. Chem. Soc., Dalton Trans.* **(1993)** 1013.
274 Hung, Y.; Martin, L. Y.; Jackels, S. C.; Tait, A. M.; Busch, D. H. *J. Am. Chem. Soc.* **(1977)** *99,* 4029.
275 Hancock, R. D., McDougall, G. J. *J. Am. Chem. Soc.* **(1980)** *102,* 6551.
276 Cram, D. J.; T. Kamda; Helgeson, R. L.; Liu, G. M. *J. Am. Chem. Soc.* **(1979)** *99,* 948.
277 Cram, D. J.; Liu, G. M. *J. Am. Chem. Soc.* **(1985)** *107,* 3657.
278 Cooper, S. R. In: *Crown compounds: Toward future applications*; VCH: Weinheim, New York, 1992.
279 Vallee, B. L.; Williams, R. J. P. *Proc. Natl. Acad. Sci. USA* **(1968)** *59,* 498.
280 Williams, R. J. P. *Eur. J. Biochem.* **(1995)** *234,* 363.
281 Comba, P. *Coord. Chem. Rev.* **(1999)** *182,* 343.
282 Comba, P. *Coord. Chem. Rev.* **(2000)** *200–202,* 217.
283 Lockhart, J. C.; Tomkinson, N. P. *J. Chem. Soc., Perkin Trans. 2* **(1992)** 533.
284 Lockhart, J. C.; Mousley, D. P.; Hill, M. N. S.; Tomkinson, N. P.; Teixidor, F.; Almajano, M. P.; Escriche, L.; Casabo, J. F.; Sillanpaeae, R.; Kivekaes, R. *J. Chem. Soc., Dalton Trans.* **(1992)** 2889.
285 Forsyth, G. A.; Lockhart, J. C. *J. Chem. Soc., Dalton Trans.* **(1994)** 2243.
286 Hay, B. P.; Zhang, D.; Rustad, J. R. *Inorg. Chem.* **(1996)** *35,* 2650.
287 Comba, P.; Fath, A.; Kühner, A.; Nuber, B. *J. Chem. Soc., Dalton Trans.* **(1997)** 1889.
288 Hansch, C. *Acc. Chem. Res.* **(1969)** *2,* 232.
289 Hancock, R. D.; Martell, A. E. *Chem. Rev.* **(1989)** *89,* 1875.
290 Hay, B. P.; Rustad, J. R.; Hostetler, C. J. *J. Am. Chem. Soc.* **(1993)** *115,* 11158.
291 Comba, P.; Goll, W.; Nuber, B.; Várnágy, K. *Eur. J. Inorg. Chem.* **(1998)** 2041.
292 Yoshizuka, K.; Inoue, K.; Ohto, K.; Gloe, K.; Stephan, H.; Rambusch, T.; Comba, P. *Solvent Extraction, in press.*
293 Comba, P. In: *Intermolecular Interactions*; Gans, W., Boeyens, J. C. A., Eds.; Plenum Press: New York, 1998.
294 Anichini, A.; Fabbrizzi, L.; Paoletti, P.; Clay, R. *J. Chem. Soc., Dalton Trans.* **(1978)** 577.
295 Fabbrizzi, L. *J. Chem. Soc., Dalton Trans.* **(1979)** 1857.
296 Thöm, V. J.; Fox, C. C.; Boeyens, J. C. A.; Hancock, R. D. *J. Am. Chem. Soc.* **(1984)** *106,* 5947.
297 Minomura, S.; Drickamer, H. G. *J. Chem. Phys.* **(1961)** *35,* 903.
298 Drickamer, H. G. *J. Chem. Phys.* **(1967)** *47,* 1880.
299 Smith, D. W. *J. Chem. Phys.* **(1969)** *50,* 2784.
300 Bermejo, M.; Pueyo, L. *J. Chem. Phys.* **(1983)** *78,* 854.
301 Hancock, R. D.; Pattrick, G.; Wade, P. W.; Hosken, G. D. *Pure Appl. Chem.* **(1993)** *65,* 473.
302 Comba, P.; Börzel, H.; Pritzkow, H.; Sickmüller, A. *Inorg. Chem.* **(1998)** *37,* 3853.
303 Schäffer, C. E.; Jorgensen, C. K. *Mol. Phys.* **(1965)** *9,* 401.
304 Schäffer, C. E. *Theoret. Chim. Acta* **(1974)** *34,* 237.
305 Larsen, E.; La Mar, G. N. *J. Chem. Educ.* **(1974)** *51,* 633.
306 Hitchman, M. A. *Trans. Met. Chem.* **(1985)** *9,* 1.

307  Gerloch, M. *Magnetism and ligand-field analysis*; Cambridge University Press: Cambridge, London, 1983.

308  Schäffer, C. E. *Inorg. Chim. Acta* **(2000)** *300–302*, 1035.

309  Comba, P.; Hilfenhaus, P.; Nuber, B. *Helv. Chim. Acta* **(1997)** *80*, 1831.

310  Comba, P.; Sickmüller, A. F. *Angew. Chem. Int. Ed. Engl.* **(1997)** *36*, 2006.

311  Bridgeman, A. J.; Gerloch, M. *Prog. Inorg. Chem.* **(1997)** *45*, 179.

312  Vanquickenborne, L. G.; Coussens, B.; Postelmans, D.; Ceulemans, A.; Pierloot, K. *Inorg. Chem.* **(1991)** *30*, 2978.

313  Edwards, W. D.; Weiner, B.; Zerner, M. C. *J. Am. Chem. Soc.* **(1986)** *108*, 2196.

314  Benedix, R.; Henning, H.; Nieke, C. *Inorg. Chim. Acta* **(1990)** *172*, 109.

315  Zarbrodsky, H.; Avnir, D. *Adv. Molec. Struct. Res.* **(1995)** *1*, 1.

316  Avnir, D.; Comba, P.; Keinan, S. *unpublished results*.

317  Gerloch, M. *Comm. Inorg. Chem.* **(1996)** *18*, 101.

318  Duer, M. J.; Essex, S. J.; Gerloch, M. *Mol. Phys.* **(1993)** *79*, 1167.

319  Brown, C. A.; Gerloch, M.; McMeeking, R. F. *Mol. Phys.* **(1988)** *64*, 771.

320  Duer, M. J.; Essex, S. J.; Gerloch, M.; Jupp, K. M. *Mol. Phys.* **(1993)** *79*, 1147.

321  Bridgeman, A. J.; Essex, S. J.; Gerloch, M. *Inorg. Chem.* **(1994)** 5411.

322  Bridgeman, A. J.; Gerloch, S. J. *Mol. Phys.* **(1993)** *79*, 1195.

323  Stratemeier, H.; Hitchman, M. A.; Comba, P.; Bernhardt, P. V.; Riley, M. J. *Inorg.Chem.* **(1991)** *30*, 4088.

324  Smith, T. D.; Pilbrow, J. R. *Coord. Chem. Rev.* **(1974)** *13*, 173.

325  Wang, D.; G.R., H. *J. Magn. Res. Ser. A* **(1995)** *117*, 1.

326  Wang, D.; Hanson, G. R. *Appl. Magn. Res.* **(1996)** *11*, 401.

327  Rosokha, S. V.; Lampeka, V. D.; Maloshtan, I. M. *J. Chem. Soc., Dalton Trans.* **(1993)** 631.

328  Comba, P.; Hambley, T. W.; Hilfenhaus, P.; Richens, D. T. *J. Chem. Soc., Dalton Trans.* **(1996)** 533.

329  Brudenell, S. J.; Bond, A. M.; Spiccia, L.; Comba, P.; Hockless, D. C. R. *Inorg. Chem.* **(1998)** *37*, 3705.

330  Wüthrich, K. *Science* **(1989)** *243*, 45.

331  Abraham, R. J.; Marsden, I.; Xiujing, L. *Magn. Reson. Chem.* **(1990)** *28*, .

332  Banci, L.; Carloni, P.; Orioli, P. L. *Proteins: Struct. Funct. Gen.* **(1994)** *18*, 216.

333  Bertini, I.; Capozzi, F.; Luchinat, C.; Piccioli, M.; Vila, A. J. *J. Am. Chem. Soc.* **(1994)** *116*, 651.

334  Banci, L.; Bertini, I.; Luchinat, C. *Nuclear and Electronic Relexation: The Magnetic Nucleus-Unpaired Electron Coupling in Solution*; VCH: Weinheim, 1991.

335  Bertini, I.; Rosato, A. In: *Molecular Modeling and Dynamics of Bioinorganic Systems*; Banci, L., Comba, P., Eds.; Kluwer: Dordrecht, 1997.

336  Summers, M. F.; South, T. L.; Kim, B.; Hare, D. R. *Biochemistry* **(1991)** *30*, 6342.

337  Bernhardt, P. V.; Jones, L. A.; Sharpe, P. C. *Inorg. Chem.* **(1997)** *36*, 2420.

338  Xie, B.; Elder, T.; Wilson, L. J.; Stanbury, D. M. *Inorg. Chem.* **(1999)** *38*, 12.

339  Lay, P. A. *J. Phys. Chem.* **(1986)** *90*, 878.

340  Golub, G.; Cohen, H.; Meyerstein, D. *J. Chem. Soc., Chem. Commun.* **(1992)** 397.

341  Comba, P.; Jakob, H. *Helv. Chim. Acta* **(1997)** *80*, 1983.

342  Ambundo, E. A.; Deydier, M.-V.; Grall, A. J.; Agnera-Vega, N.; Dressel, L. T.; Cooper, T. H.; Heeg, N. J.; Ochrymowycz, L. A.; Rorabacher , D. B. *Inorg. Chem.* **(1999)** *38*, 4233.

343  Bond, A. M.; Oldham, K. B. *J. Phys. Chem.* **(1983)** *87*, 2492.

344  Bond, A. M.; Oldham, K. B. *J. Phys. Chem.* **(1985)** *89*, 3739.

345  Bond, A. M.; Hambley, T. W.; Mann, D. R.; Snow, M. R. *Inorg. Chem.* **(1987)** *26*, 2257.

346  Geue, R. J.; Hanna, J.; Höhn, A.; Qin, C. J.; Ralph, S. F.; Sargeson, A. M.; Willis, A. C. In: *Electron Transfer Reactions*; Issied, S. S., Ed.; American Chemical Society: New York, 1997.

347  Sutin, N. *Acc. Chem. Res.* **(1982)** *15*, 275.

348  Sutin, N. *Prog. Inorg. Chem.* **(1983)** *30*, 441.

349  Marcus, R. A.; Sutin, N. *Biochem. Biophys. Acta* **(1985)** *811*, 265.

350  Creutz, C. *Prog. Inorg. Chem.* **(1983)** *30*, 1.

351  Geselowitz, D. *Inorg. Chem.* **(1981)** *20*, 4457.

352  Geue, R. J.; Pizer, R.; Sargeson, A. M. *Abstr. 183rd National Meeting of the American Chemical Society, Las Vegas, NV, April 1982*; INOR 62: Washington DC, 1982.

353  Gao, Y. D.; Lipkowitz, K. B.; Schultz, F. A. *J. Am. Chem. Soc.* **(1995)** *117*, 11932.

354 Buning, C.; Canters, G. W.; Comba, P.; Dennison, C.; Jeuken, L.; Melter, M.; Sanders-Loehr, J. *J. Am. Chem. Soc.* **(2000)** *122,* 204.

355 Kunz, R. W. *Molcular Modelling für Anwender*; Teubner: Stuttgart, 1991.

356 Maseras, F.; Morokuma, K. *J. Comput. Chem.* **(1995)** *16,* 1170.

357 Burton, V. J.; Deeth, R. J.; Kemp, C. M.; Gilbert, P. J. *J. Am. Chem. Soc.* **(1995)** *117,* 8407.

358 Brunner, H.; Bürgi, H.-B.; Ludi, A. *unpublished results.*

359 Vezzosi, I. M.; Benedetti, A.; Sladani, M.; Bataglia, L. P.; Corradi, A. B. *Inorg. Chim. Acta* **(1985)** *97,* .

360 Saburi, M.; Miyamura, K.; Morita, M.; Yoshikawa, S.; Tsuboyama, S.; Sakurai, T.; Yamazaki, H.; Tsuboyama, K. *Bull. Chem. Soc. Jpn.* **(1987)** *60,* 2581.

361 Huheey, J. E. *Inorganic chemistry – Principles of structure and reactivity*; 2nd ed.; Harper & Row: New York, 1978.

362 Jahn, H. A.; Teller, E. *Proc. R. Soc.* **(1937)** *A161,* 220.

363 Reinen, D.; Atanasov, M. *Magn. Res. Rev.* **(1991)** *15,* 167.

364 Spiro, T. G. *Copper proteins*; Wiley: Chichester, New York, 1981.

365 Hay, R. W. *Bio-inorganic chemistry*; Ellis Horwood: London, New York, 1984.

366 Reedijk, J. *Bioinorganic Catalysis*; Marcel Dekker Inc.: New York, Basel, 1993.

367 Karlin, D. K. *Bioinorganic chemistry of copper*; Chapman & Hall: London, New York, 1993.

368 Brubaker, G. R.; Johnson, D. W. *Inorg. Chem.* **(1984)** *23,* 1591.

369 Wade, P. W.; Hancock, R. D. *J. Chem. Soc., Dalton* **(1990)** 1323.

370 Wiesemann, F.; Teipel, S.; Krebs, B.; Hoeweler, U. *Inorg. Chem.* **(1994)** *33,* 1891.

371 Paulic, N.; Raos, N. *J. Coord. Chem.* **(1994)** *31,* 135.

372 Comba, P.; Zimmer, M. *Inorg. Chem.* **(1994)** *33,* 5368.

373 Deeth, R. J.; Hitchman, M. A. *Inorg. Chem.* **(1986)** *25,* 1225.

374 Ammeter, J. H.; Buergi, H. B.; Gamp, E.; Meyer-Sandrin, V.; Jensen, W. P. *Inorg. Chem.* **(1979)** *18,* 733.

375 Anderson, O. P. *J. Chem. Soc., Dalton* **(1972)** 2597.

376 Anderson, O. P. *J. Chem. Soc., Dalton* **(1973)** 1237.

377 Allmann, R.; Henke, W.; Reinen, D. *Inorg. Chem.* **(1978)** *17,* 378.

378 Bertini, I.; Dapporto, P.; Gatteschi, D.; Scozzafava, A. *J. Chem. Soc., Dalton Trans.* **(1979)** 1409.

379 Chaudhuri, P.; Oder, K.; Wieghardt, K.; Weiss, J.; Reedijk, J.; Hinrichs, W.; Wood, J.; Ozarowski, A.; Stratemaier, H.; Reinen, D. *Inorg. Chem.* **(1986)** *25,* 2951.

380 Astley, T.; Canty, A. J.; Hitchman, M. A.; Rowbottom, G. L.; Skelton, B. W.; White, A. H. *J. Chem. Soc., Dalton Trans.* **(1991)** 1981.

381 Faus, J.; Julve, M.; Amigo, J. M.; Debaerdemaeker, T. *J. Chem. Soc., Dalton Trans.* **(1989)** 1681.

382 McFadden, D. L.; McPhail, A. T.; Gross, P. M.; Garner, C. D.; Mabbs, F. E. *J. Chem. Soc., Dalton Trans.* **(1975)** 263.

383 Veidis, M. V.; Schreiber, G. H.; Gough, T. E.; Palenik, G. J. *J. Am. Chem. Soc.* **(1969)** *91,* 1859.

384 *Molecular Modeling and Dynamics of Bioinorganic Compounds*; Banci, L.; Comba, P., Eds.; Kluwer Academic Publishers: Dordrecht, 1997.

385 Freeman, H. C. *Adv. Protein Chem.* **(1967)** *22,* 257.

386 Hambley, T. W. *Acta Crystallogr.* **(1988)** *B44,* 601.

387 Kaitner, B.; Ferguson, G.; Paulic, N.; Raos, N. *J. Coord. Chem.* **(1993)** *30,* 233.

388 Colman, P. A.; Freeman, H. C.; Guss, J. M.; Murata, M.; Norris, V. A.; Ramshaw, J. A.; Venkatappa, M. P. *Nature (London)* **(1978)** *272,* 319.

389 Kim, J.; Rees, D. C. *Science* **(1992)** *257,* 1677.

390 Fields, B. A.; Guss, J. M.; Freeman, H. C. *J. Mol. Biol.* **(1991)** *222,* 1053.

391 Comba, P. In: *Molecular modeling and dynamics of bioinorganic systems*; Banci, L., Comba, P., Eds.; Kluwer: Dorderecht, Boston, 1997.

392 Vedani, A.; Dobler, M.; Dunitz, J. D. *J. Comput. Chem.* **(1986)** *7,* 701.

393 Vedani, A.; Huhta, D. W.; Jacober, S. P. *J. Am. Chem. Soc.* **(1989)** *111,* 4075.

394 Vedani, A. *J. Comput. Chem.* **(1988)** *9,* 269.

395 Vedani, A.; Huhta, D. W. *J. Am. Chem. Soc.* **(1990)** *112,* 4759.

396 Comba, P.; Hilfenhaus, P.; Karlin, K. D. *Inorg. Chem.* **(1997)** *36,* 2309.

397 Börzel, H.; Comba, P.; Katsichtis, C.; Kiefer, W.; Lienke, A.; Nagel, V.; Pritzkow, H. *Chem. Eur. J.* **(1999)** *5,* 1716.

398  Hancock, R. D.; Weaving, J. S.; Marques, H. M. *J. Chem. Soc., Chem. Commun.* **(1989)** 1176.
399  Collins, J. R.; Du, P.; Loew, G. H. *Biochemistry* **(1992)** *31,* 11 166.
400  Kuczera, K.; Kuriyan, J.; Karplus, M. *J. Mol. Biol.* **(1990)** *213,* 351.
401  Munro, O. Q.; Bradley, J. C.; Hancock, R. D.; Marques, H. M.; Marsicano, F. W. *J. Am. Chem. Soc.* **(1992)** *114,* 7218.
402  Shelnutt, J. A.; Medforth, C. J.; Berber, M. D.; Barkigia, K. M.; Smith, K. M. *J. Am. Chem. Soc.* **(1991)** *113,* 4077.
403  Medforth, C. J.; Senge, M. O.; Smith, K. M.; Sparks, L. D.; Shelnutt, J. A. *J. Am. Chem. Soc.* **(1992)** *114,* 9859.
404  Senge, M. O.; Medforth, C. J.; Sparks, L. D.; Shelnutt, J. A.; Smith, K. M. *Inorg. Chem.* **(1993)** *32,* 1716.
405  Zimmer, M. *J. Biomol. Struct. Dyn.* **(1993)** *11,* 203.
406  Kaplan, W. A.; Suslick, K. S.; Scott, R. A. *J. Am. Chem. Soc.* **(1991)** *113,* 9824.
407  Marques, H. M.; Brown, K. L. *Coord. Chem. Rev.* **(1999)** *190–192,* 127.
408  Kozelka, J.; Savinelli, R.; Berthier, G.; Flament, J.-P.; Lavery, R. *J. Comput. Chem.* **(1993)** *14,* 45.
409  Kozelka, J.; Petsko, G. A.; Lippard, S. J.; Quigley, G. J. *J. Am. Chem. Soc.* **(1985)** *107,* 4079.
410  Kozelka, J.; Petsko, G. A.; Quigley, G. J.; Lippard, S. J. *Inorg. Chem.* **(1986)** *25,* 1075.
411  Kozelka, J.; Archer, S.; Petsko, G. A.; Lippard, S. J.; Quigley, G. J. *Biopolymers* **(1987)** *26,* 1245.
412  McCarthy, S. L.; Hinde, R. J.; Miller, K. J.; Anderson, J. S.; Basch, H.; Krauss, M. *Biopolymers* **(1990)** *29,* 785.
413  Herman, F.; Kozelka, J.; Stoven, V.; Guittet, E.; Girault, J.-P.; Huynh-Dinh, T.; Igolen, J.; Lallemand, J.-Y.; Chottard, J.-C. *Eur. J. Biochem.* **(1990)** *194,* 119.
414  Hambley, T. W. *J. Chem. Soc., Chem. Commun.* **(1988)** 221.
415  Sip, M.; Schwartz, A.; Vovelle, F.; Ptak, M.; Leng, M. *Biochemistry* **(1992)** *31,* 2508.
416  Ling, E. C. H.; Allen, G. W.; Hambley, T. W. *J. Am. Chem. Soc.* **(1994)** *116,* 2673.
417  Fenton, R. R.; Easdale, W. J.; Er, H. M.; O'Mara, S. M.; McKeage, M. J.; Russell, P. J.; Hambley, T. W. *J. Med. Chem.* **(1997)** *40,* 1090.
418  Lepre, C. A.; Chassot, L.; Costello, C. E.; Lippard, S. J. *Biochemistry* **(1990)** *29,* 811.
419  Prevost, C.; Boudivillain, M.; Beudaert, P.; Leng, M.; Vovelle, F. *J. Biomol. Struct. Dynam.* **(1997)** *14,* 703.
420  Hambley, T. W. *Inorg. Chim. Acta* **(1987)** *137,* 15.
421  Yuriev, E.; Orbell, J. D. *J. Comput.-Aided Mol. Des.* **(1996)** *10,* 589.
422  Yuriev, E.; Orbell, J. D. *Inorg. Chem.* **(1996)** *35,* 7914.
423  Yuriev, E.; Orbell, J. D. *Inorg. Chem.* **(1998)** *37,* 6269.
424  Qu, Y.; Bloemink, M. J.; Reedijk, J.; Hambley, T. W.; Farrell, N. P. *J. Am. Chem. Soc.* **(1996)** *118,* 9307.
425  Brabec, V.; Kasparkova, J.; Vrana, O.; Novakova, O.; Cox, J. W.; Qu, Y.; Farrell, N. P. *Biochemistry* **(1999)** *38,* 6781.
426  Zakovska, A.; Novakova, O.; Balcarova, Z.; Bierbach, U.; Farrell, N. P.; Brabec, V. *Eur. J. Biochem.* **(1998)** *254,* 547.
427  Hambley, T. W.; Ling, E. C. H.; Messerle, B. A. *Inorg. Chem.* **(1996)** *35,* 4663.
428  Gelasco, A.; Lippard, S. J. *Biochemistry* **(1998)** *37,* 9230.
429  Dunham, S. U.; Dunham, S. U.; Turner, C. J.; Lippard, S. J. *J. Am. Chem. Soc.* **(1998)** *120,* 5395.
430  Fouchet, M.-H.; Guittet, E.; Cognet, J. A. H.; Kozelka, J.; Gauthier, C.; Bret, J. L.; Zimmermann, K.; Chottard, J.-C. *J. Biol. Inorg. Chem.* **(1997)** *2,* 83.
431  Murdoch, P.; Guo, Z.; Parkinson, J. A.; Sadler, P. J. *J. Biol. Inorg. Chem.* **(1999)** *4,* 32.
432  Huang, H.; Zhu, L.; Reid, B. R.; Drobny, G. P.; Hopking, P. B. *Science* **(1995)** *270,* 1842.
433  Paquet, F.; Perez, C.; Leng, M.; Lancelot, G.; Malinge, J. M. *J. Biomol. Struc. Dynam.* **(1996)** *4,* 67.
434  Rehmann, J. P.; Barton, J. K. *Biochemistry* **(1990)** *29,* 1710.
435  Shanzer, A.; Libman, J.; Lifson, S.; Felder, C. E. *J. Am. Chem. Soc.* **(1986)** *108,* 7609.
436  Wu, Y.-D.; Houk, K. N.; J.S., V.; Nam, W. *Inorg. Chem.* **(1992)** *31,* 718.
437  Lin, W.; Welsh, W. J.; Harris, W. R. *Inorg. Chem.* **(1994)** *33,* 884.
438  Fröhlich, N.; Frenking, G. In: *Solid State Organometallic Chemistry: Methods and Applications*; Gielen, M., Willem, R., Wrackmeyer, B., Eds.; Wiley: New York, 1999, p. 173.

439 Boeyens, J. C. A.; Levendis, D. C. *S. Afr. J. Chem.* **(1982)** *35,* 144.
440 Shubina, E. S.; Epstein, L. M.; Timofeeva, T. V.; Struchkov, Y. T.; Kreindlin, A. Z.; Fadeeva, S. S.; Rybinskaya, M. I. *J. Orgamet. Chem.* **(1988)** *346,* 56.
441 Menger, F. M.; Sherrod, M. J. *J. Am. Chem. Soc.* **(1988)** *110,* 8606.
442 Thiem, H.-J.; Brandl, M.; Breslow, R. *J. Am. Chem. Soc.* **(1988)** *110,* 8612.
443 du Plooy, K. E.; Marais, C. F.; Carlton, L.; Hunter, R.; Boeyens, J. C. A.; Coville, N. J. *Inorg. Chem.* **(1989)** *28,* 3855.
444 Beer, P. D.; Tite, E. L.; Drew, M. G. B.; Ibbotson, A. *J. Chem. Soc., Dalton Trans.* **(1990)** 2543.
445 Mackie, S. C.; Park, Y.-S.; Shurvell, H. F.; Baird, M. C. *Organometallics* **(1991)** *10,* 2993.
446 Davies, S. G.; Derome, A. E.; McNally, J. P. *J. Am. Chem. Soc.* **(1991)** *113,* 2854.
447 Rudzinski, J. M.; Osawa, E. *J. Phys. Org. Chem.* **(1992)** *5,* 382.
448 Rudzinski, J. M.; Osawa, E. *J. Phys. Org. Chem.* **(1993)** *6,* 107.
449 Blom, R.; Hammel, A.; Haaland, A.; Weidlein, J. T., T. V.; Struchkov, Y. T. *J. Organomet. Chem.* **(1993)** *462,* 131.
450 Leek, Y. v. d.; Angermund, K.; Reffke, M.; Kleinschmidt, R.; Goretzki, R.; Fink, G. *Chem. Eur. J.* **(1997)** *3,* 585.
451 Gajewski, J. J.; Gilbert, K. E.; Kreek, T. W. *J. Comput. Chem.* **(1998)** *19,* 1167.
452 Brintzinger, H.-H.; Prosenc, M.-H.; Schaper, F.; Weeber, A.; Wieser, U. *J. Mol. Struct.* **(1999)** *485–486,* 409.
453 Oslob, J. D.; Akermark, B.; Helquist, P.; Norrby, P.-O. *Organometallics* **(1997)** *16,* 3015.
454 Hagelin, H.; Akermark, B.; Norrby, P.-O. *Organometallics* **(1999)** *18,* 2884.
455 Castonguay, L. A.; Rappe, A. K.; Casewit, C. J. *J. Am. Chem. Soc.* **(1991)** *113,* 7177.
456 Brown, J. M.; Evans, P. L.; Lucy, A. R. *J. Chem Soc., Perkin Trans. II* **(1987)** 1589.
457 Hunger, J.; Beyreuther, S.; Huttner, G. *J. Mol. Mod.* **(1996)** *2,* 257.
458 Beyreuther, S.; Hunger, J.; Huttner, G. *J. Mol. Mod.* **(1996)** *2,* 273.
459 Hunger, J.; Beyreuther, S.; Huttner, G.; Allinger, K.; Radelow, U.; Zsolnai, L. *Eur. J. Inorg. Chem.* **(1998)** 693.
460 Hunger, J.; Huttner, G. *J. Comput. Chem.* **(1999)** *20,* 455.
461 Beyreuther, S.; Frick, A.; Hunger, J.; Huttner, G.; Antelmann, B.; Schober, P.; Soltek, R. *Eur. J. Inorg. Chem.* **(2000)** 597.
462 Tolman, C. A. *J. Am. Chem. Soc.* **(1970)** *92,* 2956.
463 Tolman, C. A. *Chem. Rev.* **(1977)** *77,* 313.
464 Stahl, L.; Ernst, R. D. *J. Am. Chem. Soc.* **(1987)** *109,* 5673.
465 Xing-Fu, L.; Tian-Gi, S.; Ao-Ling, G.; Guang-Li, S.; Peng-Nian, S. *Inorg. Chim. Acta* **(1987)** *129,* 227.
466 Xing-Fu, L.; Tian-Gi, S.; Ao-Ling, G.; Guang-Li, S.; Peng-Nian, S. *Inorg. Chim. Acta* **(1987)** *129,* 235.
467 Xing-Fu, L.; Peng-Nian, S.; Ao-Ling, G.; Jing-Ci, L. *Inorg. Chim. Acta* **(1987)** *129,* 245.
468 Caffery, M. L.; Brown, T. L. *Inorg. Chem.* **(1991)** *30,* 3907.
469 Lee, K. J.; Brown, T. L. *Inorg. Chem.* **(1992)** *31,* 289.
470 Bubel, R. J.; Douglass, W.; White, D. P. *J. Comput. Chem.* **(2000)** *21,* 239.
471 Boeyens, J. C. A.; Cotton, F. A.; Han, S. *Inorg. Chem.* **(1985)** *24,* 1750.
472 O'Neill, F. M.; Boeyens, J. C. A. *Inorg. Chem.* **(1990)** *29,* 1301.
473 Boeyens, J. C. A.; O'Neill, F. M. M. *Inorg. Chem.* **(1998)** *37,* 5346.
474 Boeyens, J. C. A. *Inorg. Chem.* **(1985)** *24,* 4149.
475 Clark, R. J. H.; Hempleman, A. J. *Inorg. Chem.* **(1988)** *27,* 2225.
476 Clark, R. J. H.; Hempleman, A. J.; Flint, C. D. *J. Am. Chem. Soc.* **(1986)** *108,* 518.
477 Miskowski, V. M.; Schaefer, W. P.; Santarsiero, B. D.; Gray, H. B. *Inorg. Chem.* **(1984)** *23,* 1154.
478 Lovell, T.; McGrady, J. E.; Stranger, R.; Macgregor, S. A. *Inorg. Chem.* **(1996)** *35,* 3079.
479 McGrady, J. E.; Lovell, T.; Stranger, R. *J. Phys. Chem.* **(1997)** *101,* 6265.
480 McGrady, J. E.; Lovell, T.; Stranger, R. *Inorg. Chem.* **(1997)** 3242.
481 McGrady, J. E.; Stranger, R.; Lovell, T. *Inorg. Chem.* **(1998)** *37,* 3802.
482 Stranger, R.; McGrady, J. E.; Lovell, T. *Inorg. Chem.* **(1998)** 6795.
483 Cotton, F. A. *Prog. Inorg. Chem.* **(1976)** *21,* 1.
484 Crabtree, R. H.; Lavin, M. *Inorg. Chem.* **(1986)** *25,* 805.
485 Buergi, H. B.; Dunitz, J. D. *Acc. Chem. Res.* **(1983)** *16,* 153.

486  Wipff, G.; Weiner, P.; Kollman, P. *J. Am. Chem. Soc.* **(1982)** *104,* 3249.

487  Grootenhuis, P. D. J.; Kollman, P. A. *J. Am. Chem. Soc.* **(1989)** *111,* 2152.

488  Pretsch, E.; Badertscher, M.; Welti, M.; Maruizumi, T.; Morf, W. E.; Simon, W. *Pure Appl. Chem.* **(1988)** *60,* 567.

489  Badertscher, M.; Welti, M.; Pretsch, E.; Maruizumi, T.; Ha, T.-K.; Simon, W. *J. Comput. Chem.* **(1990)** *11,* 819.

490  Varnek, A. A.; Glebov, A. S.; Petrukhin, O. M.; Kolycheva, N. V.; Ozerov, R. P. *Koord. Khim.* **(1989)** *15,* 600.

491  Burns, J. H.; Kessler, R. M. *Inorg. Chem.* **(1987)** *26,* 1370.

492  Wipff, G.; Kollman, P. *Nouv. J. Chim.* **(1985)** *9,* 457.

493  Damu, K. V.; Hancock, R. D.; Wade, P. W.; Boeyens, J. C. A.; Billing, D. G.; Dobson, S. M. *J. Chem. Soc., Dalton Trans.* **(1991)** 293.

494  Kollman, P.; Wipff, G.; Singh, U. C. *J. Am. Chem. Soc.* **(1985)** *107,* 2212.

495  Maye, P. V.; Venanzi, C. A. *J. Comput. Chem.* **(1991)** *12,* 994.

496  Lifson, S.; Felder, C. S.; Shanzer *J. Am. Chem. Soc.* **(1983)** *105,* 3866.

497  Lifson, S.; Felder, C. S.; Shanzer *Biopolymers* **(1983)** *22,* 409.

498  Lifson, S.; Felder, C. S.; Shanzer *Biochemistry* **(1984)** *23,* 2577.

499  Lifson, S.; Felder, C. S.; Shanzer *J. Biomol. Struct. Dyn.* **(1984)** *2,* 641.

500  Hay, B. P. In: *Metal-Ion Separation and Preconcentration*; Bond, A. H., Dietz, M. L., Rogers, R. D., Eds.; American Chemical Society: Vol. 716. Washington, 1999.

501  Dayan, I.; Libman, J.; Shanzer, A.; Felder, C. E.; Lifson, S. *J. Am. Chem. Soc.* **(1991)** *113,* 3431.

502  Hounshell, W. D.; Dougherty, D. A.; Hummel, J. P.; Mislow, K. *J. Am. Chem. Soc.* **(1977)** *99,* 1916.

503  Frierson, M. R.; Imam, M. R.; Zalkow, V. B.; Allinger, N. L. *J. Org. Chem.* **(1988)** *53,* 5248.

504  Allen, G. W.; Aroney, M. J.; Hambley, T. W. *J. Molec. Struct.* **(1990)** *216,* 227.

505  Bowen, J. P.; Allinger, N. L. *J. Org. Chem.* **(1987)** *52,* 2937.

506  Allinger, N. L.; Allinger, J. A.; Yan, L. Q. *J. Mol. Struct. (Theochem)* **(1989)** *201,* 363.

507  Allinger, N. L.; Quinn, M. I.; Chen, K.; Thompson, B.; Frierson, M. R. *J. Mol. Struct.* **(1989)** *194,* 1.

508  Drew, M. G. B.; Nicholson, D. G. *J. Chem. Soc., Dalton Trans.* **(1986)** 1543.

509  Ramdas, S.; Thomas, J. M.; Betteridge, P. W.; Cheetham, A. K.; Davies, E. K. *Angew. Chemie* **(1984)** *96,* 629.

510  Wright, P. A.; Thomas, J. M.; Cheetham, A. K.; Nowak, A. K. *Nature* **(1985)** *318,* 611.

511  Henson, N. J.; Cheetham, A. K.; Gale, J. D. *Chem. Mater.* **(1994)** *6,* 1647.

512  Sun, Y.; Anderson, C. J.; Pajaeu, T. S.; Reichert, D. E.; Hancock, R. D.; Motekaitis, R. J.; Martell, A. E.; Welch, M. J. *J. Med. Chem.* **(1996)** *39,* 458.

513  Brecknell, D. J.; Raber, D. J.; Ferguson, D. M. *J. Mol. Struct.* **(1985)** *124,* 343.

514  Ferguson, D. M.; Raber, D. J. *J. Comput. Chem.* **(1990)** *11,* 1061.

515  Hay, B. P. *Inorg. Chem.* **(1991)** *30,* 2876.

516  Fossheim, R.; Dahl, S. G. *Acta Chem. Scand.* **(1990)** *44,* 698.

517  Frey, S. T.; Chang, C. A.; Carvalho, J. F.; Varadarajan, A.; Schultze, L. M.; Pounds, K. L.; Horrocks, W. D. *Inorg. Chem.* **(1994)** *33,* 2882.

518  Dillen, J. L. M. *J. Comp. Chem.* **(1992)** *13,* 257.

519  Lopis, A. S.; Glasser, L.; Marsicano, F. *QCPE Bulletin* **(1997)** *17,* 16.

520  White, D. N. J.; Ermer, O. *Chem. Phys. Lett.* **(1975)** *31,* 111.

521  Comba, P.; Luther, S. M.; Maas, O.; Pritzkow, H.; Vielfort, A. *Inorg. Chem. in press.*

522  Boyens, J. C. A.; Comba, P. *Coord. Chem. Rev. in press* **(2000)**.

523  Raber, D. J.; Guida, W. C. *J. Comput. Chem.* **(1998)** *20,* 1591.

524  Osawa, E.; Lipkowitz, K. B. In: *Reviews in Computational Chemistry*; Lipkowitz, K. B. Boyd, D. B. Eds.; Wiley-VCH: New York, Weinheim, **(1995)** *6,* 355.

525  Boyd, D. B. In: *Reviews in Computational Chemistry*; Lipkowitz, K. B. Boyd, D. B. Eds.; Wiley-VCH: New York, Weinheim, **(1996)** *7,* 303.

526  Boyd, D. B. In: *Reviews in Computational Chemistry*; Lipkowitz, K. B. Boyd, D. B. Eds.; Wiley-VCH: New York, Weinheim, **(1997)** *11,* 373.

527  Jalaie, M.; Lipkowitz, K. B. In: *Reviews in Computational Chemistry*; Lipkowitz, K. B. Boyd, D. B. Eds.; Wiley-VCH: New York, Weinheim, **(2000)** *14,* 441.

# Index